The Ingenious Mind of Nature

Deciphering the Patterns of Man, Society, and the Universe

The Ingenious Mind of Nature

Deciphering the Patterns of Man, Society, and the Universe

GEORGE M. HALL

A Member of the Perseus Books Group
New York

Library of Congress Cataloging-in-Publication Data

Hall, George M.
   The ingenious mind of nature   deciphering the patterns of man,
society, and the universe / George M. Hall.
      p.   cm.
   Includes bibliographical references and index.
   ISBN 0-306-45571-4
   1. Science--Miscellanea. 2. Social sciences--Miscellanea.
I. Title.
Q173.H185 1997
003--dc21   96-45673
                                                             CIP

Figure 1 and image used on all chapter-opening pages, Negative #320446, courtesy of the Department of Library Services, American Museum of Natural History.

Figures 8, 33 (nuclear cloud), 52, and 57 courtesy of the National Archives.

Figure 19, Photo #61608-H, courtesy of the Smithsonian Institution.

Figure 29 courtesy of the Bettmann Archive.

Figure 35 courtesy of American Airlines and the Tucson Air Museum Foundation of Pima County.

Figure 38 courtesy of Dr. Sherwood Casjens, University of Utah Medical Center.

The passage on p. 150 from *What Is Life?: The Physical Aspect of the Living Cell* by Erwin Schrödinger and Issac Newton's notes in Figure 36 from a photograph in *Never at Rest: A Biography of Issac Newton* by Richard S. Westfall are reprinted with the permission of Cambridge University Press.

The passage on p. 63 from *Never Call Retreat* by Bruce Catton (©1965 Bruce Catton) is reprinted with the permission of Doubleday Dell Publishing Group, Inc.

Ada is a registered trademark of the United States Department of Defense.

IBM is a registered trademark of International Business Machines Inc.

Tinkertoy is a registered trademark of Hasbro, Inc.

ISBN 0-7382-0584-2

© 1997 George M. Hall

Basic Books is a Member of the Perseus Books Group.
Visit us on the World Wide Web at www.basicbooks.com

10 9 8 7 6 5 4 3

Printed in the United States of America

For my teachers,
especially Sister Mary Frances,
Order of Saint Benedict,
Linton Hall, Virginia

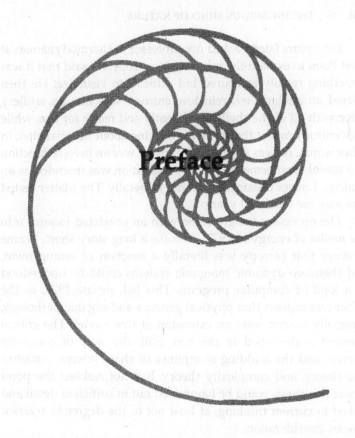

Preface

The source of this book goes back 46 years to a biology class at
Lowell High School in San Francisco. The first and second
semesters were divided into botanical and zoological studies,
respectively. In the second semester, the course was organized by
phyla to emphasize the evolution of various organs and physiological processes. It became apparent to me that by the time we
reached the humble earthworm, all of the human organs and
processes, except the skeleton (and liver), had been developed, at
least in rudimentary form. Evolution from that point forward
was an improvement on these mechanics rather than a radical
departure from them. I had no way of figuring out why, but the
question lodged itself in my mind and would not let loose.

Four years later, I asked my professor of thermodynamics at West Point to explain the meaning of entropy. He said that it was something readily measured but difficult to visualize. He then offered an analogy wherein low entropy was like an artillery piece with all of the shells piled neatly and ready for use, while high entropy meant the shells were laying about helter-skelter. In other words, he was saying that entropy was an inverse function of a useful arrangement. As this explanation was intended as an analogy, I never dreamed of taking it literally. The matter rested this way for nearly 20 years.

The necessary insight came with an unrelated inquiry into the nature of energy in 1973. To make a long story short, I came to sense that entropy was literally a function of arrangement, and therefore dynamic inorganic systems could be understood as a kind of computer program. This led, in late 1973, to the wider assumption that physical genetics and organic pathology, especially cancer, were an extension of this model. The critical moment is described in the text. Still, the state of computer science, and the budding disciplines of chaos science, catastrophe theory, and complexity theory, had not reached the point where this thesis could be hammered out in sufficient detail and linked to current thinking, at least not to the degree to warrant serious consideration.

During the next 15 years, I published a number of studies in military and defense theory and became involved in computer science on a full-time basis, publishing books and articles in that field too. This work provided the necessary background to proceed with this book. In preparing an early manuscript, it dawned on me that the principles governing genetics and evolution applied figuratively to all systems, and so the argument expanded from the biological to a more comprehensive reach, though it took five years to write it.

This leaves the question of just how much of the argument of this book is mine, and how much is the property of others. I can truthfully say that the only original idea is the thesis itself, which is stated in a single sentence near the beginning of the book. At that, Sir Isaac Newton postulated the same idea, at least in part, and published it in 1704 as part of his *Opticks*. Everything else is a matter of linking that central idea to known phenomena as a matter of logical derivation. Also, because of differences among definitions in various disciplines that basically described the same idea, I was more or less forced to develop a modified glossary, then point out how it could be linked to other terms extant. That, too, has a hint of originality to it, yet only as mere semantics.

One last point. The presentation of a new concept always raises the question of just how far it should be developed before running it up the flagpole, so to speak. There is no easy answer to that question. The goal is to present a convincing argument, but different readers will react at different rates. Closely related to this question is the matter of proportionality between abstract and physical systems on the one hand, and psychical and sociological systems on the other. The focus of the book, as the name of the concept—physiogenesis—suggests, is on physical systems. The purposes of including other types of systems is to demonstrate the symmetry or universality of the concept, not to apply it with equal depth. Moreover, such application would easily quadruple the length of the book on the grounds that sociological systems, and the mind, are considerably more complex than physical systems.

□ □ □

I am indebted to Professor Sherwood Casjens at the University of Utah Medical Center for providing the micrograph of the T-4 bacteriophage, and to Mrs. Kirsten Oftedahl, the curator at the Pima Air & Space Museum (Tucson, Arizona), for the photograph of the DC-3 aircraft. I also wish to thank Mrs. Polin Lei, at the University of Arizona Medical College library, for locating selected material related to physiology.

At Plenum Press, I wish to thank my editor Melicca McCormick for bearing with what perforce was a difficult manuscript. Lastly, I acknowledge the detailed, often overlooked work of the production editor, Arun Das, for seeing the manuscript through to print.

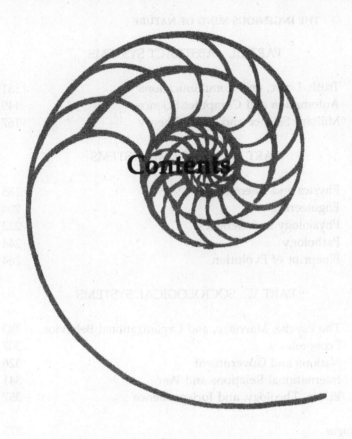

Contents

PART I.   LAY OF THE LAND

1. Concept   3
2. Categories, Structure, and Factors   27
3. Mechanics   47
4. Derivations and Applications   71

PART II.   HISTORY

5. Roots among the Physical Sciences   93
6. Insight from the Social Sciences   109

PART III. ABSTRACT SYSTEMS

7. Truth, Logic, and Communications 131
8. Automation and Computer Science 149
9. Military Science and Game Theory 167

PART IV. PHYSICAL SYSTEMS

10. Physics and Chemistry 185
11. Engineering 204
12. Physiology and Genetics 222
13. Pathology 244
14. Blueprint of Evolution 264

PART V. SOCIOLOGICAL SYSTEMS

15. The Psyche, Marriage, and Organizational Behavior 283
16. Economics 307
17. Nations and Government 326
18. International Relations and War 341
19. Ethics, Theology, and Jurisprudence 357

Epilogue 375

Glossary 379

Appendix A. Experiments, Models, Meta-Analyses 387

Appendix B. Comparative Systems 397

Notes 407

Selected Reading 431

Index 435

PART I

Lay of the Land

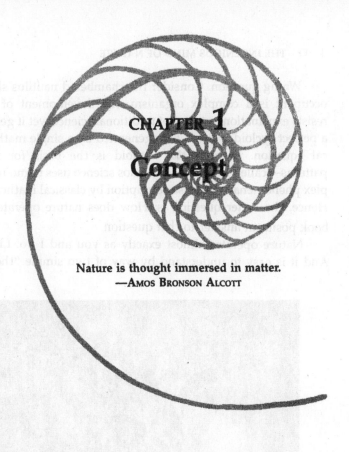

CHAPTER 1
Concept

Nature is thought immersed in matter.
—Amos Bronson Alcott

The ascendancy of the science of chaos, complexity theory, catastrophe theory, and general systems theory in the latter half of this century is traceable to a very old question: Are the laws of nature fundamentally continuous and orderly, with occasional lapses into disorder? Or are these laws fundamentally discontinuous and chaotic, from which nature evokes order on occasion by virtue of patterns that lead to such order? Traditional science points out the ability to predict solar eclipses 50,000 years from now. Chaos science smiles, then asks what the weather will be in Topeka, Kansas, at 9:00 AM local time exactly 10 years hence. Or, for that matter, 10 *days* hence. In short, is nature *inherently* orderly or chaotic?

3

Wrong question. Consider the chambered nautilus shell. Its occupant is a complex organism, the development of which resists explanation in terms of traditional science, yet it generates a perfect cycloid that can be regenerated by a single mathematical equation. In turn, that cycloid is the basis for several patterns—called fractals—that chaos science uses to model complex phenomena that resist description by classical mathematics. Hence, the better question is: How does nature operate? This book posits an answer to that question.

Nature operates almost exactly as you and I do. Literally. And it is easy to understand by way of two simple "thought"

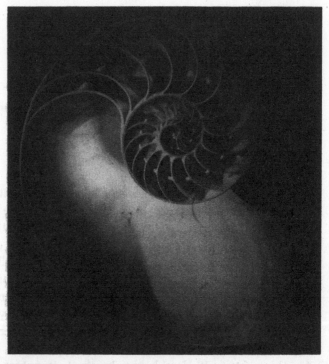

1. Chambered nautilus (*Nautilus pomphius*). This well-known seashell exemplifies the ability of a complex, nonlinear organism to produce a simple, orderly structure, which strongly suggests that apparent chaos can still wield decisive influence.

experiments. First experiment: Keep an imaginary diary for the next 24 hours. You will find that everything you see or do consists of moving or rearranging objects or the equivalent in mental tasks: working on an assembly line, writing a letter, playing a sport, grocery shopping, or cooking. That is, living consists of a continuous stream of *purposeful* configurations or patterns of behavior—in a word, *systems*. In each event, however, you exert personal initiative. You intervene, and that intervention makes the difference between disordered resources and orderly products or outcomes. Hopefully. But no such personal intervention on the part of nature has ever been detected in science, at least not this side of the big bang. So if nature does not intervene, she had to use a substitute. Matter had to be programmed *in advance* in order to evolve into solar systems and organisms, including you and me.

Understanding *how* leads to the second "thought" experiment. Consider a common billiard table expanded into a three-dimensional vacuum in a satellite wherein the centrifugal force exactly offsets the pull of gravity (a common situation encountered by astronauts). Once set in motion, the billiard balls will tumble about for a long time. Then if the spherical configurations are modified with hooks or magnets, some collisions will result in constructs. True enough, nature is not a gravity-free set of billiard balls with hooks. But atoms with various binding indexes, leading to molecules, move freely in random *patterns* when submerged in swirling water. The hydraulic pressure of the water exactly offsets the pull of gravity, and we are certain that some organic molecules first formed in tidal pools and the like. In turn, these molecules joined to form more complex molecules, which joined into microorganisms, then into non-nucleated cells, lastly into nucleated cells, which initiated the evolution of the species via multicellular organisms of ever-increasing complexity.

Still, a complex molecule is one phenomenon; an organism with a quadrillion cells, highly complex organs, and self-directed locomotion is another. The billiard-balls-in-space model lacks the heft to account for the generation of *Homo sapiens*. But then, that model uses only clumsy elements and is severely limited in space and time. Nature is more sophisticated on *two* counts. First, she ranges over the entire universe and took billions of years to unfold her designs. In human terms, that span equals 200,000,000 generations (at 20 years each). By analogy, a single penny invested at 6 percent interest, compounded quarterly, will grow to $221,649,930 in a mere 20 generations. In 100 generations, the balance will be:

$53,488,051,100,000,000,000,000,000,000,000,000,000,000,000,000

In 200 million generations, the number would fill this book. The idea is that with time, the cumulative effect of compounding—money or molecules—produces startling effects.

Second, nature relies heavily on subassemblies and variations of construction themes, and so do we, of course. An automobile contains about 5000 identifiable parts, but in practice assembly lines link only 300 subassemblies manufactured elsewhere. Similarly, nature builds organisms solely by dividing and multiplying cells (binary fission) which, in any given specimen, contain identical chromosomes (DNA). Furthermore, the physiology of all zoological organisms evinces a gradual evolution of common organs and processes. Too, consider this statement by the physicist Philippe Le Corbeiller made almost 50 years ago:

> In the third stage of scientific knowledge, which we might call deductive or axiomatic, the natural laws obtained by observation are shown to be necessary logical consequences of a few hypotheses or assumptions. The surprising thing about examples of deductive knowledge which we know today is *the extreme simplicity of the assumptions and how rich and far-removed are their consequences.*[1]

Still not convinced? Well, in that case, take note of *cellular automata.* Cellular automata is the subscience that demonstrates how infinite varieties of production can be formed from a *single* type-element.[2] In the case of the physical universe—at or above the level of atoms—we have not one element but a rich buffet of 92 naturally occurring elements, each of which comes with two or more variations in the form of isotopes. Lastly, no less a physicist than Sir Isaac Newton put the case this way:

It seems probable to me that God in the beginning formed matter in solid, massy, hard, impenetrable, moveable particles, of such sizes and figures, and with such other properties, and in such proportion to space, as most conduced to the end for which he formed them. . . . therefore that Nature may be lasting, the changes to corporeal things are to be placed only in the various separations and new associations and motions of these permanent particles.[3]

The expression "various separations and new associations and motions of . . . particles [in space]" implies that the patterns alone have the inherent ability to form into constructs, that is, without any intervention on the part of nature. So now let us take this distressingly simple concept and begin the work of recasting it into the formal terms that science demands.

PHYSIOGENESIS

Patterns of related elements in space—systems—combined with the attributes of the elements, constitute programs by which those patterns efficaciously devolve into order or disorder as the case may be. Efficacious, because the patterns or systems *inherently* possess the power and capacity to achieve results. What you see— literally—is what you will eventually get. For the physical universe, nature requires nothing else whatsoever, under any circumstances, at any time, *provided* that elements may first combine into subassemblies as a means of staging the development of more complex phenomena. This thesis, and its implication of ingenious patterns, is arbitrarily named *physiogenesis.*[4]

The first step in transitioning an idea into a disciplined theory is to define terms. The crucial terms used in this book are, for the most part, drawn from various scientific disciplines, but it must be recognized that different disciplines often define the same term in somewhat different ways or with different connotations, to include: *element, pattern, program, linear,* and *system.* The approach taken in this book is to precisely, if arbitrarily, define each relevant term using the criterion of editorial simplicity, then carefully discuss any differences at appropriate points in the text.

Element is defined as any discrete unit or construct existing in space that can be identified apart from other elements. *Configuration* is arrangement of elements (atoms, molecules, cells, and so forth), in space, combined with the properties or attributes of each element, such as velocity, mass, or shape. As such, configuration is the equivalent of the German word *gestalt*. Without considering the attributes of the elements, except for their location in space, a configuration reduces to *arrangement*. *Pattern* means the change to a configuration over a period of time, or what is the same, a configuration is a cross section of a dynamic pattern at any moment of time. To be sure, some scientists define pattern as configuration and changes to configuration as dynamics. The difference is purely semantic.

Any configuration that serves, or is intended to serve, a purpose or meet some need is also a *system*, which implies that its elements are physically or logically related. Some reference works include the attribute of purpose; others don't.[5] This book assumes inclusion of purpose, though it is recognized that an unintended configuration may nevertheless serve a purpose (herein defined as *serendipity*). In the absence of purpose, a configuration remains a configuration, but it is not a system. For example, billiard balls accidentally dropped on the floor constitute a configuration but not a system. By contrast, when billiard balls are set up on a pool table, they are part of a system, in this case a game.

Linear means that the elements of a system are so linked or related that the pattern of their operation may be described by means of traditional mathematical models. By contrast, *nonlinear* means that the elements under consideration operate—fully or in part—independent of one another, hence can be better described by techniques other than those of traditional mathematics, e.g., fractals common to the science of chaos. These definitions differ somewhat from usage in various disciplines, but the intent is roughly the same, namely, to indicate the relative mutual dependence or its lack among elements in a system.

The remaining crucial term is *program* (and its verb form *programming*). When you set elements into a configuration in order to achieve a purpose, you create a system by definition. You anticipate that system will function as you designed it to function because you know—or think you know—how those elements will interact and thus lead to the desired outcome. Alternatively, you can superimpose another system to control the first system (for example, a thermostat), which may include a set of coded written instructions, typically in the form of a computer program, that tells the system what to do. Any formal set of instructions of this kind is universally defined as a program.[6]

In the absence of a supplementary control system, you have no choice but to rely on the configuration alone to evoke the desired outcome. Colloquially, you line up the ducks and hope for the best. This reliance on configuration alone substitutes for formal programming, and in this book that technique is also called a program. Not an equivalent; a *program*. The dynamic pattern of the ensuing configurations *is* the program. This is a generalization of the fact that computer-hardware manufacturers design chips to operate in certain ways solely by virtue of configuration of the electronic components (called *firmware* or *hard-wired programming*). Too, instructions written to control a system are almost always transformed into a configuration of symbolic zeros and ones on punch cards, electromagnetic disks, or optical disks that the system "reads" and follows.

The glossary near the end of this book repeats these definitions, along with many other terms, most of which are also defined in the text when they first appear.

EVIDENCE

Valid theories must cogently explain, with consistent logic, the behavior of all phenomena to which they pertain, and they should also predict as-yet-undiscovered phenomena. The classic example of predictability was Mendeleyev's periodic table of elements, which identified "new" elements *before* they were discovered. Also, theories that integrate other theories should operate like a mental octopus, with each arm tenaciously grasping each subsumed theory, showing how they are all sired by a parent symmetry. Appendix A of this book describes 14 projects to fulfill the prophecy criterion, as it were. A subsequent section in this chapter addresses the symmetry criterion, to include the special problem of reconciling the simplicity of the physiogenetic model with the complexities of atomic physics. The balance of this section looks at the evidence—more accurately, the lack of evidence—to support any other thesis.

Aside from the psychical attributes of higher-order zoological organisms, the only things that fundamentally or ontologically exist *at or above the level of the atom* are atoms, gravity, photons, and a range of free-agent subatomic particles, all of which have attributes of momentum, force, and/or energy. Molecules, of course, are constructs of various atoms but lose their molecular properties whenever they dissemble into their constituent atoms.

The strong nuclear binding force and the weak nuclear decay force operate only within the atom, and, at that, the weaker force has been linked with the electromagnetic force.[7] Moreover, the electromagnetic force is a system or field of photons. Protons, neutrons, and electrons comprise the primary building blocks of atomic nuclei and often exist outside of nuclei. Furthermore, the neutron is a critical operator in nuclear reactions, and free-agent electrons serve many useful purposes, such as electrical current. But the free-agency activity is transient because their utility is a function of atoms sort of playing catch with them.

As for gravity, it operates like the electromagnetic "field," albeit attractive only and much, much weaker, and is commonly presumed to consist of particles called gravitons.[8] Lastly, a host of other particles exists but, with the exception of the neutrino, they decay or otherwise self-destruct rapidly.[9] In any and all events, particles constitute units in space.

Elsewhere, at least above the level of atoms, waves are an *effect*, not an ontological reality. That is, waves are the *effect* or *pattern* of a sequence of particles acting on other particles. A flag waves, but only because particles of air impinge on the strands of its cloth. A tidal wave smashes into a shoreline, but only because it is a pattern of interactions among molecules of water. That fine distinction would be small comfort to a person struck by a tidal wave, but the point remains valid. If you rearrange the particles, the wave disappears without a trace, which is exactly what occurs in an automobile muffler. In short, you cannot create a universe with huffs and puffs, though once formed it may huff and puff indefinitely.

As for energy, it too is defined only as effect—the capacity to do work (potential energy) or the process of doing work (kinetic energy). Without a doubt, energy is indispensable to the operation of dynamic systems, but, as discussed shortly, the harnessing of energy to do work is a function of configuration of elements or subassemblies in space.

Now it is said that the formation of different atomic elements introduces new properties into the physical world, and indeed it does. But similar to molecules, such properties are also effects. If you break up an atom into its constituent particles or lesser constructs (with or without conversion of mass into radiant photons), the particles remain but the properties of the atom disappear. Similarly, pure coal, fullerite, graphite, and diamonds all consist solely of carbon atoms. They differ in appearance (and perceived value), but the only fundamental difference among them is different configurations—crystals in some cases—of carbon atoms. Similarly, the old assumption of a vital force that differentiated organic constructs from inorganic molecules has been thoroughly debunked.

True enough, some biologists point out the existence of stereoisomers that are more or less unique to organic compounds. Yet these isomers are merely different configurations of the same number and type of elements, for example, ethyl alcohol and dimethyl ether (both C_2H_6O). This is no different than the spectrum of coal, fullerite, graphite, and diamonds.

Lastly, no experiment has ever found the slightest trace of a *discrete*, identifiable program, or set of instructions, or anything remotely similar, residing in atoms. It is true that programming equivalents seem to appear in various brain cells, such as synapses and neurons.[10] Yet, when you break down a brain cell into its chemical elements, that "program" also disappears. Furthermore, it took a billion years for nature to evolve cells, and it is ludicrous to think that such "programs" could operate *ex post facto.*

In short, if all that physically exists in space are particles in various guises (some perhaps with wave characteristics), combined with their respective attributes and energy content, what explanation, other than the physiogenetic model posited herein, can there be? Evidence, ladies and gentlemen; evidence! In the absence of any other evidence, we must start with what little there is and retrace the building of the universe from that. Synergy—the utility of the whole not available from constituent parts acting alone—is wonderful, but it too is an effect. And keep in mind that cosmologists have deduced that atoms probably formed within the first *second* after the primordial "big bang."[11] That is to say, the physiogenetic argument of this book *may* gloss over some aspects of the first second of time, but not the next 10 to 15 billion years.

LINEARITY VERSUS NONLINEARITY

The terms *linearity* and *nonlinearity* were defined above. In practice, only a few systems operate with near-perfect linearity, for example, gravity. Any and all disruptions shift the system in the direction of nonlinearity, and this general picture is presented in Figure 2. The important thing to note is that classical mathematics encounters increasing difficulty dealing with systems to the extent they are nonlinear. The reason is that equations are based on definitive relationships, and in the absence of such relationships, equations can only approximate relationships. This is especially true when three or more elements are related but remain independent of one another. Physicists call this the *three-body problem.*[12] However, when physically independent elements operate in tandem, for whatever reason (like members of a drill team), then as long as that order prevails, equations suffice to describe the system.

Interestingly, mathematics itself confronts nonlinearity in the form of prime numbers. A prime number is an integer that can be divided evenly only by itself and the quantity 1. All other numbers are composite, meaning they are the product of two or more prime numbers (e.g., $14 = 2 \times 7$). In 1742, Christian Goldbach, sensing an elegant simplicity here, postulated that *every even number larger than 2 is the sum of two prime numbers*. Simple, indeed. After 255 years and thousands of attempts, Goldbach's conjecture remains unproven.

Why? About 2300 years ago, Euclid proved that there was no last prime number, but no one has ever developed an equation that predicts exactly when a prime number will next occur. Only approximations are possible, and many of them are not much better than Eratosthenes' sieve. The sieve consisted of a table of

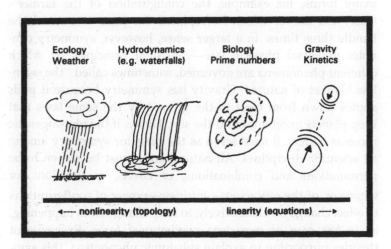

| Ecology Weather | Hydrodynamics (e.g. waterfalls) | Biology Prime numbers | Gravity Kinetics |

◄---- nonlinearity (topology)   linearity (equations) ----►

2. **Linear versus nonlinear systems.** The elements of linear systems are sufficiently linked or otherwise operate in tandem to lend themselves to description by classic mathematical technique. By contrast, nonlinear systems lack those attributes and therefore resist mathematical analysis except by approximation.

numbers, on which Eratosthenes drew lines through the multiples of each succeeding prime number. The numbers that were not crossed with these lines were (and remain forever) primes.

The significance of all this is that in the absence of neat mathematical reduction, nature's systems defy understanding in classical terms. Historically, this difficulty has spawned a substantial number of alternative assumptions, including the "vital force" doctrine mentioned above. But one by one, they have all been debunked. They lacked symmetry.

SYMMETRY

The popular meaning of *symmetry* (more at *symmetrical*) is that of an object in which the left and right halves are mirror images of one another, like a valentine heart. In science, symmetry means the properties or attributes of a system that endure despite operations performed on or with it.[13] This symmetry can take many forms, for example, the configuration of the farmer's trusty old axe of which he replaced the head twice and the handle three times. In a larger sense, however, symmetry connotes universal physical law—consistent principles by which different phenomena are governed, sometimes called "the seamless blanket of nature." Gravity has symmetry because it pulls apples down from trees to the ground by the same laws that keep planets in orbit around the sun. Hence, if the physiogenetic thesis is correct, it *must* serve as the basis for symmetry among all scientific disciplines. All natural laws must be shown to be permutations and combinations, or effects, or functions, or whatever, of the efficacious—intrinsic—power of configurations to effect change or, alternatively, to keep change from happening, as in the case of persistent gravitational force. Exclusive of theories purporting to explain subatomic phenomena, this symmetry is shown in Figure 3.

Physical *systems theory* and the *science of chaos* are based on patterns of elements operating in space. *Complexity theory* is a stepchild of chaos, holding that growth occurs only on the cusp

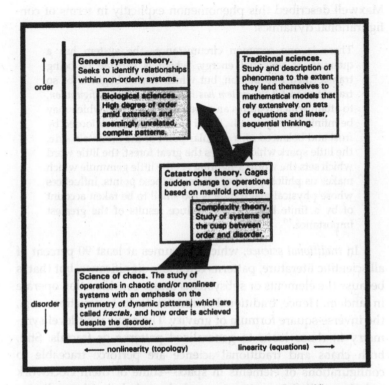

3. Relationships among scientific perspectives. Scientific perspectives differ among themselves primarily to the extent the phenomena they encompass are linear or nonlinear and, similarly, to the extent they are orderly or disorderly. The two criteria are correlated but order is not necessarily a precise function of physical linearity.

between orderly and disorderly patterns.[14] Too much order and structure prevents growth, while excessive disorder leads to disaster. That, incidentally, is the rub with systems theory; it often focuses on hierarchical order instead of equilibrative balancing.[15]

Catastrophe theory is linked to complexity theory because it describes a system that generates output out of proportion to input, at least over a defined range, for example, an amplifier. The theory also covers drastic, radical, sudden, or uncontrollable changes to output as a result of a minor output. James Clerk

Maxwell described this phenomenon explicitly in terms of configurational dynamics:

> There is one common circumstance—the system has a quantity of potential energy, which is capable of being transformed into motion, but which cannot begin to be so transformed *till the system has reached a certain configuration,* to attain which requires an expenditure of work, which may be infinitesimally small. . . . For example, the rock loosed by frost and balanced on a singular point of the mountain side, the little spark which kindles the great forest, the little word which sets the world a-fighting . . . the little gemmule which makes us philosophers or idiots. At these points, influences whose physical magnitude is too small to be taken account of by a finite being, may produce results of the greatest importance.[16]

In *traditional science,* which consumes at least 90 percent of all scientific literature, patterns often lose significance, but that is because the elements or subsystems under consideration operate in tandem. Hence, traditional symmetry looks to equations, e.g., the inverse-square formula of gravity. The equivalent-level symmetry in chaos looks to generalized patterns or fractals. Still, both chaos and traditional science are perforce traceable to configurations of elements in space—some of them clockwork orderly and linear, some not so orderly and obviously nonlinear, like weather.

The gaps in this symmetry are: (1) whether the physiogenetic model applies to abstract and sociological systems, in which *physical* configuration does not always exist, and (2) the potential discontinuity with some of the notions of current subatomic physics. The first gap is easy to address by substituting figurative relationships for geometric disposition. Fully eight of the thirteen application chapters in this book address abstract and sociological systems, and the primordial differences between physical and abstract elements are described *et passim* in the first four didactic chapters. Lastly, one of the two chapters on history addresses the development of concepts pertaining to sociological systems.

The potential gap with subatomic physics is a harder nut to crack, and in part must be postponed until such time as theoretical physics can produce a widely accepted unified symmetry of its own. Until such time, we will probably continue to hear the likes of hypothesized ten-dimensional universes, the analytical model for which requires an apparatus the size of the solar system.[17] Still, opposing symmetries would go against the grain of the prevailing notion of as-yet-undiscovered grand symmetry.[18] That would be like a kiss without a hug. Therefore, only three outcomes seem plausible. First, the physiogenetic model arises from an even simpler model operating within the atom and its fields of force. Or, second, the complexities of atomic physics may wash out, so to speak, at larger scale, like grains in a photographic image. Or, third, the physiogenetic model will be demolished. Stay tuned.

Several of the proposed experiments and projects in Appendix A address this discrepancy. In the interim, the following observation will suffice. Subatomic physics consists *primarily* of dynamic patterns of particles in space, even if some particles are also thought to be waves and even if we cannot determine the exact location and momentum of any particle at a particular moment in time (the Heisenberg uncertainty principle). That is to say, even if 10 million different subatomic particles existed, they would all exist in space and this perforce results in patterns. Need more be said?

The last item of the symmetry agenda is so-called *fuzzy logic*. Fuzzy logic is not so much a separate discipline as a technique for assessing complex situations that defy classical analysis or arise from discrete, controllable elements that blend into one another (gradations) to the point where their contributions can no longer be isolated and measured. For example, in *Goldilocks and the Three Bears*, the heroine finds the father bear's large bowl of porridge is too hot; the mother bear's medium bowl too cold; and the baby bear's little bowl "just right." Presuming the porridge was dished out at roughly the same instant, and the bowls were made of the same material, problems like this give thermodynamicists fits.

Enter fuzzy logic, which points out that the table on which the bowls rested was situated with respect to an open window, such that a draft flowed directly around the mother bear's bowl, cooling it at a much faster rate than the other two. The baby bear's smaller bowl, of course, had a higher surface-to-volume ratio, and therefore cooled faster than the father bear's bowl but not nearly as fast as the mother's bowl. Fuzzy logic, of course, has many other applications, certainly of more significance than the warm-fuzzy problem above, and we will see them at various points of this book.

ENTROPY AND ENERGY

The physiogenetic model appears to conflict with the second law of thermodynamics. This law states that in closed systems, energy becomes increasingly *less* available, hence the enclosed system will slow down and stop, as it were. How, then, can a system that draws solely on its own intrinsic power evolve into a more complex system? The key is the expression *closed system*. In a closed system, no energy of significance can either enter or leave. The only way to bypass this obstacle is to open the system, and there are two ways to do so. The first way is for an external system to resupply the system with the energy it needs. The second way is for the system itself *to reach out and grasp what it needs from its immediate environment*. The latter presumes that the system has the internal programming to do so robotically, and that the environment possesses, and is willing to release, the energy.

In more detail, the second law of thermodynamics is often called the *law of entropy*. Entropy is the measure of the *unavailability* of energy, and in practice means the *increasing* unavailability of energy. The first law of thermodynamics states that in a *closed* system (nothing of relevance enters or exits), the energy level remains constant. The second law of thermodynamics states that while the energy level remains constant, that energy will become increasingly *unavailable* to do work.[19] This seeming contradiction can be demonstrated by a system of two lakes at different elevations connected by a channel.[20] Gravity will cause water in the higher lake to flow into the lower lake until both lakes reach the same elevation. Thereafter, no further energy can be harnessed even though the two lakes collectively retain the same amount of energy with which they began. The initial lower entropy (available energy) devolves into higher entropy (unavailable energy).

Over the last century, the second law of thermodynamics has evolved into a general dogma that condemns all unattended systems to degeneration, decay, rot, or death as the case may be. Sidewalks crack, homes fill with dirt, translations suffer, empires decline and fall, organisms die. Even the Great Pyramids will eventually crumble into dust. Entropy happens! Yet clearly, a fertilized egg cell (zygote) evolves into a complex organism without any conscious *direct* effort or intervention on the part of the mother. That is, nature routinely *reverses entropy*, or, to be supertechnical, she reverses the increasing unavailability of energy. In plain English, more energy is made available.

Moreover, a woman seldom realizes that she is pregnant until the fertilized egg has multiplied about 1000-fold, and afterwards she is limited to keeping the fetus out of harm's way. But nothing in organic chemistry has ever been shown to differ fundamentally from simpler, inorganic processes. Hence, either the environment must be programmed to force-feed evolving embryos exactly what they need, or embryos must be programmed to take what they need from that environment (or both). In the case of a human embryo, both methods operate, but *any* evidence of internal-reach-out-and-grasp programming suffices to verify how the physiogenetic model works. Keep in mind that when push comes to shove, an embryo is programmed to starve the mother in an attempt to survive.

Now if the physiogenetic model is valid for all physical systems, and there is no fundamental difference between organic and inorganic systems, then man must be able to replicate this process inorganically. He has, especially in the form of a so-called computer virus. This "virus" is a segment of computer program programmed to replicate and expand at the expense of other data and programming resident on one or more memory devices.[21] The effect is destructive to the user, but technically the self-powered growth is quite orderly.

Lastly, let us assume that energy is a discrete, ontological entity apart from particles, despite the fact that it can only be described as an effect (the capacity to do work). Does that tear at the fabric of the physiogenetic model? No. Energy is transmitted only via interaction among particles and elements, be they radiant or electromagnetic field photons (which may have complementary, wavelike characteristics), or protons, or neutrons, or electrons, or gravitons, or whatever. Energy doesn't sail through space of its own accord. Hence the transfer of energy (kinetic) is unavoidably a function of configuration, and therefore so is entropy.

But one point must be made clear. Reversing entropy is *not* perpetual motion (a machine that operates eternally). An embryo processes nutrients found in its vicinity; it does not create growth from nothing. No free lunch. Ever. Furthermore and without exception, all organisms die. The difference between this efficacious growth and perpetual motion can be cleared up by reference to the common heat pump. *It reaches out and grasps.* The heat pump extracts more energy from the air than it expends in the process of that extraction, hence it gives the *appearance* of being more than 100 percent efficient. Not so. The pump merely shifts energy that exists outdoors to indoors. In mild climates it takes less energy to make the shift than is taken from the outdoor environment. And in time, the pump will break down from wear and tear, so it is not a perpetual motion machine.

CAUSALITY, DETERMINISM, AND INDETERMINACY

Causality means that the operation of a system or phenomenon can be described as a predictable—deterministic—sequence of so-called causes and effects, each "effect" being the cause of the next effect. Indeterminacy veers from this, and it has two perspectives. The first perspective states that in very small zones and in the presence of distributed outcomes resulting from apparently the same input, the internal process is impossible to describe.[22] This does *not* imply an absence of deterministic causality, only that it cannot be observed and therefore we can only deal with statistical distributions.

The second perspective assumes that distribution is ontologically indeterminate, i.e., some form of causality may operate but it is always a matter of chance. Rolling the dice seems like a matter of chance, but an evaluation of the kinetics demonstrates a deterministic causality. Not so with this second perspective, even if we could see every aspect of the system operating in ultimate detail.[23] And a variation of this second perspective eliminates the concept of causality altogether.[24] Interestingly, Albert Einstein did *not* buy into the jettisoning of causality. He defended causality with a famous statement: "God does not play dice with the universe."[25] Niels Bohr, his equally distinguished colleague at the Institute for Advanced Studies, countered: "Oh, stop telling God what to do."[26] A student of this debate compromised: "God does play dice with the universe but He loads them."[27] This book not being a theological treatise, the reader must take an assumption on faith. The book assumes that deterministic causality governs all interactions *at or above the level of an atom*, hence any pattern of these interactions is perforce equally deterministic.

Yet it doesn't really matter if the assumption is correct. Patterns of interactions exist, regardless of the nature of the gestation. Hence, the programming inherent in efficacious patterns must cope with those distributions. In other words, the programming may or may not operate as intended on each attempt, which therefore requires a larger number of attempts. We will revisit this matter at some length in the chapter on evolution. Meanwhile, let the terms *determinism* and *deterministic* refer to discrete, causal interactions whereby the outcome adheres with absolute predictability as a result of the operation of various physical laws, extended to include all patterns of interactions. *Statistical determinism*, therefore, describes statistical distributions of outcomes despite apparently identical inputs. And *actuarial determinism* is statistical determinism occurring over a period of time, typically measured in years, e.g., the actuarial studies conducted by insurance companies in order to establish policy premiums.

TELEOLOGICAL AND MIND–BODY ISSUES

The physiogenetic model also confronts two subjective issues, both of which date back to ancient times. These issues concern the role of a Creator in the design and development of the universe (teleology), and the mind–body issue. The polemics of the teleological issue are: (a) a Creator designed and set the universe in motion to achieve purposes inherent in the design, or (b) no such Creator exists and all of what we see somehow arose serendipitously. If the first viewpoint is correct, all attempts to develop a theory based on serendipity would be inimical to science. And vice versa. Fortunately, the issue is immaterial to physiogenesis because the model starts with the big bang, not behind it. However, we note that the great physicists in history believed in a Creator in some form, including Copernicus, Galileo, Kepler, Newton, Maxwell, and Einstein.[28] Newton went so far as to insert a brief theological treatise in the *Principia*.[29] Still, strictly speaking, we need concern ourselves only with the configuration of the big bang—*anthropy*—at the instant it began to expand.[30] The anthropic principle states that the elements of the universe had to be designed with the evolution of solar systems and species clearly in mind, but it does not dwell on theological issues.[31]

The mind–body issue is not so easily dispensed with. If all physical systems are fundamentally reducible to physical patterns, it follows that the psychical aspects of man—mind, emotions, feelings, and so forth—are ultimately a function of biochemical or biological mechanics. The upshot is that we are all robots. Marvin Minsky said that the mind was "a computer made out of meat."[32] Actually, that's not such a bad idea. As computers sink into obsolescence, they can be barbequed.

The body may be an immensely complex organic machine, but its movements are directed and controlled by the central nervous system (in conjunction with other biological systems), which originates, at least geometrically, in the brain. Regardless of the countless quadrillions of molecules operating in the brain, the dynamics of molecules comprise a pattern. That pattern is affected by: (1) sensations received though the senses, (2) data in the brain's memory, and (3) physiological signals generated within the body. Because one configuration perforce leads to the next, in the absence of a discrete psyche that can *originate* a change to patterns, we are all reduced to biological automatons or robots. The supposed compromise is indeterminacy, whereby changes to configurations are not deterministic but a matter of random probability arising from quantum mechanics operating within atoms. Perhaps so, but the logical upshot of that variation is to make us into biological pinball machines. *Indeterminacy is not the same thing as a purposeful decision.*

Figure 4 outlines the four contending perspectives on the mind–body issue.[33] Of these schools, the ultimate battle is arguably between *physical monism*—psyche as a product of physiology—and *interactionalism*—psyche as a discrete phenomenon extensively interacting with its physiological host. Resolution is possible only if the physical sciences achieve closure *and* that closure lacks any fundamental understanding of consciousness, emotions, and so forth. In that case, the interactionalist perspective seems to be the only plausible alternative. In the interim, this book assumes the interactionalist perspective is correct—discrete body and psyche with mutually affecting interactions. Without that assumption, all history and human behavior is entirely deterministic or actuarially deterministic as the case may be. Either a human being can make and implement purposive decisions, or he cannot do so. *There is no middle ground when it comes to the attribute of human initiative.* Still, many human behaviors result from emotional reactions, and as such may be physiologically deterministic.

In support of this assumption, the reader is asked to consider several pragmatic items of evidence. Psychoanalytic training has been opened to professionals with the appropriate doctorate in psychology; the candidates no longer need be psychiatrists (who are *always* MDs).[34] The significance is that this approach to psychotherapy is entirely mental—no physiological middleman.

Perspective	Applications	Consequences
Physical monism. The mind or psyche is a function of biology combined with environmental interactions. Varies from absolute determinism (epiphenomenalism) to statistical indeterminacy.	All psychiatric therapy must be surgical or otherwise operate as if it were a pharmaceutical.	Man is reduced either to a biological robot/automaton, or, if indeterminacy prevails, into a biological pinball machine, with or without discrete emotions.
Interactionalism. The psyche is a discrete, ontological reality. But it can manifest itself only in a physiological host and is severely constrained by physiologically based propensities.	Most psychologists try to merge these two logically incompatible schools of thought. Some behavior is influenced by psychical interaction alone without regard for biology.	Man has the potential to rise above biological determinism but physiologically induced attributes and other matters of biological circuitry usually pose major challenges.
Psychophysical parallelism. This is like interactionalism except that the psyche is independent of biology (except perhaps for the senses and the feeling of pain).	Though this perspective defies the bulk of the evidence, at times the mind can operate independent of biology (except sensory input). This can be seen, for example, in musicians with crippling arthritis who nevertheless transcend this disability completely when playing their chosen instrument.	
Neutral monism. The mind or psyche is the *only* reality. The body and the rest of the physical universe exists only in the mind, whatever that means.	This notion defies all evidence whatsoever, but there is a grain of truth to it, as when an inventor envisions a product before it is assembled, or a musician first composes a work in his mind, or the programmed physiogenesis of nature that foreordains the evolution of the species.	

Physiology developed steadily through various phyla, and organs were well-defined midway in the evolutionary chain. But the psyche arose only at the end of the process. At that, man has incomparable mental prowess vis-a-vis the chimpanzee, with whom he shares about 98 percent of the same anatomy, physiology, and biochemistry.	

4. **The mind–body issue. Four perspectives address relationships between mind and body. The primary conflict is between monism and interactionism, but there are obvious grains of truth to the other two perspectives. And keep in mind that the *full* development of the psyche was a more-or-less last-minute phenomenon on the evolutionary scale of time.**

Then, too, our systems of law clearly recognize the difference between uncontrollable behavior and malicious intent, e.g., manslaughter versus murder. But we don't send a chimpanzee—man's closest biological relative extant—to the chair for methodically

killing and eating a baby chimp.[35] Also, consider the following list
of 25 words, then attempt to make any connection whatsoever
with biological mechanics beyond facial expressions and the like.
You just cannot do it.

care	disappoint	grief	humor	pride
cheerfulness	eloquence	hate	justice	querulous
courage	envy	honor	kindness	resentment
desire	fear	hope	love	respect
dignity	forgive	humility	pity	wonder

Lastly, note that the human spirit seems indestructible at times,
clearly transcending the dull world of physical and biological
law. In a 1915 hanging in Tennessee, the culprit was asked if he
had any last words. He replied, "This will surely teach me a
lesson." Then, too, recall the well-known news photograph of
the brave individual confronting a tank in Tiananmen Square,
Beijing, China. The laws of physical mechanics dictate that he
would have been crushed, but courage can outweigh steel
despite the lack of a mathematical model to prove it.

More poignantly, on the death of her betrothed on Flanders
field in the Great War, the poet Amy Lowell penned the line
"Christ! What are patterns for?" Though many analysts have
before and since cogitated on the patterns of military tactics to
which she referred, none have attempted to assuage this lament
in scientific terms. Too, consider what Louis Pasteur said as his
young daughter lay dying:

> I gave myself up to those feelings of eternity, which come
> naturally at the bedside of a cherished child drawing its last
> breath. At those supreme moments there is something in the
> depths of our souls that tells us the world is more than a
> mere combination of phenomena appropriate for a mechani-
> cal equilibrium wrought from the chaos of nature solely by
> way of the gradual actions of the forces of nature.[36]

Or consider this excerpt from Emerson's most noted address, *The
American Scholar*:

> The scholar of the first age received into him the world
> around; brooded thereon; gave it the new arrangement of

his own mind, and uttered it again. It came into him life; it went out from him truth. It came to him short-lived actions; it went out from him immortal thoughts. It came to him business; it went out from him poetry. It was dead fact; now, it is quick thought. It can stand, and it can go. It now endures, it now flies, it now inspires.[37]

For this too is the mind of nature; that she gives to us the power to re-create if not create. The laws of nature are inviolate, but we have the power, and the volition, to realign patterns and thereby change what would otherwise occur. On the other hand, this does not imply a transcendental, hedonistic free-for-all. Psychological maturity is marked by good judgment, integrity, a sense of justice, kindness (at least up to a point), and many other like attributes. Those attributes do not arise from sloppy logic or waving some kind of magic wand. On the contrary, attainment of maturity—a kind of psychological system—is often a lifelong struggle. Ask any psychiatrist. Enough said.

□   □   □

This concludes the introduction. The intent has been to state the physiogenetic thesis unambiguously, tie it in with the gloss, at least, of accumulated evidence, demonstrate its parent symmetry with both linear and nonlinear scientific disciplines, resolve conflicts with existing doctrines, or, that failing, demonstrate that the conflict is immaterial, and conclude by confronting the mind–body issue head-on.

The next chapter describes the categories of systems, the various internal structure of systems, elemental factors in general, and the attributes of elements in systems that affect interactions. The third chapter classifies and describes the mechanical aspects of systems. The fourth chapter then shoehorns all of this discussion by recapitulation into a gross-level Euclidean-format derivation, drawing on the terms developed in previous chapters without repeating the definitions and descriptions. The reader can study or scan that section; it makes no new points. That chapter also outlines how the book applies the physiogenetic model—often figuratively— to the operation of representative systems from all categories.

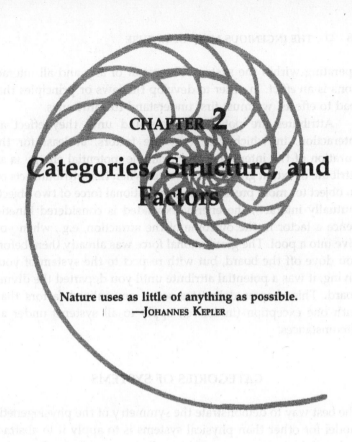

CHAPTER 2
Categories, Structure, and Factors

Nature uses as little of anything as possible.
—JOHANNES KEPLER

This chapter begins with a section on the three categories of systems—abstract, physical, and sociological—and then compares those categories with the three forms of internal control: relational, hierarchical, and integral. These categories and forms are then used consistently throughout the book. The balance of this chapter covers attributes and factors.

By definition, the elements of a system must have interrelationships, and except for a few purely static systems those elements must interact (with or without an external observer). Else, the system will do nothing except exist. *How* those elements interact is a function of their properties or attributes. *If* and *when* they interact depends on the dynamics (or patterns)

27

operating within the system. The result of any and all interactions is an *effect*. In order to develop the laws or principles that lead to effects, we must first understand the attributes.

Attributes are potential unless and until they effect an interaction, in which case they are factors, at least for the duration of the interaction. For example, potential energy is an attribute of extant gravitational force. When that attribute acts on an object (or more precisely, the gravitational force of two objects mutually interact), the energy expended is considered kinetic, hence a factor in the outcome of the attraction, e.g., when you dive into a pool. The gravitational force was already there before you dove off the board, but with respect to the system of your diving, it was a potential attribute until you departed the diving board. This chapter delineates seven generalized factors that, with one exception (initiative), apply to all systems under all circumstances.

CATEGORIES OF SYSTEMS

The best way to demonstrate the symmetry of the physiogenetic model for other than physical systems is to apply it to abstract and sociological systems, making allowance only for figurative variations of crucial terms, and introducing the factor of human initiative. To this end, the book formally categorizes all known systems into the three implied groups: *physical*, *abstract*, and *sociological*.[1]

Physical systems have already been described. Abstract systems—for example, mathematics—differ from physical systems in two major aspects. First, no matter, mass, or material is involved; the elements are ideas, statements of facts, laws, and concepts. Second, physical arrangement is not relevant (geometry and kinematics excepted). Moreover, the queen of the sciences is believed by some observers to exist apart from human existence, because virtually all mathematicians, in all ages, in all languages, will agree if a purported proof of a mathematical theorem is valid or invalid.

In practice, most abstract systems are dependent on the actions and decisions of human players, for example, law, recipes, and games. Still, even when an abstract system superintends a physical system, the abstract relationships can be understood without continuous reference to the physical media on or over which the abstract system is superimposed. Even theoreticians develop theories by intuitively sensing relationships among concepts rather than the physical phenomena that gave rise to those concepts.

Sociological systems range from the human psyche itself, through groups and institutions, coming full circle to social abstractions developed from experience, such as systems of law, ethics, and theology. As implied, the overwhelming difference is that sociological systems include the factor of initiative, or what is sometimes called volition or "free will." This means that each element—a person—in a sociological system can and will at times react unpredictably. Consequently, attempts to impose rigorous order over sociological systems may and frequently do fail. History bears witness to this in the tens of thousands of revolutions, revolts, mutinies, and rebellions. Still, individuals do not always act unpredictably, and when they do, the cumulative effect of this unpredictability sometimes falls into predictable patterns. Accordingly, some sociological systems work as if they were mechanical. Otherwise, the human race would experience continuous anarchy.

INTERNAL STRUCTURE OF SYSTEMS

Systems may also be subdivided in terms of their degree or form of internal control, and this applies to all three categories. These forms are arbitrarily labeled as: *relational*, *hierarchical*, and *integral*. In practice, few systems hew solely to a single shape or form, but instead exhibit gradations or mixtures for the maw of fuzzy set theory. Still, in most cases one form usually predominates in each system. Figure 5 compares the three internal forms of relationships and control with the three categories of systems.

	RELATIONAL	HIERARCHICAL	INTEGRAL
A B S T R A C T	Games and the conduct of war are the most well-known examples of the relational form. Effective control must take crucial elemental attributes (at least) fully into account.	Derivations and systems of logical thought, such as Euclid's *Elements*, provide examples of abstract systems with clearly hierarchical (and indispensable) sinews of control.	Works of literature and art are systems that integrate constituent elements to the point whereby those elements no longer possess their own identity, or cannot use it if they do.
P H Y S I C A L	Ecology is the classic case here. Organisms exercise considerable freedom of operations until the breaking point of the regional ecology is reached. Then all hell can break loose.	Atoms, most machines, and the solar system all require this form. Without hierarchical control, each such system will disintegrate or, as a minimum, cease to function.	Cells and organisms are common examples. Their many functions are so intertwined and interlaced that only an integral form can aptly manage the enveloping system and its growth.
S o c i o l o g i c a l	Democracies exemplify this form. The form is also common in foreign affairs, geopolitics, and, for that matter, many economic perspectives.	Most nonconstitutional monarchies (typically for worse) and legal systems (usually for better) are excellent examples of this form.	This form applies to corporate economics when the bottom line takes precedence over all departmental goals and budgets.
	Education commonly recognizes the uniqueness of each student, the necessity to relate and structure knowledge, and the crucial importance of personal integrity.		

5. **Categories versus forms of systems.** Systems can be categorized as abstract, physical, or sociological. All categories can in turn be divided into relational, hierarchical, or integral forms. This matrix depicts the nine possible combinations.

In the *relational* form of system, the elements are loosely related or linked, which permits the system to achieve its objective while allowing significant independence for constituent elements. Theoretical democracy is an obvious example, and relational forms are also found in games, war, and ecology.

The *hierarchical* form of system starts with the relational form, then strengthens or intensifies the linkage among elements to the point whereby the system objective takes precedence over most elemental goals. Many if not most human organizations depend on this form. Except at the top, every employee of every company or institution has a boss, and every boss has his or her boss. And one of the fastest ways to earn a "pink slip" is to give priority to personal goals over organizational objectives. Similarly, many machines link various subassemblies into a hierarchical form, for example, computers. The central processing unit controls major subsystems (e.g., printers, keyboards), which in turn control their own minutiae.

The *integral* form of system is the hierarchical form carried to the point where the constituent elements no longer retain their individual characteristics or ability to make decisions, or if they do, the effect is negligible with respect to the encompassing system. A molecule is a clear-cut example of the integral form with respect to its constituent atoms. Hydrogen burns fiercely. Oxygen feeds fires. But when combined as water, the molecules quench most fires. Also, a thought is an integral system in which the words that comprise the idea no longer have a separate identity (with respect to that idea). Mark Twain once told a speaker that he had a book containing every word of the speaker's remarks. This allegation made the speaker quite uncomfortable—until Clemens revealed a dictionary.[2]

FACTORS IN GENERAL

Now let us turn to the factors, first by considering this question: Why did the supposedly "unsinkable" *Titanic* in fact sink? We briefly review here the facts of its sinking in terms of the two systems operating in confluence at the time. You are asked to keep this case in mind as we proceed through the description of factors affecting the operation of systems. At the end of the chapter, try to deduce the reason for the failure of the two systems within which the *Titanic* operated. The answer is *not* simple. As an aid, Figure 6 provides a matrix of the factors compared with the three categories of systems.

The parent shipping line that owned the *Titanic* profited by transporting passengers across the Atlantic Ocean *safely*. Safety, in this situation, is primarily a matter of avoiding violent storms and collisions with shoals, other ships, and icebergs. The physical system consisted of the ship, one iceberg, the ocean in the immediate area, and the local weather. This physical system was

	Abstract	Physical	Sociological
Purpose and design	Abstract designs are variants of game theory that impose order on logical chaos.	Most physical systems have clear goals or at least goals that can be readily deduced.	The aim is to achieve goals or impose order beyond the pale of individual choice.
Elements and attributes	The elements are data, knowledge, and rules. Most elemental attributes are typically details.	Elements have static and dynamic attributes. The dynamic attributes usually relate to motion.	The elements are the rules and individuals. Attributes are logical and psychical.
Quantity and variance	Abstract systems tolerate variance and quantity only if the elements can be logically linked.	Organic systems amply demonstrate variance; inorganic systems, much less so.	Quantity ranges from two to a billion, often with exponentially increasing variance.
Initiative	Initiative operates in game choices and in the creation of most abstract systems.	Initiative does not occur in physical systems except in higher forms of animal life.	Initiative almost always works to the detriment of the intent of social systems.
Degrees of freedom	Logical systems tend to eliminate degrees of freedom while game theory depends on it.	Most physical systems eschew degrees of freedom. Ecology and weather are exceptions.	In this category, degrees of freedom strongly correlate with initiative.
Arrangement and configuration	Logical systems arrange theorems. Games arrange players or tokens according to rules.	Arrangement is always geometric and can usually be drawn to scale on ordinary media.	Arrangement tends to be schematic in terms of lines of influence and communication.
Dominant objects and focal points	Axioms dominate in logical systems while players and instruments are the key in games.	Dominance is achieved by mass, by momentum, or by leverage gained from position.	Strong leaders tend to dominate and depend heavily on interior lines of authority.

6. **Factors versus categories of systems.** Although elements in systems can be described by a mere seven aspects, the proportionality and significance among them, especially initiative, differs markedly in different categories of systems.

encompassed by a sociological system consisting of the ship's captain, the crew, and the passengers. Note that the iceberg vastly outweighed the ship, moved at a much slower velocity, and was in the process of gradually melting. The melting is part of the larger physical system of the earth's environment, in this case nature's way of preventing a rerun of the Ice Age—but that is not relevant here.

Also note that the *Titanic* had a double hull divided crossways into 16 watertight compartments (like thick slices of bread), any *four* of which could be flooded without sinking the ship. Late on the night of April 14, 1912, the *Titanic* struck an iceberg, ruptured *one* or *two* of the compartments, and sunk in less than 5 hours. Only 706 of the 2224 people on board were saved, mostly women and children. Lookouts had been posted, and the watch on the bridge of the ship was aware of the iceberg before the collision. Lastly, presume that the ship *could* sustain itself despite four flooded compartments.

PURPOSE AND DESIGN

A system, by definition, consists of elements organized or configured to "stand together" to effect a purpose, or achieve a goal, or equivalent. That purpose may be intended by way of an initial configuration, in which case it equates with *design*. In the case of a serendipitous system—an unintended configuration arising from the operation of other systems that nevertheless proves useful—the configuration at the time the system first proves useful suffices for a design. In either case, when we say that such-and-such system is designed to do so-and-so, we mean that the system's elements or components have been placed into a configuration that leads to so-and-so—its *efficacy*.

Hence, the proper way to evaluate a system is to compare its operation or outcome with its purpose. In London some years ago, a half-filled bus sped past some people waiting to board. When one of the would-be passengers complained to the Ministry of Transportation, the response was that the bus was already late, and that if it had stopped to pick up any more passengers, it would have been even more late arriving at its final destination.[3] The bus line, as a purely physical system, worked fine (as designed by the manufacturing engineer). As part of a sociological system, it obviously failed of purpose.

A design does not ensure that a system will operate as intended. An old adage advises that the road to hell is paved with good intentions. The most notorious case of this was the communist theory developed by Marx and Engels in 1848. The

Communist Manifesto had little impact on the attempted European revolutions of that year, but as modified by Lenin and Mao Zedong earlier in *this* century, the doctrine led to tyrannies visiting death on 90 million people.[4] When Lady Astor visited Stalin in 1931, she asked the dictator, "How long are you going to go on killing people?" Stalin calmly answered, "As long as necessary."[5] The intentions of Marx and Engels may have been good, but their system contributed to a slaughter that exceeded the Holocaust by a factor of 15, that is, the ratio of 90 million deaths ultimately traceable to Marx and Engels to the 6 million of the Holocaust.

In a few cases, systems are presumed to exist that really don't but which are criticized for failing to achieve their supposed purpose. For example, the provision of healthcare in the United States is often regarded as a system. It is not a system in any strict sense of the word because it was never designed to operate as such, nor does it so operate serendipitously. At best, it is a conglomeration of providers, institutions, and insurance plans lacking a unifying objective or goal, at least in terms of user-observers. Elsewhere, a system may continue in operation after its purpose loses validity. For example, war can evolve into a brawl in which the principals no longer have a clear idea of what they are trying to achieve, e.g., Vietnam.[6]

In yet other cases, purpose is a mystery, especially in the human psyche, i.e., what is the meaning of life? Philosophers and psychologists remain puzzled as to the purpose of the conflicting attributes of the psyche. Spinoza lamented this problem, saying that man's impotence "to govern or restrain the affects" he called bondage. "A man who is under their control is not his own master, but is mastered by fortune, in whose power he is, so that he is often forced to follow the worse, although he sees the better before him."[7]

The point is, in the absence of a known purpose, the evaluation of a system is limited to describing its relationships. For example, Rube Goldberg's fictional contraptions always had an obvious purpose, hence his systems lent themselves to evaluation in terms of their ability to achieve that purpose (however comically inefficient they may have been). By contrast, it is not possible to evaluate a configuration of doodles on a piece of paper *as a system*, though those doodles, and their relationships if any, may be described at great length.

ELEMENTS AND THEIR ATTRIBUTES

By definition, all systems have two or more elements in some form: matter, ideas, or people (or a combination of these). Furthermore, and with the exception of a static point in geometry, every element must have one or more attributes, be it simple like velocity or subtle as eloquence in a statesman. Do not underestimate the importance of attributes. Eloquence in a statesman may contribute significantly to a nation's survival, witness Sir Winston Churchill during World War II: "We shall fight on the beaches, we shall fight on the landing grounds, we shall fight in . . . the streets . . . . We shall never surrender."[8]

Nor should the variety of attributes per element be underestimated. A simple spherical game ball has mass, diameter (hence density), variable velocity (when in play), composition, and surface characteristics. These spheres may be hollow or solid, and different spheres are used for basketball, soccer, volleyball, lacrosse, baseball, softball, racquetball, tennis, ping-pong, golf, and billiards. Furthermore, although different manufacturers produce identical balls, the products have slightly different characteristics. Lastly, those characteristics change with time. For example, a tennis ball goes "flat" when it loses its resiliency.

Not all attributes are equally important in a system. Alternatively, they may have different effects at different times. For example, ladders can be made of wood, steel, aluminum, magnesium, or other material, and that material does not make much difference for simple jobs. The choice *does* matter when the ladder is a 40-foot extension model and the user must raise it by himself.

Additionally, elements may be systems in their own right, i.e., *subsystems, subassemblies,* or *subunits.* A corporate computer system may have a payroll subsystem, a customer database subsystem, and an inventory subsystem. An automobile has dozens of subassemblies. Governments have multitudinous agencies and directorates. Subsystems (or equivalents) are indispensable in many systems, especially physiology and genetics. The most obvious example is that *all* complex organisms develop by replicating and linking cells. And complex organisms are further organized into a higher level of subunits that we call organs.

QUANTITY OF ELEMENTS AND VARIANCE OF ATTRIBUTES

Quantity is an inventory of elements in a system on hand at any given moment. *Variance* is the measure of how the attributes of those elements differ from one another, insofar as those differences affect performance of the system. The clearest examples of quantity are dollars in a bank account and players on a sports team. Dollars have no variance; by contrast, team sports require different specialties (in football: quarterback, linebacker, and others).

As much as possible, variance is measured in statistical terms, and those measurements have significance only with respect to variance in related systems (or to an external standard). For example, let us say that the diameter of a widget is 0.500 inch. That attribute tells us something about widgets that stands as a fact. Then let us say that the standard deviation of the diameter in various samples of widgets coming off the assembly line is plus or minus 0.004 inch. That also tells us something, but the reported magnitude immediately raises questions as to significance. Does that variance exceed the allowable tolerance of the machine into which the widget will be fitted?

INITIATIVE

William James liked to tell the story about the man who tried to join his local Determinist Club. When asked why he wanted to join, he said he chose to and was promptly shown the door. He then went across the street to seek acceptance in the League of Free Will, where he was asked the same question. When he replied that he had no other choice, his application there was rejected. Whether the reader prefers to think of making a choice as "free will" or "volition" or "initiative" is immaterial. The important point is that James's anecdote illustrates a key issue in psychology: Are the decisions made by an individual entirely the consequence of stimuli and bio-mechanical brain processing (to include reference to stored information)? Or can the individual exercise at least some degree of initiative? Can he or she choose a new course of action that is *not* foreordained by operation of his or her biochemistry? As discussed in the first chapter, if man lacks the ability to exercise initiative, then we are all machines, and the doctrine of guilt and intent in judicial systems is futile. But if he does possess this ability, then all sociological systems must take initiative into account whereas physical systems (and abstract systems apart from human behavior) need not do so. That is to say, the human element in any sociological system is *inherently* unpredictable.

Virtually everything a person does affects one or more systems or, alternatively, prevents or modifies how the stimuli of external systems affect him or her. It follows that initiative causes the behavior of one or more systems (to include one's own psyche or body) to change from an otherwise deterministic course. But without the ability to exercise initiative, everything that happens is determined by the dynamics of all previous relevant configurations. It does not matter if umpteen quintillion atoms and molecules are involved; the *pattern* of changes to those countless elements is a predictable consequence of previous configurations.

The potential for initiative is not necessarily limited to *Homo sapiens*. On the contrary, we must ask to what extent it occurs in lower-order animals, and, for that matter, whether a machine can be made with "a mind of its own," so to speak (sometimes called a *Frankenstein* or a *golem*). These are hard questions to answer, but to date there is no evidence of bona fide initiative working in any machine and little evidence of it in animals beyond a few rudimentary *choices*. Man's nearest biological relative on the evolutionary scale is the chimpanzee. Chimpanzees have existed on earth for tens of millions of years. Yet their behavior today does not seem to differ significantly from that at any time before, at least when compared to the differences between early cavemen and modern civilization. That is to say, repetitive animal behavior extending over millions of years may be associated with emotions but it hardly exemplifies the exercise of initiative.

DEGREES OF FREEDOM

Degrees of freedom measure the physical, mathematical, or logical independence among elements in a system—the extent or degree to which they remain unlinked and thus "free" to operate regardless of other elements. For example, elements in a physical system may have a logical or taxonomic relationship, but if they are not physically linked or mutually subject to situationally dominant force (e.g., electromagnetism), those elements can move or otherwise operate independent of one another. Any interactions depend entirely on configurational dynamics. By contrast, if the elements are linked, then such linkage sets the configuration and more or less forces them to interact.

In practice, most man-made physical systems have few degrees of freedom (though they typically have controls to modify machine behavior). By contrast, nature's physical systems are marked by many degrees of freedom, especially weather and ecology. Because degrees of freedom correlate with increasing disorder or colloquial chaos, man-made machines evince order while nature (gravity excepted) *appears* to be chaotic. Appears. There is nothing more orderly than the gestation of a fertilized egg cell into a full-grown, complex organism. Hence, we may conclude that order can be evoked from disorder regardless of degrees of freedom, but as those degrees increase, the task grows exponentially more difficult and must rely on an increasing period of trial and error.

Note also that degrees of freedom are distinct from variance. Variance gauges differences among the attributes of elements. Degrees of freedom measures maneuver room, as it were, and, for humans and perhaps other animals, the *opportunity* to exercise initiative. This difference is illustrated by numerous cartoons of prisoners chained to the walls in ancient dungeons planning how to escape. In real life, chain gangs physically link prisoners and prisons restrict movement. These measures *reduce* both degrees of freedom and the *opportunity* to exercise initiative. Paratroopers offer a clear case of the opposite situation. While descending, the troopers have maximum degrees of freedom with respect to military purpose because a unit leader cannot manage or lead them. Then when they land and re-form into units, the degrees of freedom effectively diminish because that leader can effectively exercise initiative by directing their employment as a unit.

In general, then, the implied task of a system maker is to remove or negate degrees of freedom in order to evoke more order out of disorder or colloquial chaos—to make elements work together that would be difficult if not impossible to do in the absence of a system. The task is easiest to understand in physical systems: Assemble the elements into a system by some means of linkage, be it physical or by fields of force. In abstract systems, logical relationships must be identified or developed. But in sociological systems, the task runs headlong into the attribute of initiative. Where excess initiative is potentially detrimental, the key is to configure the system so that its constituents engage each other constructively, e.g., the checks and balances written into the U.S. Constitution.

7. Initiative versus degrees of freedom. Massed paratroopers in the air illustrate maximum degrees of freedom without significant initiative. Initiative is regained immediately on landing, but vulnerability is high until the troopers link up into coherent units.

ARRANGEMENT AND CONFIGURATION

To review, arrangement in physical systems is the geometric relationship of elements, and configuration is arrangement combined with the attributes of those elements. Hence, arrangement is visually obvious in physical systems. In abstract and sociological systems, arrangement can be visualized graphically by symbolically plotting the relationships, e.g., a sociogram or organization chart. But whether (and how) two or more elements in a physical system will interact depends on their respective position and attributes. A baseball hit "over the fence" with a clearance of 1 millimeter results in a home run. One millimeter less, and the team in the field has a chance to make an out.

In abstract and sociological systems, logical relationships usually replace the geometric arrangement common to physical systems. That is to say, physical distance and configuration are not relevant. For example, two individuals can converse over a telephone line across the street or halfway around the globe without being conscious of the difference. Similarly, though data in a network computer system may be distributed worldwide, software and communications can link all of it as if it were stored in a nearby filing cabinet.

The significance, efficiency, or synergism (advantages gained from a system impossible to achieve by the elements operating independently) of various configurations is another matter. For example, poetry is composed of words and music is composed of notes, both of which are arranged in various sequences that can be recorded on paper objectively. Yet the effect of poetry and music on the psyche is in part subjective.

Another point. Arrangement subsumes *density*. Density is the ratio of elements to the space they occupy and it has two variants: (1) the ratio of elements per unit of space and (2) how those elements are differentially configured within that space, which is usually called *concentration*. A common example of a simple ratio is population density: the number of people living in a stated geographical area or region. An example of concentration is the population density of cities compared to rural areas within a state or region. Concentration is very important in war and games, because success often arises not so much from numerical density as from how that density is congregated and maneuvered. Much more on this in Chapter 9.

DOMINANT ELEMENTS AND FOCAL POINTS

If a professional football team were to play a high school counterpart, the outcome would not be in doubt. The force of a professional team would significantly "outweigh" the high schoolers. As such, the pros would retain a dominant status and win every time. By contrast, in a game between two evenly matched professional teams, dominance alternates between the teams based on alternating "possession." The rules of football have been intentionally written to give the team with possession more advantages than the defending team, for example, penalties for sacking the quarterback and pass interference.

In sociological systems, psychical force must also be considered, for example, Tiananmen Square, where a lone rebel confronted a column of tanks (1989). In this transitory sociological system, the tanks were intended to suppress a rebellion by threatening to kill the participants. The man on foot obviously did not have the physical force to halt the tanks, but his courage had the same effect on the driver of the lead tank.

Dominance may be temporary in any system. This was the fate of dinosaurs and, to this day, the rule applies equally to entertainers and nations. In the 19th century, the sun did not set on the British Empire because it spanned the globe. That hegemony no longer prevails, though the English language now dominates international communications, and English jurisprudence may yet prove to be the ultimate contribution of Western civilization.

Dominance may also be situational. That is, an element may exercise extraordinary influence by virtue of its position with respect to other elements in a system, such as an incumbent in a position of authority. Emperor Peter III of Russia (1762) was incompetent, a profligate, feeble (if not half insane), yet he wielded considerable authority to the detriment of his country. But when his estranged wife engineered a coup that forced him to abdicate the throne, then imprisoned (and later, by proxy, murdered) him, placing herself on the throne, Peter III remained the man he was but no longer exerted authority.[9] The queen, a remarkable woman who became known as Catherine the Great, also exemplifies the idea of focal point. If she hadn't been Peter's wife, she would not have gained the throne.

For an interesting fictional example of combined temporary and situational dominance, consider the role of the butler in *The Admirable Crichton*. A wealthy English family is stranded on a desert island. Finding themselves unable to cope with nature, Crichton, their butler, takes over as head of the family but returns to his servitude after they are rescued.

Physical and psychical dominance has a geometric equivalent in *focal points*, a term that comes from optics. In a camera the focal point is situated somewhere within the lens. The paths of light reflected or radiated from the object in view converge at this point before expanding again to form an image on the film (called the *focal plane*). Elsewhere, focal points go by different names, especially *center of gravity*. In a physical center of gravity, the entire weight of an object may be considered to be concentrated at one point. In theory, any object can be balanced when perched on a rod at this point.

Except in abstract mathematical systems, however, a focal point cannot exert dominance other than in partnership with some element. Unless an element or external system exists to take advantage of that position, the potential dominance remains moot. Referring to mechanical advantage obtained by leverage, Archimedes said, "Give me somewhere to stand, and I will move the earth." On a lesser scale, the proverbial little old lady in tennis shoes can lift a 20-ton block using a very long lever or, better yet, a differential chain hoist.

For an example in a sociological system, consider the convention that wrote the U.S. Constitution during the summer of 1787. Even a cursory review of the proceedings clearly demonstrates that several key issues were the focal points of power— the division of power between the two houses, the powers of the President, and the franchise.

Another aspect of dominance arises from skewed distribution of resources and attributes in many systems. A comparative handful of elements possess superior ability or capability to the point where those elements dominate the system. The significance is that while many political systems attempt to impose an equality among citizens, a few individuals are inherently dominant. Dominance equates with power, and power corrupts.

In the physical universe, a mere two of the 92 naturally occurring elements account for 75 percent of terrestrial weight, namely, oxygen and silicon, and it takes only eight elements to exceed the 98th percentile of that weight.[10]

8. Constitutional Convention of 1787. Catherine Bowen hailed this event as a miracle, but in reality it was a confluence of great minds confronting an immediate problem, guided by an extraordinary man. Given such factors, a sociological system can produce profound results.

Further, a mere six countries (of more than 200) control roughly half of the world's land mass. Fewer than 10 countries control about 90 percent of available natural resources that aren't consumed by mere survival.[11] Five countries claim nearly half the world's population, and 20 account for about 90 percent of it. In the United States, eight states have more than half of the country's population, which means that, in theory, eight states could control the House of Representatives. As a corollary, the 26 states with the lowest population (about 10 percent of the total U.S. population) could control the Senate.

Elsewhere, fewer than 10 countries generate 90 percent of the world's gross domestic product. That distribution becomes even more skewed when the numbers are factored to reflect only wealth above the subsistence level. Then, too, about 5 percent of corporations in the United States account for 85 percent of the profits (and an even higher percentage of corporate income taxes). Eighteen percent of individuals, and family units filing jointly, foot 74 percent of the bill.[12]

□ □ □

Now let us return to the sinking-of-the-*Titanic* case. The design of the ship was an obviously important factor because of its presumed "unsinkability," perhaps diluting the captain's sensitivity to risks. The key elements and attributes were the iceberg and its momentum, the ship and both its velocity and its engineered ability to withstand collision, the captain and his professional competence and judgment, and the weather and its uncertainties. Neither quantity nor variance among elements is significant here, except possibly in the sense of how the *Titanic* was presumed to differ from previous ships. Initiative resided almost exclusively with the captain, for under the circumstances only he had the authority to change the ship's direction or speed. Hence, there were no psychological degrees of freedom— the iceberg had no mind of its own. The physical degrees of freedom were two: the ship and iceberg. The key configuration was the distance of the iceberg from the ship when it was first spotted. And on balance, the ship was arguably the dominant object because it retained the initiative to safely avoid or, that failing, to minimize the damage of a collision. When that initiative was lost or misspent, the iceberg defaulted into dominance and won the argument.

The technical explanation is that while the impact of the iceberg ruptured only one or two of the watertight compartments, the force of that impact reverberated throughout the ship, causing internal ruptures between bulkheads (walls) and the hull, which permitted water entering into the iceberg-ruptured compartment to flow internally into a total of at least five compartments.

The sociological issue focuses on the responsibility of the captain to avoid a fatal collision. As any collision with an iceberg is potentially fatal, he was obliged to reduce the ship's velocity in foul, foggy weather to minimize the damage from any impact, and preferably to avoid it altogether by steering clear of the iceberg. Note that there is an optimum speed here, which reduces the force of any impact yet retains enough momentum to permit the ship to be steered clear if that more favorable option is possible. Lastly, the ship's captain likely grew overconfident about the ship's supposed "unsinkability" and in all probability did not realize the semi-chain-reaction potential of internal rupture. We will never know what his thinking was, of course, but obviously it failed and that failure cost 1518 people their lives.

Now let us proceed to decipher how attributes and factors emanating from various elements lead to the mechanical or operational aspects of their respective encompassing systems. The verb used here—*decipher*—was chosen advisedly because the dynamics of changing configurations—patterns—are in fact a program written in the only code available to nature.

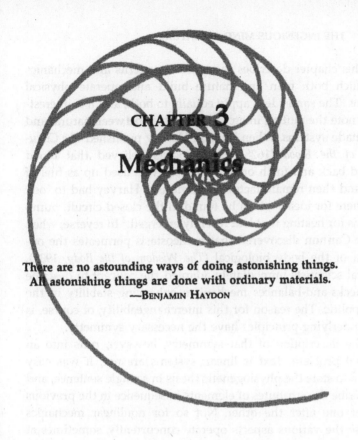

CHAPTER 3
Mechanics

There are no astounding ways of doing astonishing things.
All astonishing things are done with ordinary materials.
—BENJAMIN HAYDON

In 1649, Queen Christina persuaded the aging Descartes to move to Sweden in order to become her tutor in science. Once there, Descartes harped on his major theme that all natural systems, including animals, reduced to mechanical, clockwork principles. Christina expressed her doubt by noting that she had never seen her watch give birth to little baby watches.[1] In point of fact, though, most watches today are manufactured by programmed, mechanical assembly lines controlled by clocks. Nature, if the physiogenetic thesis is valid, uses similar processes in working her assembly line. The only difference is that it takes more time—by a factor of billions—when programming must rely on unlinked elements in space.

47

This chapter describes the various patterns and mechanics by which both man and nature build and operate physical systems. The same ideas apply equally to both, and it is interesting to note the mutual interchange of ideas between natural and man-made systems. When William Harvey published *The Circulation of the Blood* (1628), physiologists believed that blood sloshed back and forth or, alternatively, was used up as fuel of sorts and then remanufactured. Obviously, Harvey had to look elsewhere for ideas, which he found in the closed-circuit, pump systems for heating that had recently emerged.[2] In reverse, when Walter Cannon discovered that homeostasis permeates the operation of the body biological (*The Wisdom of the Body*, 1932), political scientists immediately adapted the concept to describe the checks-and-balances mechanism enhancing stability in the body politic. The reason for this interchangeability, of course, is that underlying principles have the necessary symmetry.

The description of that symmetry, however, runs into an editorial problem. Text is linear; systems are not. It was easy enough to state the physiogenetic thesis in a single sentence, and to describe the attributes of elements in sequence in the previous chapter one after the other. Not so for nonlinear mechanics because the various aspects operate concurrently, sometimes at cross-purposes. Operational facet "A" helps explain operational facet "B," and vice versa. The sequence chosen was based on the fewest cross-references. Also, bear in mind that much of the terminology is arbitrarily defined to avoid confusion with conflicting usage among various scientific disciplines.

STABILIZATION

Loose elements in space are inherently unstable and chaotic. Systems can be formed from such elements, either by imposing direct linkage or by confining and arranging them in such a way that they continue to operate in tandem. For the latter, the ensuing pattern is defined as *equilibrium*. However, some of the elements involved may grow out of synch, go astray, or exert untoward dominance.

The operation of a supplementary control or regulatory system superimposed over, or inherent as a subsystem within, the main system to throttle wayward elements is defined as *homeostasis*. Lastly, if the controlling system or subsystem is itself programmed to improve its own performance, that programming and its effects are defined as *cybernetics*. Information fed into these controlling systems, be they homeostatic or cybernetic, is defined as *feedback*. Corrective instructions transmitted to a system out of whack is defined as an *adjustment*. And any dynamic pattern created by the elements in equilibrium is defined as a *cycle*.

A simple system that illustrates equilibrium consists of two springs mounted on a rod, the length of which is less than the combined length of the two springs before they are compressed, as shown in Figure 9. To fit them on that rod, you must compress at least one of the springs. The compressed spring will immediately decompress, and the force of its momentum will compress the other spring. The decompressing spring decelerates because of the decreasing ratio of its force with respect to the increasing tension of the compressing spring. At a certain point the velocity of the decompressing spring reaches zero, whereupon the compressed spring begins to accelerate in the opposite direction, then decelerates, then stops, and so on, causing the process to operate cyclically. Incidentally, the leading or dominant force in this model is whichever spring is decompressing; hence, homeostasis may also be thought of as keeping dominant element(s) or force(s) in check.

Friction—primarily internal within the springs—will slowly tamp this equilibrium to the point where the two springs equally compress sans momentum and therefore stop, which is called static equilibrium. In order to keep the equilibrium dynamic, it is necessary to supply additional energy to the springs—sort of a push during each cycle to ensure that at least one of the springs compresses to the same length as on the first cycle. This is homeostasis.

	CONCEPTUAL MECHANICS	OPERATING CHARACTERISTICS
EQUILIBRIUM	Self-sustaining equilibrium	*Prerequisite configuration.* The springs must have nearly equal length and elasticity in order to exercise give-and-take equilibrium. *Breakdown.* This simple equilibrium winds down because the available energy is used up by friction. *Application.* Many atoms and/or molecules in a chemical reaction continue to alternate between former and new configurations because energy transfers among them. This is called stabilized equilibrium.
HOMEOSTASIS	FEEDBACK CONTROL DEVICE TO MAINTAIN EQUILIBRIUM RATE.	*Prerequisite configuration.* The control device must be able to "resupply" the energy lost to friction *and/or* stop (or divert) any excess energy gained by either spring. *Breakdown.* The controlled stabilization will break down whenever the springs gain more energy than the control can thwart, or, alternatively, lose more than it can resupply. *Application.* Complex organisms depend on sympathetic/parasympathetic controls to regulate physiological equilibrium, but severe trauma or excesses overtax those controls.
CYBERNETICS	DEVICE IS PROGRAMMED TO MODIFY ITSELF AFTER EACH USE.	*Prerequisite configuration.* The control device senses unusual increases/decreases to the give-and-take of the springs, then reprograms itself to react more efficiently the next time. *Breakdown.* The device's complexity itself is prone to malfunction and/or a condition may arise that causes the device to reprogram itself to self-destruct or destroy the system. *Application.* The organic autoimmune system can overreact and begin to destroy itself or a related biochemical mechanism, mistaking it (or its operation) for an invading antigen.

9. **Equilibrium, homeostasis, and cybernetics.** Equilibrium, or self-sustainment, results from the give-and-take among elements of a system. When the give-and-take fails, supplementary mechanisms—homeostasis—can be activated to restore balance. Cybernetics programs these homeostatic mechanisms to improve performance based on experience.

Now let us say that a human operator replaces the springs in this system with other springs of different temper. This exchange might exceed the effective operating limits of the homeostatic control mechanism and thus cause the system to stop or to operate erratically. The solutions are: build a more versatile control or, alteratively, program the control to change its operating characteristics based on failures. The second solution is cybernetic, and in this case is extraneous. If you know that differently tempered springs will be inserted, why bother to wait for failure? But in very complex, long-term-use systems, the range of potential problems can be so wide and unpredictable that it is simpler to employ a control device that learns from its experience. This is what happens in the body's autoimmune system, though the sophistication of its cybernetic programming is moot if the first experience combating a specific antigen proves fatal. We may care about individuals; nature does not.

To put the case another way, cybernetics substitutes internal programming for external control, and the underlying process is the same for inorganic and organic phenomena alike. Furthermore, cybernetic programming has two distinct levels. The first level is feedback that causes a system to react to stimuli according to fixed rules. As such, the system will react the same way each time the stimulus occurs (which, semantically speaking, also makes it a special case of homeostasis). A speed governor uses feedback to control speed. Increases to speed mechanically reduce the flow of fuel, cutting it off altogether at a maximum allowable speed. The second level consists of programming that modifies internal controls in reaction to experience, for example, if the speed governor were programmed to change the maximum allowable speed as a consequence of various patterns of usage.

CONTROLLABILITY, SUSTAINABILITY, AND RESILIENCY

The compressed-springs model provides a good opportunity to define some related terms. *Controllability* means the extent to which external influence or internal feedback mechanisms control the operation of a system. The purpose may be to prolong the life of the system, but more likely it lets the system perform specific work that cannot be achieved without this control.

Controllability is not a major issue with simple machines and, furthermore, the objective of some systems is to control an external system, e.g., a thermostat. Controllability *is* a major issue for sociological systems and, for that matter, abstract systems, especially computer setups and war. The more degrees of freedom, the more that control is needed to prevent disorder chaos.

The extent to which a system resists degradation—its ability to resist or postpone increasing entropy—is the measure of its *sustainability*. The system perseveres. This perseverance may be achieved by periodic external maintenance, by refueling in some form, by internal mechanics, or by sheer mass. Regardless of method, the design of the system cannot let too many of its constituent elements get out of control, leave the premises, or coalesce into constructs that hamper the operation. The ability of a system to achieve this control without collapsing or coagulating into a useless lump constitutes its *resiliency*.

Restated, resiliency is the ability of a system to parry actions (intentional or natural) directed at its demise. Colloquially, this ability is known as "give-and-take" and "rolling with the punches." Formal adjectival synonyms include: *compensatory*, *counterbalancing*, and *offsetting*. The system stretches, like a rubber band, at least up to a breaking point.

The corollary to sustainability and resiliency is that most systems resist change. This is desirable in physical apparatus but a detriment in most social institutions, hence the pejorative connotation of the word *bureaucracy*. Bureaucracies are resilient in the sense that they thwart or parry actions that threaten their existence, but few of them are flexible. Perhaps the best illustration of this inertia may be gleaned from a talk by General Creighton Abrams at the U.S. Army War College in 1974. General Abrams, as Chief of Staff of the Army, had been trying for years to make the reserve forces a more integral part of the total force structure and was at the moment recounting the difficulties of his task. His biographer, Lewis Sorley, described the scene (noting that the general was dying from cancer):

"Gentlemen, do you know what it takes to move the US Army"—and here he put his hand up like a clock's hand at noon—"just one degree?"—and he moved his hand ever so slightly off the vertical. Then silently, he put his head down and shook it.[3]

There are two approaches to changing or altering the behavior of systems. Colloquially, the first approach may be called brute force—the imposing of a massive, counteracting force (which may result in destruction rather than alteration). The worst case of this approach in recent times occurred during the Vietnam conflict: "We must destroy such and such village in order to save it." By contrast, the second approach mandates that the system be designed in such a way that permits a signal to activate a major change. This approach is clearly evident in organisms, whereby minute biochemical signals from within the body signal the brain to initiate major corrective actions, sometimes to excess as in the case of allergies. Also, compare the two approaches used to stop an 18-wheeler barreling downhill at 70 miles per hour. The design approach relies on a pedal or lever that actuates the brake system. If the brakes fail, the driver must rely on gravity and tons of stone piled on an incline, i.e., a runaway-truck ramp on a mountain road.

Few systems beyond simple mechanical apparatuses facilitate change. It is difficult to steer institutions away from disaster or, for that matter, to make diamonds from amorphous carbon. Instead, most configurations tend to defy modification short of destruction. The result is that complex phenomena rarely respond in direct proportion to the force applied to change them. On the contrary, it usually takes a concerted effort or a strategic approach, whereby force is applied to *focal points*. Focal points are points in a system that cause the system to react more strongly to a stimulus than if that same stimulus were applied elsewhere, e.g., acupuncture. And sometimes the only way to change a system is to separate the elements and start over (though there is no guarantee the new configuration will work). Compare the successful American Revolution with the contemporaneous failure of the French Revolution.

TURNING POINTS AND POWER CURVES

Efficiency is the means by which maximum output or production is obtained from the least possible input or effort—colloquially, "to get the mostest out of the leastest." Ideally, minor increases to this input should generate a handsome return on investment. This is called a *power curve*. When an organization sputters, or a political campaign falters, leaders often say that they are "behind the power curve" (a variation of "behind the eight ball"). When the opposite is true, they may say they are "riding the power curve."

You can easily graph a power curve, for example, the range of an aircraft with respect to its fuel load, as shown in Figure 10. First, the aircraft takes considerable fuel just to take off. Second, it requires an exponentially increasing amount of fuel to carry an additional *usable* supply until that supply is needed. Third, at some point any further loading reduces the range, culminating in the inability of the plane even to get off the ground.

The segment of the curve representing a significant increase in range as a function of increased fuel load is the *power curve* (or *power curve zone*) and begins at the aptly named *upswing point*. Within this zone, the aircraft's range extends significantly with small increases to fuel load (at least compared to the amount of fuel required to get off the ground). The leveling of the slope represents the *point of diminishing returns*. This means that further increases to fuel load will not significantly extend the aircraft's range, because most of that additional fuel is expended in carrying itself to the point where it can be used. Lastly, the turn of the curve downwards represents the *point of negative returns*. At this point, any further increase to fuel load *decreases* the range of the aircraft (because more fuel is used lifting and transporting that additional fuel than it adds to the flying range). Points of upswing, diminishing returns, and negative returns are called *turning points*, because they all represent a point in time or operation that signals a major change to efficiency, up or down.

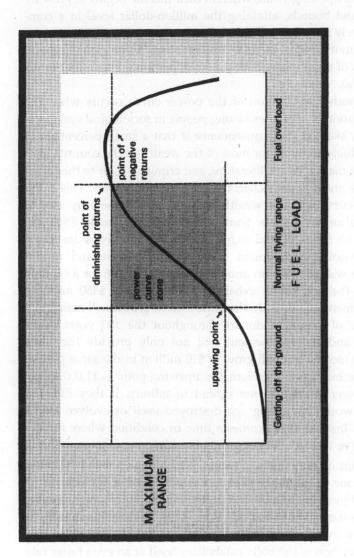

10. Power curves and turning points. Some systems generate significant output or influence as a result of marginal input or change, at least over a range. This includes the range of an aircraft as a function of its fuel load.

Elsewhere, consider that self-made millionaires often struggle for years on little money and long hours. Then they reach an upswing point, wherein their income begins to grow by leaps and bounds, attaining the million-dollar level in a comparatively short time. Granted, becoming a millionaire may seem uncommon, but in fact the United States alone claims several million of them (though only a small percentage earn that much *annually*).

Another application of the power curve occurs when the distribution of attributes among people in sociological systems is heavily skewed. The consequence is that a small percentage of individuals accounts for most of the wealth in the country, and, for that matter, its art, literature, and criminals. Left to their own devices, the rich get richer; the poor become poorer. This is the power-curve effect of wealth. The old joke is the easy way to make a million dollars: Start with $900,000. To make $900,000, start with $800,000, and so on. In a similar vein, consider that a million-dollar endowment at 8 percent interest and capital growth will generate an annual income of $80,036 for a century before the account is exhausted. Add a mere $450 and the endowment will provide the same income plus it will maintain a balance of a million dollars throughout the 100 years.[4] Add $4500, and the endowment will not only provide the same annual income but will grow to $10 million in the same period. In these examples, therefore, the upswing point is $1,000,450.

Power curves do not extend to infinity. If they did, the world would have long ago destroyed itself or evolved into a utopia. Instead, there comes a time or condition where further change or increase no longer yields a high rate of return. This is the point of diminishing returns, which is sometimes followed by a zone of negative returns whereby further investment or expenditure *reduces* total output. Examples of this process extend even to starvation. A starving person can improve his chances of survival by avoiding protein-rich food. Proteins intensify metabolism; hence, the body catabolizes itself at an even faster rate than would occur by fasting. In practice, this means that more people would have survived the famed Donner Pass tragedy (1846–1847) if they had avoided cannibalization.[5]

CRUCIAL CONFIGURATIONS AND THE REVERSAL OF ENTROPY

The power curve measures efficiency, but the implication is that no system can be 100 percent efficient. Losses related to friction or whatever will eventually cause each system to break down or otherwise stop running—entropy. This is easy to visualize. Drop a handful of marbles into a circle drawn on a perfectly level surface. More of them will move outward than inward because at the circumference of the circle the percentage of possible vectors "escaping" the circle is higher that the percentage facing inward. Alternatively, interactions among elements may cause them to coalesce into detrimental constructs: pipes and arteries become clogged; bureaucracies grow ever more sluggish.

The reversal of this fate, as briefly discussed in the first chapter, is the creation of a pattern that causes the system to reach beyond the circle, so to speak, and fetch what it needs without loss of what it already has. The configuration at the instant a pattern first does so is defined as a *crucial configuration*. The consequence of attaining a crucial configuration is herein defined as *catastrophe*, be it constructive or destructive.[6]

An offspring of the science of chaos describes crucial configurations as *complexity*.[7] This concept posits that autonomic growth from simplicity to complexity occurs only on the cusp between order and disorder. Highly ordered systems—bureaucracies for example—are too self-encumbered to tolerate sophisticated growth or much human initiative. By contrast, most disorderly situations—anarchy for example—rarely evolve into order, much less into growth. Hence, it is only on the cusp between order and disorder that a configuration of elements has just the right mix of control and energy to evoke growth in some form. Orson Welles put this phenomenon into colorful terms with his celebrated observation that three decades of war and bloodshed under the Borgias in Italy also produced Leonardo da Vinci, Michelangelo, and the Renaissance; whereas more than a half-millennium of peace, law, and order in Switzerland yielded the cuckoo clock.[8]

Crucial configurations are rare. Less than 1 percent of organic molecules evolve into microorganisms.[9] And it is equally rare for a nation to create a new system of government that continues intact after more than two centuries. The Constitutional Convention of 1787 in the United States was exceptional. Rare or not, we can graph autonomic growth and catastrophe, as depicted in Figure 11. Because it transcends efficiency by virtue of its own fetching, the power curve turns vertical. The point at which a crucial configuration occurs is herein called a *critical point*. It is a generalization of *critical mass*—the mass of radioactive material necessary to trigger a chain-reaction release of nuclear energy.

To be sure, the reader may find it difficult to place destructive catastrophes and constructive growth in the same basket. Yet both processes reverse entropy, and the same phenomenon can be simultaneously constructive and destructive. For example, we commonly view the morbidity of infectious diseases as destructive, and indeed they are. But from a larger perspective, the deaths of bearers prevent the symptoms from triggering an epidemic that would eliminate mankind. As a worst-case scenario, the Black Death is commonly thought to have killed a third of Europe's population. Yet if the victims had lived longer than they did, the disease might very well have claimed virtually the entire population.

However, constructive catastrophes—reversals of entropy— *tend* to differ from destructive counterparts in terms of speed. Destructive catastrophes often release stored or potential energy in great amounts in seconds, be they chemical or nuclear. In nuclear reactions, the release converts mass into energy, yet once past that fine point of distinction the effect is the same. Moreover, the volcanic eruption on Krakatoa in August 1883 is said to have dispersed more chemical energy than convertible by the

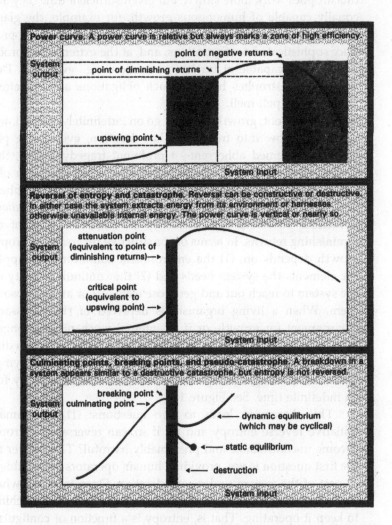

Power curve. A power curve is relative but always marks a zone of high efficiency.

System output

point of negative returns

point of diminishing returns

upswing point

System input

Reversal of entropy and catastrophe. Reversal can be constructive or destructive. In either case the system extracts energy from its environment or harnesses otherwise unavailable internal energy. The power curve is vertical or nearly so.

System output

attenuation point (equivalent to point of diminishing returns)→

critical point (equivalent to upswing point)→

System input

Culminating points, breaking points, and pseudo-catastrophe. A breakdown in a system appears similar to a destructive catastrophe, but entropy is not reversed.

System output

breaking point
culminating point→

dynamic equilibrium (which may be cyclical)

static equilibrium

destruction

System input

11. **Power curves and catastrophes.** When a crucial configuration reverses entropy, the power curve goes vertical because no other input is necessary to maintain the reaction. The system is programmed to get what it needs on its own steam. But true catastrophes differ from breakdowns, which can be thought of as pseudocatastrophes.

most powerful thermonuclear bomb in existence.[10] Constructive catastrophes work more slowly, but given sufficient time they are equally capable of humongous growth, for example, the giant sequoia tree. The reason is that constructive catastrophes produce sophisticated configurations, and, at the extreme, the product must be capable of further re-creation, e.g., evolution. Destructive catastrophes have no such obligations and therefore may proceed pell-mell.

In any event, growth does not go on indefinitely. If it did, we would all grow into freaks. In that situation, even apple pie would be deemed abhorrent—a profound tragedy. Hence, the vertical power curve of a catastrophe must eventually taper off, analogous to a point of diminishing returns. For catastrophes, however, this turning point is called the *point of attenuation*. Epidemics attenuate or abate; they do not reach a point of diminishing returns. In terms of constructive reversal of entropy, growth depends on: (1) the environment continuing to supply the elements the system needs and (2) the continuing ability of the system to reach out and get those elements or at least absorb them. When a living organism is deprived of the necessary environment for growth, or if its internal mechanics no longer function, the organism dies. In time, of course, all organisms die. However, most of them replicate themselves before dying; hence, from a larger perspective nature can reverse entropy for an indefinite time. See Figure 12.

This discussion leads to two questions: (1) can human initiative reverse entropy and (2) if so, can reversal of entropy become uncontrollable and presumably harmful? The answer to the first question is yes, providing human operators are included as part of the system under consideration. Operators take what is needed from beyond the machine and add it to that machine to keep it operating. That is, entropy is a function of configuration, and most configurations devolve from order to disorder. The exercise of human initiative restores order by way of resupply or reconfiguring. Every reader of this book eats several times a day (resupply) and *may* clean off his or her desktop occasionally (reconfiguring).

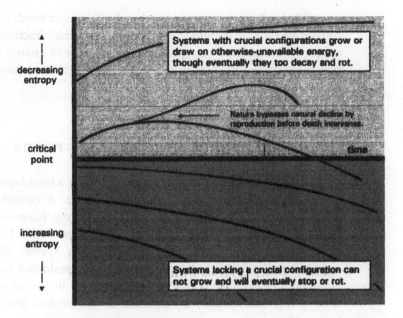

12. Entropy reversal. Systems decay except in those few situations where a crucial configuration causes the system to expand or otherwise improve by virtue of its ability to take what it needs from its environment. No single organism can reverse entropy permanently, but it can spawn new organisms to carry on the process indefinitely.

The second question is profound, but again the answer is yes. Nature often reverses entropy in a destructive, uncontrollable catastrophe, at least until the dynamics attenuate or burn themselves out. A metastasized cancer is usually inoperable and therefore fatal. On a larger scale, an epidemic may rage out of control for years. The bubonic plague in the 14th century claimed roughly a third of Europe's population. As for sociological systems, recall the Third Reich and World War II. The cost of bringing that political malignancy to an end was 60 million lives.

But can man engineer a machine or system that escapes control? Yes. An example is the potential for creating a destructive strain of a virus in a biogenetic experiment. As for a mechanical Frankenstein, consider a thermonuclear device that is used to ignite some sort of superbomb. This superbomb would convert its *entire* mass into energy, not just the small fraction that occurs with current weapons. A sudden release of energy with that magnitude could literally fracture the earth and ruin the entire day.

BREAKDOWNS AND CULMINATING POINTS

A "mild" catastrophe is sometimes referred to as a *breakdown*, but technically speaking the two are poles apart. A catastrophe either triggers autonomous growth or, alternately, releases otherwise unavailable energy, both without further external intervention. In a breakdown, neither occurs. The system simply stops operating; hence, a breakdown may be considered to be a *pseudocatastrophe*. The graph of a breakdown, in juxtaposition with actual catastrophe, is depicted in the bottom frame of Figure 11.

In most cases, a breakdown is preceded by a *culminating point* (or point of *overstretch*). A culminating point means the system is headed for breakdown, perhaps irretrievably so. That is to say, the system may continue to operate but it is doomed. Hence, a *physical* turning point is not the same thing as a culminating point. For example, consider an object thrown up into the air. The culminating point occurs when the object is first thrown because it started to decelerate immediately, *not* when it stopped and began its descent. Had the object carried its own propellent, it would have used that fuel to overcome gravity, and therefore the culminating point would not have been reached until the fuel ran out. If the fuel was sufficient to escape the earth's gravitational pull, then the culminating point would have never been reached. Pam Shriver once made all this clear at a quarterfinal tennis match at Wimbledon. After being thoroughly trounced, she was asked to identify the turning point of the match. Her reply was that it occurred when she walked out on the court.

It is difficult to find a better description of the dynamics of passing a culminating point than Bruce Catton's analysis of the Confederacy's fatal weakness during the Civil War:

> The problems were interlocking. One thing that made it impossible to enlarge the armies was the number of men exempt and on detail; one thing that made it impossible to feed and clothe the armies properly was the fact that not enough men were exempt and on detail, so that railroads and wagon lines and processing plants and mills operated badly; and one thing that caused desertions was the fact that the armies were poorly fed and clothed.[11]

In summary here, a culminating point marks the configuration of a system that is on the verge of a fatal mistake. The system is overcommitted and therefore *subject* to eventual loss if not collapse. A small effort by an external force will likely cause the system to break, like a balloon one breath away from bursting. Culminating and breaking points underwrite the maxim "death is nature's way of telling you to slow down."

CYCLES AND PERIODICITY

Dynamic equilibrium almost always evinces a cycle, and this cycle is often harnessed for useful purposes, for example, oscillation in electronic circuits and vibrations in crystals, because of which quartz has become the standard controller for timepieces. In other cases, dynamic equilibrium can *increase* in intensity, often causing the system to reach its breaking point. The collapse of the Tacoma Narrows Bridge in Puget Sound, Washington, in 1940 is a good illustration of this overstretch (though wind action rather than excess live loading was responsible for the collapse). Similarly, when military troops march across a bridge, they are given a route-step command. This prevents harmonic vibration (a form of equilibrium) from weakening if not destroying the bridge.

Cycles may be periodic or aperiodic. *Periodic* means that the cycle is linear and hence the system behavior is predictable by way of comparatively simple equations. *Aperiodicity* means the ability of a system to cycle through a seemingly random pattern of configurations, and then repeat those patterns, sort of. Admittedly, this is a sloppy definition, but it is clearly evident in the reproduction of any higher-order species. Moreover, no two people are exactly alike, not even identical twins, but they all cycle through fertilized egg, fetus, baby, child, teenager, adult, senior, and death, though death for some occurs earlier in the cycle.

Aperiodicity is always a matter of degree, and no agreement on when a formerly periodic process becomes aperiodic is possible. Nor is there agreement on when aperiodicity devolves into sheer disorder. Look at the desktop of any married writer. He may see it as a model of order, while his wife recoils in horror from the obvious disaster. Perhaps the easiest way to visualize aperiodicity is to compare an ordinary billiard table with one that has curved railings, especially when the curve itself changes shape with time. This is depicted in Figure 13. If the table is square, the trace of the movement of the billiard ball repeats itself. On a rectangular table, the trace generates a series of parallelograms that eventually repeat themselves, hence this constitutes a highly controlled or limited aperiodicity. If the railings are curved, the resulting trace may or may not exhibit aperiodicity. If it does, the repeating pattern will be intricate and, at that, never repeats itself in exactly the same way.

METAMORPHOSIS AMONG FORMS OF SYSTEMS

As described in the previous chapter, systems take one of three primary forms: relational, hierarchical, or integral (despite overlapping features in specific systems). The relational form retains the identity and purpose of its elements while linking or relating them to achieve a necessary purpose, e.g., a democracy. The hierarchical form intensifies that linkage to the point whereby the elements are subordinated to the overarching system, i.e., most corporations and their organization charts. The integral

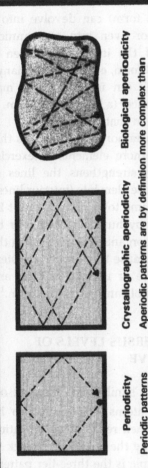

Periodicity

Periodic patterns clone themselves exactly. As such, the pattern cannot replicate itself or otherwise grow. Any apparent replication requires an external agent, for example, the honeycomb pattern in beehives.

Crystallographic aperiodicity

Aperiodic patterns are by definition more complex than periodic counterparts, but they are able to replicate themselves, be it exactly or approximately. The greater the complexity, the less approximate replication is likely to be. In simple cases, replication is effected by the *moving* object exiting a port into a similar shaped arena. The equivalent in nature is to let a budding system shift into a similar paradigm of forces.

Biological aperiodicity

13. Periodicity versus aperiodicity. Aperiodicity is periodicity in which old configurations reappear only after a long sequence of interactions and then only by approximation. The reproduction of organisms is a pervasive application of aperiodicity.

form is hierarchical to the point where the elements are no longer recognizable as such or at least have little utility outside the system, for example, words in a paragraph with respect to the central idea expressed therein.

As depicted in Figure 14, any of the three forms of systems can metamorphose into either of the other two forms. For example, a democracy (a relational form) can devolve into a dictatorship (a hierarchical form) or even into a tyrannical system where the individuality of the individual citizen is submerged into the state (an integral form, e.g., Nazi Germany). Elsewhere, a healthy biological cell (an integral form) may suddenly transform into a virulent cancer (a hierarchical form, at least with respect to the host organism that it consumes).

The primary reason for most metamorphic changes is that many configurations allow one or more elements to exercise dominant influence, which in turn strengthens the lines of communication or linkage among those elements (interior lines), hence intensifying control. By the same token, any attempt by one element to wrest control—be it by human initiative or by mechanical determinism, e.g., a microorganism—can weaken the overall control within the system. Because the form of a system is a function primarily of its degree of internal control, any significant major change to that control can change the form.[12]

SCALED REPLICATION VERSUS LEVELS OF PERSPECTIVE

The term *fractal* was coined to label similar patterns that occurred at different scales, notably atoms that appear to be subminiature solar systems (though their respectively operating characteristics differ sharply, especially the ability or its lack to bind with neighbors). Another example is the three-tier pattern of cell, organism, and the earth, despite each level emphasizing a different morphology. All organisms above one-celled models are built from cells and, collectively, those cells operate as a system that mimics cellular mechanics, to include homeostasis.

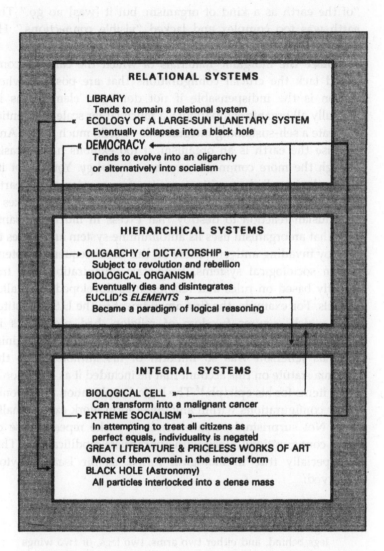

RELATIONAL SYSTEMS

LIBRARY
Tends to remain a relational system
ECOLOGY OF A LARGE-SUN PLANETARY SYSTEM
Eventually collapses into a black hole
DEMOCRACY
Tends to evolve into an oligarchy
or alternatively into socialism

HIERARCHICAL SYSTEMS

OLIGARCHY or DICTATORSHIP »
Subject to revolution and rebellion
BIOLOGICAL ORGANISM
Eventually dies and disintegrates
EUCLID'S *ELEMENTS* »
Became a paradigm of logical reasoning

INTEGRAL SYSTEMS

BIOLOGICAL CELL »
Can transform into a malignant cancer
EXTREME SOCIALISM »
In attempting to treat all citizens as
perfect equals, individuality is negated
GREAT LITERATURE & PRICELESS WORKS OF ART
Most of them remain in the integral form
BLACK HOLE (Astronomy)
All particles interlocked into a dense mass

14. Morphological dynamics. The internal relationships in a system can change because of the dynamics of their constituent elements. At various turning points, these changes can amount to a metamorphosis from one form to another in any direction.

As for the earth, Lewis Thomas remarked that he tried thinking "of the earth as a kind of organism, but it [was] no go." The earth was too complex and lacked "visible connections." He concluded that it was *"most* like a single cell."[13]

Since the inorganic materials of which the earth is composed lack the crucial configurations that are possible when carbon is the indispensable if not dominant element (as in virtually all organic compounds), a much larger scale is essential to create a self-sustaining integral unit—much, much larger. And because the earth is an organism, it too exhibits homeostasis, though the more common name here is *ecology*. You doubt it? When the so-called carrying capacity of any region of the earth is exceeded by population—animal and/or human—it evokes a Malthusian reaction to destroy that excess in much the same way that an organism uses its autoimmune-system antibodies to destroy invading antigens. More on this in subsequent chapters.

In sociological systems, large-scale organizations are frequently based on rules and relationships developed in smaller models. For example, the first amendment to the U.S. Constitution, which ensures freedom of religion, had precedents in various state laws, especially in Rhode Island and Virginia. Thomas Jefferson was so mindful of his authorship of the Virginia statute on this account that he included it as one of only three items for his epitaph.[14] The logical explanation is that only a few configurations within any given form work exceptionally well. Not surprisingly, nature tends to make repeated use of these configurations, albeit with appropriate modifications. This is especially true in biological forms, as Sir Isaac Newton observed:

> The uniformity in the bodies of animals [have] generally a right and left side shaped alike, and on either side ... two legs behind, and either two arms, two legs, or two wings before upon their shoulders, and between their shoulders a neck running down into a backbone, and a head upon it; and in the head two ears, two eyes, a nose, a mouth, and a tongue, alike situated.[15]

While this similarity is not found throughout all phyla, a clearly discernible symmetry does exist from which all species seem to have been templated with ever increasing complexity. More on this in Chapter 14.

Now let us consider the ability to view the same system from different perspectives. Here, we are not looking at comparable systems, but the same system from different viewpoints. For example, the automobile may be considered as a purely mechanical system, a negative contributor to ecology, a positive contributor to the national economy, or a major item in a household budget system. Then, too, a painting can be viewed as so many dabs of paint on canvas, or as a masterpiece of art, or as an item of inventory in a museum.

In warfare, we can identify five discrete levels of perspective. These levels are: (1) heroism, (2) tactics, (3) operations, (4) theaters of war, and (5) national purpose. Of these levels, *heroism* is the easiest to understand. In *The Ballad of East and West*, Rudyard Kipling wrote: "But there is neither East nor West, border, nor breed, nor birth, when two strong men stand face to face, though they come from the ends of the earth." At the next level, *tactics* comprise techniques and procedures for immediate use in battle. But as wars are seldom resolved by single battles, at least not in modern times, the sequence of battles must be managed. That task is called an *operation* (or *campaign*). On the other hand, when wars are widespread most fighting is subdivided into distinct geographical arenas called *theaters*. Finally, war should be conducted in light of explicit *national purpose*. Unfortunately, wars sometimes take on a life of their own to the point where tens of thousands of servicemen may be sacrificed without any perceivable gain to national purpose, especially Vietnam.

Differing perspectives complicate the study and management of systems. Even when each level is fully understood, the conflicts can generate dilemmas and pose difficult ethical issues. For example, our hearts go out to a child with a rare dysfunction whose life can be saved only by a million-dollar series of operations. But if that money were spent on prenatal care for hundreds of poor families, statistical analysis tells us that several children will escape an early death. In an era of limited resources, which option rates priority?

□ □ □

In sum, we have seen how the factors and attributes of elements in a system operate on a more-or-less mechanical basis, with figurative equivalents for abstract and sociological systems. These mechanics do not always reduce to neat, linear equations, but they are ruthlessly logical. In the next chapter, we will tighten up that logic in the form of a gross-level Euclidean-format derivation.

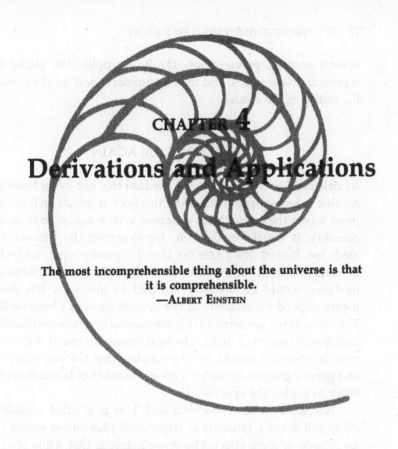

CHAPTER 4

Derivations and Applications

> The most incomprehensible thing about the universe is that
> it is comprehensible.
> —ALBERT EINSTEIN

This chapter draws on the material developed in the first three chapters and, in part, recasts it into a semiformal set of axioms and theorems, beginning with another look at the causality issue because it is so central to the physiogenetic model. This is followed by three sections on assumptions, axioms, and a transitional axiom, the last dealing with the problem of starting at or above the level of an atom. The theorems come next, divided into two sections: primary and secondary theorems. Secondary theorems are more descriptive than derivative. The concluding sections extend the physiogenetic model to abstract and sociological systems (in part, figuratively), present a graph showing relationships among systems that can and cannot reverse entropy, preview how the book applies the model to representative systems, and offer a summary view on the potential utility of the model.

CAUSALITY ISSUE AGAIN

As discussed in the first chapter, the idea that one event leads to another is known as *causality*, and this book depends on it to the point where the entire thesis collapses in its absence. To be sure, causality is clearly evident in, for example, the practice of medicine. Indeed, it's a safe bet that if causality were suddenly to evaporate from the medical literature, the entire profession of medicine would be instantly reduced to quackery. But does nature depend on causality at the deepest roots of phenomena? This issue is not concerned with deterministic versus statistically distributed outcomes under identical circumstances. If the latter view is correct, it would merely mandate that the evolutionary and genetic processes employ a greater number of interactions in order to evolve the species.

No, what we are concerned with here is whether causality *exists*. If it doesn't, causality in larger-scale phenomena would be an illusion or mere effect. The consequence is that while physiogenesis might be a workable, computerlike model, it would not be a fundamental truth. Obviously, this book cannot resolve that issue, nor need it do so. If, over the next century, the physiogenetic model proves its utility, then the philosophical debate would devolve into sophistry. In the interim, we can review three plausible reasons why the no-causality idea has such a strong grip in science.

The first reason is bona fide humility carried to an extreme. When Niels Bohr postulated the complementarity principle, he meant that at the subatomic level, we could not "measure" a particle's location and momentum in the same experiment.[1] In time, this truism evolved into the tenet that if we could not measure both concurrently, we should not try to deduce what was happening.[2] In turn, this tenet evolved into the doctrine that whatever the internal mechanics were, they fundamentally operated by chance and only collectively produced a statistical distribution. Finally, this doctrine led to a coda—inaccurately called complementarity—that our whole notion of causality was faulty.[3] The humility, of course, was in not presuming to postulate more than experiments could verify. The excess was in ignoring the capability of inferred theoretical constructs to explain by deduction how the pieces fit together. James R. Newman, an eminent science writer, put the case this way:

> In this century, the professional philosophers have let the physicists get away with murder. It is a safe bet that no other group of scientists could have passed off and gained acceptance for such an extraordinary principle as complementarity, nor succeeded in elevating indeterminacy to a universal law.[4]

The second reason is the impossibility of mathematically describing the interrelationships and dynamics of countless trillions of particles, elements, and molecules moving in space independent of one another. The idea of order evolving out of this disorder seems like throwing a stick of dynamite into a piano and expecting to hear a Beethoven concerto. *Chopsticks*, maybe, but not Beethoven. Yet the science of chaos, and its offspring complexity theory, have shown that nature *can* play Beethoven by way of patterns combined with the unmistakable evolutionary sequence of increasingly complex organic subassemblies.

The third reason is the attempt to eradicate the concept of God from science, or, at the very most, to reduce God into a *Creator Emeritus*. As one Nobel laureate in physics recently put the case in a 20-million-subscriber magazine interview, we must relegate God or equivalent to a far corner of the universe (though hedging his bet by saying that once we found that corner, it could prove to be infinitely large).[5] Hence, if there was a Creator, and that Creator did design the universe to unfold the way it has, then presumably He would have the power to intervene in His own creation. There is no evidence of this, but science seems intent on eliminating even the potential. The way to eliminate this potential is to eliminate the Creator. This also requires eliminating the concept of causality, because causality forces us to trace evolution "backwards" from the "now" to the big bang (in umpteen trillion easy steps). Once there, figuratively speaking, we would be confronted with a design—what some scientists call the anthropic principle—which brings the Creator back in the picture.

We will not resolve this issue either, but suffice it to say that science has *not* dismissed causality with any kind of logic that remotely approaches scientific rigor. The dismissal is, at best, a coda of faith. By contrast, this book is based on a reasonable inference from the mass of evidence, not faith *per se*. It makes no claim to certainty; only that its model seems to hold water without significant leakage, at least at or above the level of atoms.

ASSUMPTIONS

There are two assumptions, namely, that classical mathematics (e.g., algebra, geometry, calculus) are a valid means of describing physical phenomena, and that the generally accepted laws of gravity, kinetics, thermodynamics, chemistry, and so forth, remain valid to describe the behavior of atoms, and all constructs thereof. The implication is that the esoteric notions prevalent in subatomic physics, insofar as they differ from classical kinetics and the like, have no significant bearing on phenomena at or above the level of atoms. As a matter of pragmatic evidence, we rarely see any sustained discussion of these notions in textbooks and reference works in chemistry, genetics, evolution, and so forth.

AXIOMS

There are four axioms. The first axiom is that space is an infinite void, in which matter moves, and time is "a continuum which lacks spatial dimensions and in which events succeed one another from past through present to future."[6] The passage of time cannot be reversed or slowed, but objects therein can be rearranged into the same configuration that occurred at an earlier time and that the time rate of equilibrative cycles can slow down or speed up, thus giving the appearance of time itself slowing or accelerating.

The second axiom is, in part, a derivative from the first axiom, namely, that matter in the form of atoms exists in space, and if so, then any two or more atoms (and/or constructs thereof) comprise an arrangement that is describable in terms of classical mathematics.

The third axiom is that the psyche is a discrete entity or construct of subentities apart from physical matter, but that it can evince itself only when, and only to the extent that, a physiological construct permits it to do so. This is analogous to the fact that the abstract logic of a computer program remains latent until a computer with sufficient circuitry becomes available to run that program. The implication of this axiom is that any shortcomings or malfunctions of the physiological host can severely affect the conduct of the psyche.

The fourth axiom is that the psyche, at least in higher-order animals and certainly in *Homo sapiens*, can effect a change to otherwise deterministic (or statistically deterministic) events and their consequences by initiating a signal via the cerebrum and nervous system that directs the body to effect that change (the process of which is beyond controversy).

TRANSITIONAL AXIOM

The physiogenetic model commences at or above the level of an atom, and this requires a transitional axiom, at least until such time as the differences between the model and the esoteric notions of subatomic physics are reconciled. The focal point of this transitional bridge is the matter of *effects*. When the results of physical interactions and constructs among atoms create new attributes, those attributes are transitory effects. That is, they are not ontological, presumably conserved properties. The reason is that when a construct is dissembled into its constituent elements, effects disappear without a trace while the total of mass and energy remains the same. On the other hand, velocity and momentum (which is conserved) are not effects precisely because they are conserved.

The rub comes with atoms because their chemical properties are effects. When the constituent particles are fissioned (or fused), the chemical attributes disappear without a trace. But until the physiogenetic thesis and subatomic physics are reconciled, we may treat atoms as if they existed ontologically, with the single exception of atomic fusion and fission itself.

PRIMARY THEOREMS

The axioms lead to 20 theorems linking the various factors and mechanics described in previous chapters. However, the Euclidean practice of tracing each theorem back to relevant axioms and previous theorems is, for the most part, omitted. Instead, this logical path is depicted in Figure 15. The numbers on the figure correspond with theorem numbers.

1. *All physical patterns are deterministic (or statistically deterministic as the case may be), except by the intervention of psychical initiative.* If the outcome of every interaction is deterministic (or statistically deterministic), it follows that the pattern of all such interactions must follow suit. No other factor operates to change this, save psychical initiative.

2. *Atoms may form into molecules, and both may form into constructs, which are called system configurations (systems, for short) if they have a purpose, or configurations if they don't.* The laws of chemistry assure this, and the balance is a matter of definition.

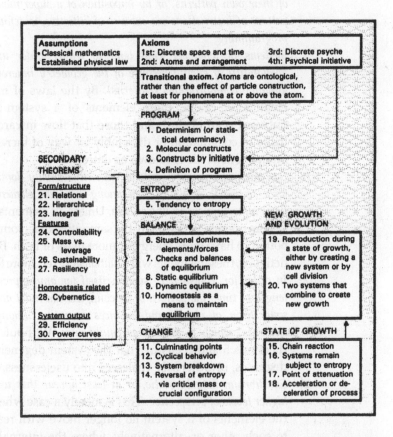

Assumptions
- Classical mathematics
- Established physical law

Axioms
1st: Discrete space and time
2nd: Atoms and arrangement
3rd: Discrete psyche
4th: Psychical initiative

Transitional axiom. Atoms are ontological, rather than the effect of particle construction, at least for phenomena at or above the atom.

PROGRAM
1. Determinism (or statistical determinacy)
2. Molecular constructs
3. Constructs by initiative
4. Definition of program

ENTROPY
5. Tendency to entropy

BALANCE
6. Situational dominant elements/forces
7. Checks and balances of equilibrium
8. Static equilibrium
9. Dynamic equilibrium
10. Homeostasis as a means to maintain equilibrium

CHANGE
11. Culminating points
12. Cyclical behavior
13. System breakdown
14. Reversal of entropy via critical mass or crucial configuration

SECONDARY THEOREMS

Form/structure
21. Relational
22. Hierarchical
23. Integral

Features
24. Controllability
25. Mass vs. leverage
26. Sustainability
27. Resiliency

Homeostasis related
28. Cybernetics

System output
29. Efficiency
30. Power curves

NEW GROWTH AND EVOLUTION
19. Reproduction during a state of growth, either by creating a new system or by cell division
20. Two systems that combine to create new growth

STATE OF GROWTH
15. Chain reaction
16. Systems remain subject to entropy
17. Point of attenuation
18. Acceleration or deceleration of process

15. Systems in a nutshell. Hamlet lamented that if he were bounded by a nutshell, he would be king of infinite space. Nature uses systems a million times smaller, some of which evolve into blue whales and giant sequoias. And the relationships among the mechanics systems can be reduced to a logical nutshell.

3. *The exercise of initiative can cause atoms to form into molecules, and both into constructs, or, alternatively, to dissemble molecules and constructs.* Axiomatic.

4. *The formation of molecules and constructs, or their dissembling, by virtue of the determinism (or statistical determinism)*

of their own patterns, or by imposition of a superintendi pattern *of instructions via exercise of initiative, is defined* programming.

5. *Systems of whatever kind tend to reduce order and* availability *of energy because of the geometry inherent* dynamic *configurations (patterns).* By the laws of mat ematics, et cetera, more elements of a system flc outward than inward, and those that flow inward a liable to coagulation or dissembly by way of increase interactions. This is known as entropy.

6. *Systems must usually contend with situationally domina* forces, *either those acting from among their own elements* those *impacting externally, or both.* Unless all elements ir system operate with exactly the same attributes, some a bound to have more energy or more mass or both. By netic law, these elements will *situationally* exert more for

7. *In order for a system to sustain itself for any appreciat* length *of time, the pattern of its configuration must achie* some *form of checks and balances operating among* elements, *which is defined as equilibrium.* Without th theorems 5 and 6 ensure that the system degenerate degrades, or otherwise coalesces into uselessness.

8. *Equilibrium may be static, or at least appear that way* higher *levels of perspective.* This is merely a case where the elements of a system no longer move with respe to each other or, alternatively, where the internal d namics are immaterial to observers.

9. *Equilibrium may also by dynamic, in which case equilibri* is *more difficult to achieve.* If the elements are not in sta relationship, they must be dynamic, and dynamics a always subject to entropy, more so than static config rations.

10. *The configuration or pattern of a system can enhance* equilibrium *by way of supplementary patterns or externa* operating *systems that keep dominant forces in check, her* defined *as homeostasis.* This is a matter of definition.

11. *In the attempt to increase entropy, most systems will reach a configuration where their fate immediately hinges on that configuration at that instant, herein defined as the culminating point.* Given the operation of entropy, a configuration must occur just before the system loses its ability to function, which is defined as the culminating point.

12. *One outcome of the culminating point is for the system to repeat (or nearly so) the previous pattern of its behavior, herein defined as a cycle.* This is the result of homeostasis.

13. *Another possible outcome of the culminating point is for the system to break down or otherwise degrade into uselessness.* When homeostasis fails, this is the only alternative.

14. *The third possible outcome of a culminating point is for the system's programming to begin to fetch what it needs from its environment and add that material or energy to itself. Alternatively, the system may begin to access internal energy or material not otherwise available. In this case, the culminating point becomes a critical point, the configuration at that instant is defined as a crucial configuration (figuratively: critical mass), and the process as a matter of reversing entropy.* This is a variation on a cyclical reaction, whereby instead of the dynamic pattern turning inward, it turns outward. However, in the case of accessing otherwise unavailable energy, it also turns inward, for example, nuclear radioactivity.

15. *The events that follow a crucial configuration, until such time as the system is no longer capable of sustaining growth or drawing on otherwise unavailable resources, are defined as a chain reaction.* In order to grow, the programming of the system must be able to effect a sustained series of interactions without further externally directed control or input.

16. *A chain reaction ends when either the system's programming can no longer sustain the mechanics to grow or the environment (or*

internal system itself) no longer offers, or makes available, t *necessary resources to grow. This is defined as the point* *attenuation.* The fact that a system reverses entropy f a period of time does not exempt it from entropy pe manently. Sooner or latter the programming brea down or the resources run out.

17. *All systems in a state of growth remain subject to the acti of entropy, both internally and from without, and in tin every system in growth will reach a point of attenuatio however long or short its tenure. This is simply a extension of theorem 16.*

18. *The process of reaching attenuation can be advanced retarded by the manner in which one or more systems in state of growth consume the finite resources of their respe tive environment. Advancement by way of excess consum tion constitutes reaching a culminating point, as well as point of attenuation, whereby the result is to end grows This is defined as a Malthusian reaction.* When consum tion of finite resources is accelerated, the point attenuation will occur sooner.

19. *While a system is in a state of growth, its programming m construct and release a new system or, alternatively, fissi into two or more new systems, herein defined as reprodu tion.* A pattern can be programmed to do anything the resources are available, hence Immanuel Kant famous observation: "Creation is never complete; forever goes on."

20. *Two or more systems in a state of growth may interact produce a new system, and that new system may be a ne replica of its forebears or it may be a different system. In tun that system may go on, with or without other systems in a st of growth, to produce yet other systems. When this process internal to an encompassing system, it is defined as geneti When the process spawns physically independent systems, it defined as reproduction or evolution. This is a variation theorem 19, and largely a matter of definition.*

SECONDARY THEOREMS

The next 10 theorems, though numbered from 21 to 30, do not follow logically from theorems 1 through 20. Instead, they are descriptive theorems that apply to one or more aspects of physiogenesis, though strictly speaking they could also be derived with equal rigor. The reason for this approach was to minimize distracting ideas among the primary theorems.

21. *A system wherein the elements have the least linkage or impelling relationships is defined as a relational system.* This is a matter of definition.

22. *A system wherein a few elements continue to exert dominant influence over other elements, usually by way of a tiered structure, is defined as a hierarchical system.* Ditto.

23. *A system wherein the elements are linked to the point of losing discrete identification, at least with respect to that system, is defined as an integral system.* Ditto.

24. *The degree to which a system can control or modify its behavior or, alternatively, permits an external system to control or modify its behavior, is defined as controllability.* This, too, is a matter of definition, but it implies a hierarchical structure, at least in part.

25. *Controllability may be exerted by mass or by leverage, but leverage requires the system to have a configuration that lends itself to a quasi-chain-reaction process.* Comparatively large mass typically constitutes a situational dominant element, hence may more easily influence the system (though this risks damage). But in order for a small force to control a comparatively larger system, the effect of that force must set off a quasi chain reaction of events, whereby each cause produces an amplified effect. This is not a matter of reversing entropy because there is not normally any growth *per se*. However, when a system does reverse entropy, it also may sustain growth by way of leverage.

26. *The extent to which a system resists entropy or, alternatively, the extent to which a system in a state of growth resists attenuation, is defined as sustainability.* Also a definition.

27. *The extent to which a system can modify the pattern of its configuration on inimical external interactions or internal mechanics, without significant loss of its ability to meet its purpose or the efficiency by which it does so, is defined as resiliency.* Ditto.

28. *The process of homeostasis can be enhanced by programming that modifies the homeostatic process itself based on experience, herein defined as cybernetics.* Again, programming can be made to do anything if the necessary resources are available.

29. *The ratio of useful work that a system performs to the amount of resources entered or consumed, be it by an external system or by virtue of reversing entropy, is defined as the measure of efficiency.* This is a matter of definition.

30. *When the efficiency of a system is such that comparatively small increases to input result in continued high output, the operation of the system for that range of input is defined as operating on a power curve.* This is a colloquial definition, but it serves as a prelude to reversing entropy, whereby autonomous growth causes the power curve to go vertical. No further input is necessary, except that which the system itself fetches from its environment.

MODIFICATION FOR ABSTRACT
AND SOCIOLOGICAL SYSTEMS

The physiogenetic model inherent in the above theorems also applies to abstract and sociological systems, *intact* insofar as those systems are manifested in physical terms, except as noted, that the exercise of human initiative can effect a change to an otherwise deterministic course of events. By contrast, the theorems apply only figuratively to pure abstract systems—like mathematics and jurisprudence—because ideas substitute for matter. Thus, physical distance is not a consideration, and time is a consideration only in the sense of the history of the development of certain ideas. Logical relationships take the place of interactions, and most logical systems are hierarchical in nature. Dominant ideas substitute for dominant elements and, except in games, those dominant ideas tend to stay dominant. Yet pure abstract systems can still have figurative power curves (in terms of utility), reversal of entropy (in terms of discovery), and continued growth (in terms of how far a theory extends). Moreover, the physiogenetic model itself is stated as a quasi-mathematical abstract model. And it seems entirely possible to develop a formal derivation of theorems that apply to pure abstract systems—in what might be called an *epistemogenetic* model (from the word for the study of truth: *epistemology*). To that end, project *Logos* (described in Appendix A) is intended as a start, but an epistemogenetic model is not essential for this book.

The parallel for sociological systems would be a *sociogenetic* model. But if sociology follows psychology, the prerequisite would be a *psychogenetic* model. Chapter 15 makes only a stab at the latter. Nevertheless, this book applies the concepts of the physiogenetic model—combined with human initiative—to a variety of sociological systems and uses the same terminology as consistently as possible. For example, the concept of a physical turning point and power curve apply both physically and figuratively in sociology, as evidenced by this well-known passage from Shakespeare's *Julius Caesar* (Act IV, scene iii, lines 218–224):

> There is a tide in the affairs of men,
> Which, taken at the flood, leads on to fortune;
> Omitted, all the voyage of their life
> Is bound in shallows and in miseries.
> On such a full sea are we now afloat,
> And we must take the current when it serves,
> Or lose our ventures.

But the exercise of initiative, which is often beset by other physical attributes, can make mincemeat of attempts to apply the physiogenetic model to human endeavor. Consider the attempt of Sir Winston Churchill to improve military performance during World War II by way of the British Chiefs of Staff system. He afterwards lamented: "You may take the most gallant sailor, the most intrepid airman, and the most audacious soldier, put them together—and what do you get? The sum of their fears."[7] Still, human initiative, despite its ability to change an otherwise deterministic outcome, often acts in predictable, nearly deterministic ways. For example, some aspects of behavior are clearly predictable under some conditions (an angry reaction on discovery of fraud). Other behavior is a complete mystery (how the pioneers survived the westward trek without microwave ovens). Without at least some predictability, all sociological systems would degenerate into anarchy.

Perhaps the best way to show the parallelism of abstract, physical, and sociological applications of the concepts underwriting the physiogenetic model is to compare examples from each category on one chart, as depicted in Figure 16. The first column describes the game of chess; the second, chromosomes (DNA); and the third, the U.S. Constitution. Incidentally, the format of this illustration is used without modification for the eight charts in Appendix B that apply these concepts to 24 different subcategories of systems.

ENTROPIC INDEX

Configurations and patterns can be numbered, like paintings in a gallery, but those numbers do not describe the art. Still, science thrives on quantification, if only by way of a scale that compares different cases of related phenomena. Along these lines, the author has developed a scale to gauge systems as they approach or veer from the critical point—hence crucial configuration—at which their operation reverses entropy. This scale is arbitrarily named the *entropic index scale* (EI scale) and is modeled after the familiar pH scale (which gauges hydrogen–ion activity: the degree of acidity or alkalinity). The pH scale ranges from 0 to 14,

	Game of Chess	Chromosomes (DNA)	U.S. Constitution
• Intent • Design	To provide a contest of skill and mental ability—especially in spatial reasoning—between two players.	To provide a biophysical engine for nature to evolve various species, and to permit extant specimens to reproduce.	To create a system of government with checks and balances among branches, and with the citizens and states.
• Elements • Quantity • Attributes • Variance	There are six different type-elements on an 8*8 square playing board, the permissible movement of which is rigidly specified.	Gene sections, despite a common build-up from common amino acids, each possess their own distinct attributes.	Elements and attributes are permissible powers and relationships among branches, and how they are controlled.
• Degrees of freedom • Initiative	The idea is for a player to retain maximum degrees of freedom among playing pieces in order to employ maximum initiative.	Initiative is not a factor, and degrees of freedom must be curtailed in order for programmed growth to proceed.	The ultimate object to minimize controlling structure and thus permit the maximum exercise of initiative.
• Crucial arrangements • Interior lines • Dominant elements	Victory arises from an arrangement whereby the opponent's king yields dominance to one or more pieces of the victor.	The double-helix design is crucial to replication, but different genes exercise dominance at different times.	The idea is to provide a set of political interior lines that thwarts any attempt by one branch to achieve dominance.
• Resiliency • Sustainability • Resistance to change	Rules of the game *almost* ensure that sustainability will fail as one player proves more able to accommodate change.	Within any organism, the DNA of each cell remains identical, even if that specimen has quadrillions of cells.	It provides for maximum sustainability and/or resiliency despite any and all political changes occurring over time.
• Equilibrium • Homeostasis • Cycles • Cybernetics	The cycle of alternating plays lets the players try to upset equilibrium, but overly aggressive moves usually beggar disaster.	Nature offers no better example of perfectly balancing these four properties in a single system.	The idea of homeostasis was observed in the Constitution the same year it was identified in biology (1932).
• Power curves • Catastrophe • Culminating points	Victory is usually signaled by one player forcing the other into a culminating-point configuration, whereby loss impends.	Binary fission is a superb example of culminating points and highly controlled, positive catastrophe.	Every constitution seeks to avoid catastrophes and culminating points, yet to provide for a power curve of growth.

16. **Sample applications.** The mechanics of physiogenesis apply to abstract and sociological systems, albeit with some figurative adaptions. This chart compares the game of chess (an abstract system), DNA chromosomes (a physical system), and the U.S. Constitution (a sociological system).

a range that the entropic index adopts intact. This index is shown in Figure 17.

However, entropy is an *inverse* relationship; the magnitude of entropy is said to *increase* as energy becomes *less* available. Hence, the value of 0 on the EI scale represents maximum availability of energy or absolute order, while 14 represents zero

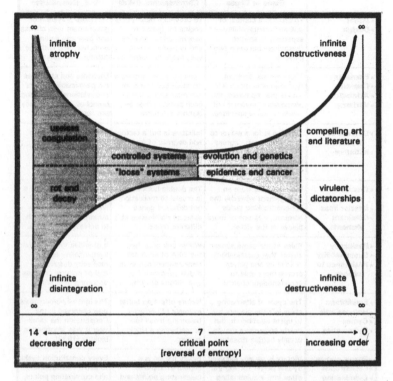

17. Entropic index. Most systems beg disorder and grow increasingly ineffective. A few systems reverse this process by using their own configuration to take what they need from their environment. The equivalent in abstract and sociological systems is a configuration that increasingly gains popular support or recognition. The entropic index scales this range.

availability of energy (maximum unavailability), disorder beyond repair, or an utterly chaotic degradation. Reversal of entropy occurs at the midpoint value of 7, i.e., the *critical point*.

On the graph, the horizontal axis represents the scale itself, and is purely mathematical. By contrast, the vertical axis is mathematical only in the sense of topology, which depicts relationships more so than quantity. The employment of infinity marks symbolizes the fact that zero entropy (maximum availability of energy or absolute order) inherently entails infinite magnitude. This situation occurs only when a system retains its utility permanently without the need for more energy or further modification. Examples include completed systems of logic (Euclid's *Elements*) and works of art. Interestingly, the figurative infinity can approach mathematical infinity, as when a classic painting is considered to be "priceless." Conversely, maximum unavailability constitutes a permanent lack of order or energy— what is sometimes called absolute death. The black hole in astronomy is a good example. We also note that the differently shaded divisions on the graph imply discrete increments. In actuality, the entropic measure of systems could be more accurately portrayed as a gradient.

Keep three points in mind. First, a system typically increases entropy by dispersing its elements or by coagulating them into useless subunits, like a ruptured artery (dispersing blood) or a cirrhotic liver (shutting down essential physiological processes). Second, the criterion for reversing entropy—colloquially—is that the system begins to take on a life of its own. This means that the system is programmed (be it a formal set of instructions or by virtue of patterns) to obtain the resources from its environment that it needs to grow *without the need for external assistance*. Third, the few systems that reverse entropy seem, not surprisingly, to be compact and efficiently organized, most notably in biological organisms.

APPLICATIONS

Following the two chapters on historical development of concepts in the physical and social sciences, this book devotes three chapters to abstract systems, five to physical systems, and five more to sociological systems. Abstract systems are placed first because they better illustrate the concepts of the physiogenetic model, and most of those concepts are widely accepted among experts in the respective fields. Indeed, the text in the chapter on automation will cure insomnia for any professional in computer science. And for the text on military science, experts in that field will probably agonize over plowing through the principles of war for the 200th time. Still, that is not the point. The point is to anchor the physiogenetic model to widely accepted theory before applying it to physical systems.

The text on physical systems begins with a chapter on basic physics and chemistry, and one on engineering. This involves less controversy because most of the phenomena are very well understood. The application of the physiogenetic model, therefore, is little more than an exercise in lexicography—a drudge harmless enough as Samuel Johnson put it. This is followed by three chapters on genetics and physiology, pathology, and evolution, and this is where the gauntlet is thrown down. The physiogenetic model is applied up front to organic phenomena in a single chart as a prelude to reducing the whole of these subject fields into a purely mechanical model (exclusive of the psyche). Nature relies on the same theme—the same symmetry—over and over again with only relatively minor variations to reproduce specimens within a species, or to use one (and usually two) species to generate a new species. From this vantage point, we can also see cancer as a rare aberration of the genetic process. To be sure, cancer is *not* rare in *Homo sapiens* as organisms, but it is rare among cells. Less than one cell in a trillion turns cancerous, and virtually all cancers originate in a single cell.

Following these chapters, the book turns to sociological systems, starting with psychology. As mentioned, the time is not ripe for a formal psychogenetic model, and much less so for a derivative sociogenetic model. But we can make a stab at the former, then demonstrate how the concepts of the physiogenetic model apply to psychology and sociology, sometimes figuratively, sometimes literally. Think preview of coming attractions. Other minds much stronger than the author's will someday tackle the matter of formal derivation.

UTILITY

The general utility of the physiogenetic thesis, and its extension to abstract and sociological systems, depends on two overwhelming factors, both beyond the control of the author: whether the theory is correct, and if correct, whether it is widely accepted as such. It is not the business of a theoretician to harp on his own theory; that would beggar egotism. The proper approach is to do as good a job as possible with developing and presenting the theory, then step back. If it is correct, it will speak for itself, and if so, then it will come eventually to find a home in science, and from that its utility. That utility, like all theories, permits researchers to more quickly and more accurately focus their work.

This having been said, Appendix A proposes 14 experiments and projects that emanate from the theory. Beyond that, however, the utility of this theory is the recognition of human initiative apart from biological determinism. If each person bears the power to change the course of events, then he or she also bears some obligation and responsibility to do so, at least under some circumstances. True, this is not a bed of roses. The editor-in-chief of one dictionary struggled for months with the definition of *existentialism:*[8]

> The doctrine that existence takes precedence over essence, and holds that man is totally free and responsible for his acts, and that this responsibility is the source of the dread and anguish that encompasses him.[9]

So there are two choices. The first is let the dread and anguish sink in even further, perhaps to the point of psychotic despair. We see a hint of this in a recent conference among British computer scientists, who concluded that human life and judgment was being taken over by machines and that there was nothing to stop it.[10] The other option is to rise above the lethargy and begin to treat the human spirit as something more than a product of mechanistic or biochemically based psychology. Charles Dickens once expressed this idea in his beloved *Christmas Carol*, wherein Ebenezer Scrooge has a one-sided conversation with the "Ghost of Christmas Yet to Come":

> Men's courses will foreshadow certain ends, to which, if persevered in, they must lead... But if the courses be departed from, the ends will change. Say it thus with what you show me!

Creator=Nature=Mind. Sort of. Suffice it to say that when the last equation of science is written, and the final pattern of chaos is mapped out, man will still be faced with himself.

PART II

History

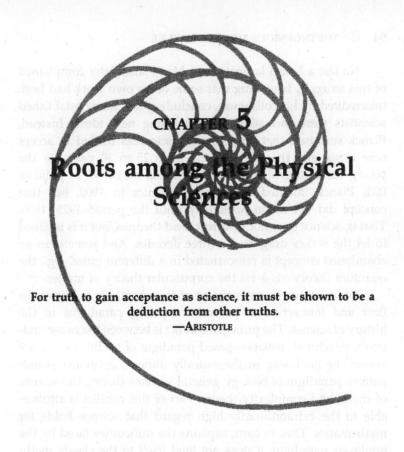

CHAPTER 5

Roots among the Physical Sciences

For truth to gain acceptance as science, it must be shown to be a deduction from other truths.

—ARISTOTLE

The common perception of progress in science is that of a steady accumulation of knowledge and understanding, highlighted by the work of a comparative handful of well-known persons. From the distant perspective of time and history, this is accurate enough for nearly the same reason that viewing land while flying over it minimizes the day-to-day tribulations of those living on it. That "steady accumulation" is a kind of smoothing out of a grand learning curve of sorts, erratically punctuated by the contentiousness of competing theories and paradigms, the enigmas of unexpected experimental findings, and the uphill struggles faced by new concepts no matter how valid they may later prove to be.

93

No less a Nobel laureate than Max Planck later complained of this struggle, lamenting that some of his own work had been miscredited by his colleagues, concluding that many established scientists were incapable of assimilating new ideas. Instead, Planck surmised that only younger scientists tended to accept new ideas, and therefore it takes from 25 to 30 years for the young Turks to evolve into the orthodox majority.[1] In point of fact, Planck initiated quantum mechanics in 1900, but that concept did not attain orthodoxy until the period 1925–1930. That is, science does not resurrect dead theories, but it is inclined to let the wakes drag on for three decades. And sometimes an abandoned concept is reincarnated in a different guise, e.g., the quantum theory vis-à-vis the corpuscular theory of matter.

Accordingly, the main theme of this chapter is to depict the flow and interaction of several competing paradigms in the history of science. The principal conflict is between the cause-and-effect, structured, equation-based paradigm of traditional science versus the nonlinear, mathematically difficult, relationship-and-pattern paradigm of biology, general systems theory, the science of chaos, and complexity theory. Part of this conflict is attributable to the extraordinarily high regard that science holds for mathematics. This, in turn, explains the difficulties faced by the nonlinear paradigm; it does not lend itself to the classic mathematical elegance of traditional science. Yet it is equally obvious that atoms, molecules, and other forms of physical elements exist and operate physically independent of one another in space.

TRADITIONAL PHYSICS

Despite its culminating section titled "System of the World," Newton's *Principia* covered only what today is called *mechanics* or *kinetics*. The import, however, went beyond mechanics by showing that at least some natural laws applied universally to all relevant phenomena at all levels. The implication of Newton's work was that other universal laws could be eventually found to explain all other phenomena. Toward that goal, Newton envisioned nature to be a system of particles in space, to include corpuscular units of light.[2]

This corpuscular-based conjecture quickly lost ground to the theories of Christiaan Huygens, a contemporary of Newton and equally famous at the time. Huygens assumed that light consisted of waves propelled through an ether in space (an idea

18. Sir Isaac Newton. Newton set a rigorous tone for science that remains undiminished to this day. Voltaire once remarked that such accomplishment could only fall to one man, and history has borne this out.

later debunked) because waves offered a much neater explanation of refraction and other optical phenomena.[3] Accordingly,
scientists over the next century and a half came to reject
corpuscle theories entirely, and subsequent studies in electromagnetism furthered the wave theory. In 1873, James Clerk
Maxwell consolidated all that was known about this subject into
his magnum opus *Treatise on Electricity and Magnetism*. However,
in rejecting corpuscle theory, he had to turn to the ether theory.
The concluding section of the treatise was titled "The ether
cannot be got rid of." His reasoning was that if waves had no
mass yet carried energy, they would have to do so by nudging
(at the speed of light) the supposed ether along the path of
travel.

Another major difference between Newtonian and Max-
wellian mechanics concerned the frame of reference for motion.
Absolute motion means that all motion can be mathematically
anchored to a set of identifiable coordinates in space (which,
today, would be the site of the so-called big bang).[4] Relative
motion means that no such frame of reference exists. Newton
recognized the difference between absolute and relative motion,
but assumed that the equations and laws governing motion and
forces must be derived only on the basis of relative motion.[5] That
is to say, while larger bodies emitting a correspondingly stronger
gravitational pull give the appearance of absolute motion, it is
equally valid *mathematically* to envision heavier bodies moving
with respect to lighter bodies. By contrast, electromagnetic
radiation propelled itself at the speed of light, hence Maxwell's
theories presumed an absolute frame of reference for motion. He
had no idea where the origin of this reference frame lay but,
logically, it had to exist. As might be expected, many scientists
could not accept the idea that nature needed a reference frame
for some phenomena but not for others.[6]

In 1887, A. A. Michelson and E. W. Morley attempted to
reconcile this dilemma with a pivotal experiment.[7] This experiment used an interferometer that split a single "segment" of a
light wave into two parts and then routed each part on round
trips at right angles to one another. This interferometer was

mounted on a large flat stone, supported by a much larger block of wood floating in a vat of mercury. If the earth was traveling through the ether (presupposing an absolute frame of reference), then light waves traveling in the path of the earth's rotation would be shortened by way of the Doppler effect. A wave traveling perpendicular to this path would not. Hence, when a ray of light was split in order to travel along these two paths and then reflected back to a common point, changes in rotation of the instrument would generate a change in the diffraction pattern. In diffraction, light waves "cancel" one another analogous to the way a muffler operates. But despite a dozen attempts with ever more sophisticated apparatus, the predicted patterns were never seen.

The implication was that the absolute frame of reference for motion did *not* exist; hence the primary assumption in Maxwell's definitive theory collapsed. Yet because his theory worked so well, two different explanations were proffered to account for the negative results of the Michelson–Morley (M-M) experiment. The first explanation was developed by Hendrik Antoon Lorentz; the second (10 years later), by Albert Einstein. Both explanations became known as relativistic theories. Both culminated with the same set of transformation equations, but the theoretical underpinnings differed radically.[8]

Lorentz postulated that objects compressed themselves as a function of their velocity that exactly compensated for the lack of an observed Doppler effect in the M-M experiment. He derived these equations from conventional mechanics, implying that the force necessary to accelerate an object was added to that object, the dynamics of which reduced its girth in the direction of its velocity. Lorentz also maintained Maxwell's absolute reference frame, which meant that all relativistic effects were to be calculated unidirectionally from that frame.

Einstein took a different tack. He posited that the reference frame was the speed of light itself. That speed was constant to all observers irrespective of their motion. From this he rederived the Lorentz transformation equations by kinematic analysis. That is, he compared traces (loci) of the path of radiation as

viewed by different observers traveling at different speeds wit
respect to each other. This meant that relativity was bidire
tional. If observers A and B were moving apart from (or towar
one another, A would measure B's mass as heavier than his ow
while to B, A's mass would appear to be heavier.

THE GREAT CONFLICT

Needless to say, bidirectional relativity raised eyebrows, and
was at this juncture that physics leaned evermore heavily towai
esoteric mathematical models. The first such model came
1908. In that year, the mathematician Hermann Minkowski sai
that the bidirectional nature of Einstein's relativity theory r
quired that space and time could not exist as separate concep
but only as some sort of continuum.[9] Einstein replied that, und
such conditions, he no longer understood his own theory.[10]

Then in 1916, Einstein published his general theory
relativity. This theory predicted that light waves "bent" whe
passing through a gravitational field. When this prediction wa
confirmed, Einstein was acknowledged to be a genius, furth
enhanced by confirmation of the famed $e = mc^2$ equation derive
from his original (special) theory of relativity. From this pi
nacle, Einstein devoted the better part of his life to developing
unified field theory. This theory sought to unify mathematical
the then-known fields of force in nature: gravity and electroma
netism. The implication was that once the fields were linked, t
derivation of all natural laws would follow.[11] In this Einste
failed, but the effort inspired many others to seek this Holy Gra
of physics, to include the subsequently discovered stro
nuclear binding force and the weak nuclear decay force.

The ultimate task was to discern the nature of these forc
Were the forces primordial? Or were they the result of a field
particles (as indeed the electromagnetic "field" consists of ph
tons)? If particles, are the particles corpuscular, or waves, or bot
or primarily corpuscular under some conditions and waves und

other conditions? These issues have not been fully resolved, and some analysts believe they will never be resolved.[12]

In any event, the corpuscular theory made a comeback in the form of quantum theory. In 1900, Max Planck was confronted with some experimental results that defied then-current theory.[13] The only way out, he reasoned, was to assume that nature at root behaved in discrete increments rather than continuous action. This became known as quantum mechanics and eventually gained widespread acceptance as science deduced the structure of atoms and molecules. The rub was that particles did not dovetail with field theory. Despite his earlier support of Planck's ideas, Einstein later wrote that "in the foundations of any consistent field theory, the particle concept must not appear in addition to the field concept." He argued that the "theory must be based solely on partial differential equations and their singularity-free solutions."[14] A pure field theory cannot cope with independent objects operating in space.

This dilemma brought the two competing paradigms—linear versus nonlinear systems—into sharp focus. Hundreds of different particles have been discovered, though most of them are short-lived. Moreover, stable protons and electrons, and less-stable neutrons (outside of a nucleus), are now presumed to be constructs of fundamental building blocks, usually called *quarks*, which themselves may have their own building blocks.[15]

CHEMISTRY AND THERMODYNAMICS

Chemistry found its Newton in the many-sided genius of Antoine Lavoisier. Lavoisier published *Elements of Chemistry* in 1789, aided immensely by his young wife, Marie, who learned English in order to translate the writings of Cavendish and Priestley for her husband's study. In the *Elements*, Lavoisier brought an end to the inconsistent phlogiston theory. Phlogiston theory held that matter first disappeared in some reactions, then reappeared in others. His explanation demonstrated that matter was preserved and thus clarified how chemical equations should be formulated. Also, Lavoisier's methods and approach to analysis were a model of prudence, and he might have carried his work further but for the French Revolution. In that madness, he was sent to the guillotine (for his role as an appointed tax collector). The next day, the mathematician Joseph Lagrange lamented that while it took only a moment to sever Lavoisier's head, it would take a century to produce another like it.

Lagrange was wrong. Measured from Lavoisier's birthdate, it took only 80 years to produce Dimitry Ivanovich Mendeleyev. Mendeleyev integrated many speculations on relationships among the various elements into the now-familiar periodic table. Using this table, he predicted the existence of three "new" elements: gallium, scandium, and germanium. When all three were found within 15 years, the world took note (except in his native Russia). The significance was the recognition that nature relied heavily on permutations and combinations of simple underlying mechanics to produce widely divergent elements. Though nuclear physics took 50 years to reveal the mechanics, scientists working in that field were incalculably aided by the periodic table.

Chemical reactions, of course, often generate or absorb heat, a process that led to the study of thermodynamics, which means the dynamics of heat transfer. Here, another many-sided French genius—Joseph Fourier—wrote a seminal treatise, titled *Analytical Theory of Heat* (1822). Among other theorems, he formalized the emerging doctrine of conservation of energy, analogous to Newton's assumption of the conservation of momentum. This law became known as the first law of thermodynamics. Meanwhile, the concept of entropy was emerging from various studies. Then as now, entropy presumed the world and all its systems tended to disorder and decreased availability of energy. In 1869, A. F. Horstmann formalized this idea in what became known as the second law of thermodynamics. To be sure, some chemical reactions are reversible, meaning that unavailable energy again becomes available. Still, both reactions consume energy in friction or its equivalent. As such, no reaction is 100 percent reversible, and therefore entropy will always increase in a closed system. Since that time, the concept of entropy has infiltrated into nearly every discipline, especially information theory and management science.

BIOLOGY, GENETICS, AND EVOLUTION

For centuries, biological phenomena were assumed to arise from a different source than nonliving material. This font was called the *vital force*. The trouble was—like the ether in physics—no one could find a trace of it. Every complex molecule was recognized as an extension of simpler molecules, and every intricate organic process was deduced as a combination of conventional reactions. This realization has been accelerated by the increasing comparison of physiological processes with man-made machinery. At one time, the heart was considered to be the seat of man's soul. Today, it is recognized as a pump, pure and simple.

The seminal discovery in all of biology was the cell. In 1665, Robert Hooke published his findings in *Micrographia*. Twenty years later, Antonie van Leeuwenhoek invented the microscope and used it to describe cells in some detail, reporting his findings to the Royal Society. It took another 150 years before the German biologists Matthias J. Schleiden and Theodor Schwann formalized the cellular theory of biology (1815). Schwann reduced it to one sentence: "The cells are organisms, and animals as well as plants are aggregates of these organisms, arranged in accordance with definite laws."[16]

About a half-century later, three momentous theories appeared. The first was Darwin's thesis on evolution (1859), which needs no comment here. The second was Gregor Mendel's laws of heredity (1865). The mathematical precision of Mendel's work strongly suggested an underlying mechanics in genetics. The third was Claude Bernard's *Introduction to the Study of Experimental Medicine* (1865). Among other conclusions, Bernard's work formalized a doctrine about organisms, which he called *constancy*. This concept was really an amalgam of two biological mechanisms that he could not differentiate at the time. The first is *equilibrium*, a process well known to chemistry. The second is *homeostasis*, which was later developed by Walter Cannon in *The Wisdom of the Body* (1932). As described earlier, equilibrium is the give-and-take between contending forces in a system, analogous to the play of a football game, or marriage for that matter. Homeostasis goes beyond this by activating counterforces to offset any force that upsets equilibrative balance in a system.

In 1944, the Nobel laureate physicist Erwin Schrödinger wrote his seminal essay *What is Life? The Physical Aspect of the Living Cell.* This essay set many minds thinking about the relationships among physics, machinery and engineering, the chromosomes, and by implication, automation. The foremost example of this thinking was Norbert Wiener's *Cybernetics: Or Control and Communication in the Animal and the Machine* (1948), which cited Schrödinger's essay. Wiener compared the ability of biological cells to learn from experience—especially the process of sensitizing the autoimmune system—with man's ability to program a machine that "learns" from its "experience." An example of this is an automated chess-playing machine that changes its strategy based on the moves of its human opponent.[17]

AUTOMATION AND COMPUTER SCIENCE

Automation is commonly, and inaccurately, equated with robots. Robots perform manual work by following a program of instructions in some form. Some of them resemble human arms and the like, but most robots are traditional machines that have been automated by way of programs. A program is a set of instructions that tells a machine or system what to do. Hence, computer science is, in effect, a generalization on the techniques essential for control of robots. Instead of doing manual work, programs are written to manipulate electronic circuits in order to produce answers to quantitative and other logical questions. However, a computer-actuated printer is, in fact, a robot; and robots can be programmed to work like a simple computer despite the obvious inefficiency.

The roots of automation trace to ancient times and made their formal debut in 1305. In that year, Ramon Lull completed his *Ars Magna* (Great Art), in which he envisioned mechanization of the structure of knowledge, using wheels and gears to relate various ideas. His prognosis was either ignored or ridiculed, but by 1550 automation was used to program displays with water gardens and fountains, some with quacking mechanical ducks.

By 1700, automation found a practical use in programmed looms to weave patterns in cloth. Most were destroyed by workers fearful of losing their jobs, but the concept persisted and by 1800 evolved into easily programmed robotic machines— Jacquard looms—that could weave any conceivable pattern. A generation later, the British government funded Charles Babbage to develop a robotic-like computer, which he called an "analytical engine." England saw its value as an aid to the computational difficulties involved in navigation—no small consideration in what was then the world's foremost maritime power. His protégé, Augusta Ada Lovelace (Lord Byron's daughter, and for whom the Ada programming language is named), wrote a treatise on how to program this computer. Drawing an analogy to the automated loom, she said it could weave "algebraic patterns," then went on to say:

> A new, a vast and powerful language is developed through which alone we can adequately express great facts of the natural world, and those unceasing changes of mutual relationships which, visibly or invisibly, consciously or unconsciously, are interminably going on in the agencies of the creation we live amidst.[18]

Like St. Joan (martyred 1431; canonized 1920), Ramon Lull was vindicated half a millennium after his death.

Intrigued by these ideas, Stanley Jervons built another computer, modeled on an ordinary piano. He called it an "automated logic machine," demonstrating it before the Royal Society in 1874. Fifteen years later, a Census Bureau employee,

Input. Early automation used punchcard input. Today, control values and programs are written as magnetized bits or their equivalent on optical surfaces, but the basic idea remains unchanged.

Processing. All processing consists of rearranging matter or energy content in some form or manner.

• *Robotics:* This is usually a step-by-step assembly or construction process.

• *Information processing (computers).* Manipulation of bit patterns using circuits symbolic of intermediate calculations.

Output. Output is the last rearrangement or reconfiguration of the processing, though it is often transmitted elsewhere.

• *Robotics.* Typically a subassembly or product, or a refinement thereof.

• *Information processing.* Output bitstrings are usually transformed into human readable language, data, graphs or other images.

Jacquard Loom

19. Automation as the basis for dynamic systems. The principles of Jacquard's loom underwrite all automation. Indeed, the first electronic computer—the *Colossus*—closely resembled an automated loom. We are beginning to realize that similar principles underwrite the ability of chromosomes to "weave" biological tapestries.

Herman Hollerith, realized that it would take 11 years to compile the data from the 1890 decennial accounting by hand. He persuaded his superiors to give him a contract to build automated tabulators nearer in design to Jacquard's loom than Babbage's engine. These machines reduced the anticipated 11-year period to 7 years. Afterward, Hollerith quit his job and formed a company to sell improved models to the business community. That company later evolved into IBM.

Meanwhile, automated computing aroused an interest in academe. At first, the work progressed slowly and unheralded. Then came World War II, at which time the British government was impelled to break the German *enigma* code and so built the world's first electronic computer. That computer was located at Bletchley Park and nicknamed *Colossus*. From that point forward, computers became the stuff of popular history.

MILITARY SCIENCE

Although some scholars regard military science as an oxymoron, in many ways that science was the forerunner of both operations research and the science of chaos. The reason is that war is an obvious nonlinear system that nevertheless mandates decisive action. Interestingly, success or failure can be explained by a handful of principles irrespective of the tactics or technology employed, though this does not mean that war operates smoothly. On the contrary, battle is marked by an uncontrollable disorder known in the trade as fog or friction. Yet, Napoleon regarded war as "the science of sciences" and an art of immense proportions that encompassed all others.[19]

That appellation may be debatable in degree, but throughout history various military analysts and theoreticians strove to find the system behind war. The oldest treatise still held in high regard is Sun Tzu's *The Art of War*, written about 500 B.C. Among other things, the ancient scholar clearly recognized culminating points by suggesting that he who knew when the odds favored him, and when they did not, would be the victor in the end. That is to say, successful military commanders correctly sense when to attack and when to back off.

Two millennia later, two Prussian theoreticians wrote books that led to the study of war as a formal discipline: Baron de Jomini's *The Art of War* (1837) and Carl von Clausewitz's *On War* (1830, published 1843). Jomini concentrated on then-current tactics, most of which have become obsolete. By contrast, Clausewitz eschewed tactics and instead identified the principles by which all wars, in all eras, could be understood.

Some of these principles eventually found their way into other disciplines, especially the *principle of mass*. Clausewitz said that the outcome of a battle was related primarily to which side had the greater mass *at the critical or focal point of that battle*. A weak force can defeat a strong force if it can concentrate what force it *does* have against its opponent's "center of gravity." We also see this in the martial arts, whereby a weak person can easily deck a towering adversary. Perhaps, then, Napoleon was nearly correct. War, if not the science of sciences, presents a clear application of the principles underwriting the dynamics of systems.

NONLINEAR SCIENCES

The nonlinear sciences include general systems theory, operations research, systems analysis, systems engineering, the science of chaos, catastrophe theory, complexity theory, and some aspects of fuzzy logic. Many of these disciplines can be traced to progenitors in mathematics, especially group theory, topography (in part), and mathematical complexity theory. What gave them impetus in this century was the inability of traditional scientific models to deal with complex phenomena, or only with unacceptable accuracy, or only by plodding through time-consuming calculations (a strong impetus for computer development).

The exigencies of war aggravated this shortcoming to the point where something had to be done, and so operations research came of age during the Battle of Britain. The problem faced by Great Britain, when Germany began her air assault, was that the *Luftwaffe* had clear superiority in quantity of aircraft, pilot skills, and possibly aircraft design. Even if the last point is not true, the British Hurricanes and Spitfires were often flown by men with only a few weeks of flight training, whereas many German pilots had several years of recent combat experience. Furthermore, supply lines to the British Isles were at the mercy of German submarine warfare, and she had no allies beyond what the United States sent in material.

The heroism of the British fighter pilots cannot be denied, but against such odds it takes more than courage to prevail. Resources on hand had to be configured, scheduled, and optimized to the nth degree. Led by Air Chief Marshal Sir Hugh Dowding, the Royal Air Force Fighter Command rationed and allocated fuel to the gallon in order to inflict the optimum damage on German aircraft. Dowding realized only too well that Germany had the wherewithal to win. The idea was that if Great Britain could resist long enough, and make the raids costly enough, then Germany might forego the campaign.[20] This is exactly what happened, and, as a result, operations research (OR) earned a good reputation. Unfortunately, OR—and its close cousin, systems analysis—later fell into disrepute, at least in the United States. This downfall occurred during the Kennedy–Johnson administrations. Secretary of Defense Robert McNamara's analysts, who meant well, tried to quantify qualitative issues such as the value of human life, exemplified by the infamous "body-count" syndrome.

Unrelated to OR but even more important was the development of the atom bomb. The phenomenon of nuclear fission had been documented in scientific circles for a decade, and the concept of a critical-mass chain reaction was openly discussed in *The New York Times* as early as 1939.[21] In turn, the geometric mechanics of chain reactions led to a general understanding of how catastrophes operated, which, in part, set the stage for the science of chaos. Also, the complexities of the hydrodynamic problems in implosion bomb design forced the physicists and mathematicians (including John von Neumann) at Los Alamos to use IBM, Hollerith-type punch-card calculating machines—forerunners of the modern computer.[22]

The irony here was that the advent of computers enabled traditional, linear science to overshadow its nonlinear competition. The computer's speed negated the objection of time-consuming calculations required by more-or-less complex linear models that roughly mimicked the behavior of nonlinear systems. Then—more irony—the computer itself indirectly spawned the science of chaos. In 1973, Edward Lorenz, a meteorologist, accidentally changed a minor parameter in a computer weather model, and that model reacted with a furious symbolic hurricane. Lorenz claimed that the minor change was mathematically equivalent to the flapping of a butterfly's wings, and so labeled it. To be sure, any analysis of the program would have shown this phenomenon to be the logical or symbolic equivalent of a catastrophic chain reaction but, serendipitously, the findings were reported as chaos.

Remarkably, this new science of chaos picked up steam in less than 10 years. The first notable, if philosophical, book on the subject was Louis J. Halle's *Out of Chaos* (1980). It was followed by James Gleick's *Chaos: Making a New Science* (1987), a tradebook reportorial account of this new discipline. The heart of this discipline is the identification and study of configurations and patterns and their efficacious dynamics—the inherent ability of one pattern to evolve into more useful patterns or, alternatively, to continue productivity or utility despite radical, ongoing change, for example, weather. These patterns are called fractals.

□ □ □

In summary, we have seen how the physical sciences became divided into two paradigms or schools of thought. One school stressed linear, mathematical models; the other, nonlinear configurations and patterns of elements or equivalents in space, which do not readily lend themselves to neat mathematical treatment. The obvious task is to integrate these two paradigms or, more accurately, to discover the underlying symmetry that links them. This book, of course, presumes to take on that task.

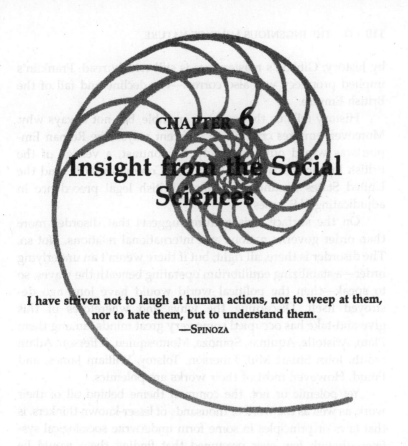

CHAPTER 6

Insight from the Social Sciences

I have striven not to laugh at human actions, nor to weep at them, nor to hate them, but to understand them.
—SPINOZA

The scene is Paris; the year, 1777. *Dramatis personae:* Benjamin Franklin and Edward Gibbon. Franklin is the United States envoy to France. Gibbon is the newly acclaimed author of the first volume of *The Decline and Fall of the Roman Empire* and a member of the British Parliament on vacation. Franklin sends a note to Gibbon, requesting that they meet socially despite their respective nations being at war. Gibbon replies that although he greatly admires Franklin as an individual, it would be inappropriate for him to converse socially with a British subject in rebellion. Franklin ripostes that he greatly admires Gibbon as a historian, and that if Gibbon ever decides to write on book on the decline and fall of the British Empire, he will provide some relevant material.[1] Franklin's estimate of Gibbon has been ratified by history; Gibbon's masterpiece is still widely read. Franklin's implied prophecy was also correct—the decline and fall of the British Empire.

History tells us that empires crumble, but not always why. Moreover, empires crumble in different ways. The Roman Empire was wiped out completely. By contrast, a vestige of the British Empire remains in the British Commonwealth. And the United States continues to follow British legal precedence in adjudicating U.S. cases.

On the surface, this picture suggests that disorder more than order governs nations and international relations. Not so. The disorder is there, all right, but if there weren't an underlying order—a stabilizing equilibrium operating beneath the waves, so to speak—then the political world would have long ago destroyed itself. Discerning the whys and wherefores of this give-and-take has occupied some very great minds, among them Plato, Aristotle, Aquinas, Spinoza, Montesquieu, Jefferson, Adam Smith, John Stuart Mill, Emerson, Tolstoy, William James, and Freud. However, most of their works are polemics.

Yet polemic or not, the common theme behind all of their work, as well as the work of thousands of lesser-known thinkers, is that laws or principles in some form underwrite sociological systems, though few ever presumed that finding them would be simple. Certainly, the repair of the most complicated machine is easy compared to the task of restoring mental health to a child driven to insanity after several years in a Nazi concentration camp. For that matter, no amount of physics will ever explain the absolute horror of the Holocaust. Yet we must have sociological systems if only to survive. If they are to be improved beyond a survival level, we need to understand how they operate and why they fail.

ISSUES

The study of sociological systems contends with many issues, most of which arise from conflicts between individuals exercising their initiative and systems that structure if not stymie that volition for the sake of organizational goals or needs. Some of these issues are:

Nature versus Nurture   This issue asks if human conduct is *primarily* influenced by genetic inheritance or *primarily* by the social (and sometimes the physical) environment. The difficulty of resolution stems from the substantial interaction of individual choices and the environmental reaction to those choices. The degree of cultural or societal isolation or immersion must also be taken into account. Though some cultures once existed in virtual isolation, that situation seldom occurs today. Every culture interlocks, so to speak, with other cultures, be it for mutual benefit, mutual detriment, or as geopolitical fodder.

Regional Factors   This issue covers weather, climate, geography, and the like. We know that some aspects of culture are similar worldwide, while others are in part a function of regional characteristics. In addition, different population sizes and densities can significantly influence culture, commonly urban versus rural communities.

Intellectual Acuity   This covers beliefs, extent of knowledge, technological prowess, and the ease or difficulty of communications. Clearly, behavior is influenced by beliefs, knowledge, and technology. But as the level of technology rises, does this weaken systems of belief? Does this potential weakening lead to moral decay? Lastly, are efficient communication networks always a blessing? Roughly 80 percent of the U.S. population fears invasion of privacy traceable to computerized databases.[2]

Economic Motivation   No serious observer denies the tendency of economic forces to shape individual and societal choices, but some regard these forces as peripheral while others view them as the overriding source of most human behavior. John Stuart Mill seldom discoursed on economics; Marx and Engels made it central to their writing.

Function versus Structure   Do social organizations mold their structure in light of function? Or does function evolve primarily from structure? Some analysts, especially in Great Britain, stress the structure of relationships—configurations—in culture as the key to understanding sociological systems. Other analysts, especially in the United States, stress the functions or purpose for which societies and states are formed. The British constitution, for example, is unwritten—a high-level common law that evolved slowly from glacial changes to social structures. By contrast, the U.S. Constitution is a singular written document that begins with a well-known list of functions: "We the people of the United States, in order to form a more perfect union, establish justice, insure domestic tranquillity, provide for the common defense, promote the general welfare, and secure the blessings of liberty. . . ."

Growth versus Decline   Do societies respond principally to challenges, as Arnold Toynbee concluded? Or is growth inevitable up to some point, followed by decline? The second viewpoint is analogous to life cycles of biological organisms, and, interestingly, when Walter Cannon first identified homeostasis in organisms (1939), political scientists immediately adopted the term to describe societies under government rule.

Collective Ego   Do sociological organizations harbor a collective ego? A collective ego (or collective conscience) is a "spirit" that supposedly exists independent of the psychical nature of individuals, or, that failing, people collectively act as if there were such a discrete "spirit." This conscience may be the artificial person created at law to represent the state or a corporation. Or it may be a collective feeling of belonging to something or some ideal above and beyond the self, for which one may be willing to die, such as nationalism.

During her prolonged heresy trial, Joan of Arc was asked why she carried her personal banner (which depicted the Last

Judgment) into the cathedral at Reims during the coronation of Charles VII of France (following the victorious battle at Orléans). Her reply suffers in translation, but we can still glimpse the effect of a collective conscience. Joan rose to her feet, looked quietly at her many judges, then calmly said: "It had bourn the burden; it had earned the honor."[3] Interestingly, history credits the Maid of Orléans with initiating the idea of nationalism, and she is now regarded as the patron saint of France.

PSYCHOLOGY AND PSYCHIATRY

The disciplines of psychology and psychiatry trace their roots to ancient Greece. Moreover, and unlike parallel writing in the physical sciences, some of these discourses remain valid. The ancients were, as we are now, confronted with the consequences of human behavior. Furthermore, some of the ancients—especially Greeks—were poets and writers who recorded this behavior in terms appealing to the modern mind. On the other hand, few of them probed the underlying reasons for behavior, and little was or could be done to ameliorate madness and other psychiatric dysfunctions. That task would have to wait for nearly two millennia.

The first comprehensive probe of the psyche appeared in Spinoza's *Ethics* (1663–1666, published posthumously in 1677). The *Ethics* is often regarded as a supreme achievement, and Franz Alexander—Freud's first student—said that Spinoza's influence on modern ideas was pervasive to the point where "many of his basic concepts became part of the general ideological climate that influenced Freud without his knowing its origin."[4]

Then in 1890, William James compiled his two-volume study *The Principles of Psychology*. Spinoza had written *Ethics* in scholastic terms and waxed philosophical, which made acceptance by science problematic. By contrast, James wrote with scientific rigor, concentrating on the more verifiable aspects of psychology. Although James's book was not honored on a par with Newton's *Principia*, the effect was nearly the same because it arguably marked the turning point of psychology from philosophy into a scientific discipline.

Shortly afterwards, Sigmund Freud merged, in effect, the fields of psychology and psychiatry. Psychiatry had existed as a branch of medicine for more than a century, but for the most part it concentrated on the palliative care of the insane. Freud changed that by pulling back the curtain on the hidden recesses of the human psyche. Or at least he tried, for many of Freud's ideas have long since been discarded or modified beyond recognition. Still, while his emphasis on the sexual drive was probably overwrought, Freud must be given credit for originating psychoanalytic therapy and making the subconscious an integral part of psychiatry. In so doing, he opened a field of study that attracted hundreds of noted scholars.

Of these scholars, Carl Jung is perhaps the best known, perhaps because Jung also sought to understand the spiritual or religious nature of man. Elsewhere, Jean Piaget nudged psychiatry further into pediatrics by way of a psychological theory on how a child develops the power to think (cognitive ontogeny). Other major contributors include Alfred Adler, Erich Fromm, Erik Erikson, Karen Horney, Kurt Lewin, Arnold Gesell, Karl Menninger, Harry Stack Sullivan, Rollo May, and Abraham Maslow.

To be sure, controversy continues to mark psychology and psychiatry. The most serious contention is the role of physical versus psychical influences on behavior. Alexander and Selesnick said (sarcastically) that "no longer a devil but a *deus ex machina*, a disturbed brain chemistry rather than a person's own life experiences, is responsible for mental illness" which leads to the conviction that "the disturbed mind can now be cured by drugs and that the patient himself as a person no longer needs to try to understand the source of his troubles and master them by improved self-knowledge."[5] So is human behavior entirely a function of biochemistry, so that all dysfunctions can be alleviated by pharmaceuticals? Or when the pharmacopeia is exhausted, will there still be a psyche to contend with?

ANTHROPOLOGY AND SOCIOLOGY

Sociology deals with relationships among individuals, how these relationships affect group norms, and how those norms affect individual behavior. Anthropology adds the dimension of how this process—usually called *culture*—developed in different regions and how that history affects current group behavior. Understandably, these fields of study stress systems more so than psychology, because once individuals band together to achieve some common purpose, the very act of doing so falls under the definition of a system. However, unlike physical or rigid abstract systems, organizations must contend with higher degrees of freedom. Except for chain gangs, people are not physically linked nor are they irretrievably bound by logical syllogisms. On the contrary, many sociological discourses speculate on the how and why of creating social order out of human anarchy and why social order sometimes degenerates into chaotic behavior, especially by way of riots, rebellions, and revolutions.

These theories have been categorized under four schools of thought: interaction, conflict, evolution, and functional.[6] Relationships and interactions, of course, are the *sine qua non* of systems, and the mass of human interactions in sociological systems are bound to create conflict. Whether such conflict leads to evolutionary or revolutionary change depends on how well the environment contains the struggle (political homeostasis). And because all systems have at least one purpose or aim, the functional school deserves equal billing.

Let us look at how perceived systems within society affected noted scholars. Of these scholars, Charles Darwin heads the list. *The Origin of the Species* (1859) and *The Descent of Man* (1871) concentrated on biological evolution, but the obvious implication was that regardless of the mind's capability of rising to sublime heights of accomplishment, the physiological host remained subject to biological instincts. His adjunct thesis (to evolution itself) was survival of the fittest, which conflicted with man's occasional desire to treat everyone as equal.

Shortly after *The Origin of the Species* appeared, Herbert Spencer published *First Principles* (1861), which he claimed to be a unified theory covering all phenomena. The primary theme of Spencer's book was that all complex phenomena found in nature—including the whole of biology—were nothing more than permutations and combinations of simple processes, just as the most complex mathematical theories are ultimately permutations and combinations of simple arithmetic.[7] However, because Spencer relied heavily on bits and pieces of questionable evidence, most of his thesis has been ignored.

About the same time, Edward Tyler, who like Darwin based his conclusions on what he saw on voyages, attempted to develop an integrated understanding of psychological, cultural, and spiritual influences on human behavior.[8] He came to the conclusion that despite major differences among cultures, the nature of the human mind was singular. In turn, this conclusion encouraged other anthropologists and sociologists to seek relevant universal laws.

At first, this synthesis emphasized the individual, but that emphasis soon changed. James Frazier devoted most of his life to studying the role of myths—or mythical powers attributed to objects—and how they unified various cultures. Émile Durkheim, in his book *Suicide* (1897), stressed the collective conscience—a group ego that could dominate if not warp the exercise of human initiative.[9] This focus on collective phenomena found another advocate in the person of Franz Boas, who moved anthropology about as close as anyone could to the physical sciences.[10] Although Boas was personally rebellious against all kinds of authority, he subordinated himself to facts, especially when those facts were susceptible to statistical analysis. He criticized any student who let feelings intrude on such analysis.

Bronislaw Malinowski backed off from this rigor by concentrating on functional motivations. He concluded that the seminal aspects of culture resulted from individuals collectively filling a need they could not achieve by striving alone.[11] Alfred Kroeber took another tack in classifying cultures by regions as if they were biological species.[12] And Ruth Benedict claimed that the study of culture as a science was futile; too many variations intruded.[13] Hence, any scientific attempt to find generalizations was doomed. A psychiatrist at heart, if not by sheepskin, her studies concentrated on how the individual fulfills personal goals without disengaging from his or her cultural environment.

In perspective, many scholars in anthropology and sociology gave impetus to viewing society and cultures as systems. These systems were especially evident in isolated cultures, like the Anasazi Indians who lived in Mesa Verde and Chaco Canyon. But modern cultures are complex. They exhibit so many overlapping subsystems—legal, economic, religious, and so forth—that it is all but impossible to isolate any one subsystem for study.

20. **Misleading anthropology.** Ruins, such as these at Mesa Verde, indicate that many ancient cultures flourished under systems in which human initiative played only a minor role. But the history of rebellion and revolution suggests that human initiative is an explosive force. To ignore it is to beg failure of understanding among the social sciences.

ORGANIZATION AND MANAGEMENT THEORY

Organization theory facilitates the study of complex social entities because most organizations are limited and have a singular focus. Furthermore, most management analysts agree that the common focus of their work is on systems. In turn, organizational systems can be viewed from many perspectives. One perspective stresses corporate or business economics. The tendency of poorly managed companies is to lose money and go out of business, often via bankruptcy. That is, too much cash flows outward, i.e., financial entropy. Business economics seeks ways to reverse this tendency in order to increase profits and cash flow. Hence, most studies from this perspective ask a common question: What does it take to make and keep a company profitable? If it is already profitable, the question is: What does it take to make it more profitable? Sometimes the answer is to increase market share. At other times the answer is a new product, or lower prices, or higher prices, or cutting expenses, or changing the advertising thrust.

A second perspective offers a comparison between leadership and management skill. Leadership has two forms. One form concentrates on motivation of individuals; the other, on leading an organization in a direction that it would not go in the absence of that leadership. The motivational form has been studied intensely by the military but, beyond the battlefield, motivation may be less important than structure, organization, and sound strategy.[14] By contrast, organizational leadership is the stuff of M.B.A. programs. Alfred P. Sloan, at General Motors, and Lee Iacocca, at Chrysler Corporation, are probably the best-known cases. Sloan's autobiographical *My Years with General Motors* (1964) is considered a classic.

By contrast, management is less concerned with heroes or motivation beyond the effect of good human relations skills. To the manager, people are primarily resources. As an extreme example, Frederick Taylor used time-motion studies that mapped out work activities as if each worker were an automaton that could be reprogrammed in order to improve efficiency. Understandably, other observers recoiled from this perspective, i.e., McGregor's Theory X versus Theory Y and the Management Grid seminars, both of which emphasize the importance of the individual within organizational settings.

Another perspective—morphology—stresses *optimum* organization, which almost by definition is a variation on operations research. Its hallmark is the familiar organization chart. These charts have spawned thousands of jokes, but a less biased view brings out several bona fide issues. One issue asks if it is better to organize along departmental lines (manufacturing, sales, and so on) or along functional lines (e.g., new consumer products). Another issue focuses on the optimum mix of centralized (hierarchical) versus decentralized (relational) control. In practice, some organizational tasks benefit from a hierarchical, top-down implementation of decisions. Other tasks thrive on the relational form, with its emphasis on creativity.

Yet another perspective compares public and institutional management. Although nonprofit organizations lack the profit motive by definition, they still must cope with other problems common to all organizations. This situation leads to two opinions. One opinion advocates separate graduate degree programs in public administration. The other opinion holds that M.B.A. programs are more useful because it is easy (in theory) to think of profit in terms of providing more services for the same dollar. Regardless of the point of view, two conclusions encounter little debate. First, nonprofit organizations typically suffer from higher entropy (read: bureaucracy). Second, institutions— especially government agencies—continue to grow in number and size, more so than proprietary firms. They take on a life of their own, as it were, arguably in excess of public need—one of the so-called Parkinson's laws.[15]

To be sure, entropy in the form of increasing sluggishness infects all organizations. Virtually all organizations grow moribund, decay, swell up out of proportion to mission, or otherwise outlive their usefulness. In a time of crisis under strong leadership this entropy can be reversed, but the reversal is seldom permanent. The solution—such as it may be—is the sociological counterpart to organic life cycles. Old systems wither; new systems emerge.

POLITICAL SCIENCE

Though it preceded organizational theory, political science may be considered as organizational theory applied to jurisdictions, especially nations. Given the extremes of human behavior, the conflicts between the state and individuals, and the lack of definitive goals (except in times of crisis), the relationships among citizens and government must be codified to preclude chaos in the form of anarchy. The basic question of political science is: How does government work, and if it isn't working well, how should it work?

All systems operate. They operate because their constituent elements relate and interact. The pattern of those interactions arises from their logical and/or physical arrangement and the attributes of elements. As such, all forms of government systems have at least one configuration and usually several overlapping configurations. The elements are: individuals, customs, laws and regulations, and the state as an entity. The state also controls natural resources: waterways, land, ores, fresh water, and so on, most of which are relatively constant from year to year.

As a result, individuals and families cannot independently provide for all of their own needs. They depend on some form of system to meet those needs while leaving them free to pursue their own interests and aspirations. That system is called the state, which in time tends to take on a life of its own, if only because the law recognizes it as an artificial person. Next, many citizens—perhaps most—participate in the state in order to further their own ambitions, regardless if their contributions enhance or denigrate the common good. Because of this, conflicts will arise between the bona fide rights of the state and those of citizens.

Lastly, a government is obliged to establish some form of political equilibrium or homeostasis to keep these conflicts within acceptable limits, neither stultifying the nation nor begging revolt and revolution. The ideal system will reverse entropy, so to speak, by ensuring continuous growth. In practice, many governments settle for the status quo, providing that "quo" has sufficient status in the eyes of the majority of citizens.

To attain either goal, virtually all governments take one of five configurations: (1) dictatorship, (2) monarchy, (3) oligarchy, (4) aristocracy, and (5) democracy, usually in the form of a constitutional monarchy or a republic.[16] These forms have been studied, dissected, resurrected, manipulated, and modified by literally millions of articles and books, many of which also expound on communism and socialism. But keep in mind that neither communism nor socialism is a form of government despite many opinions to the contrary. The reason is that they are economic systems or models that, in theory at least, can be incorporated into any form of government. In practice, virtually all national-level communistic systems devolve into dictatorships, whereas socialism, or the alternative of so-called welfare capitalism, can be found to one degree or another in all forms.

Perhaps the most intriguing aspect of governmental configurations is how they spawn rebellion and revolution—the political science equivalent of catastrophe. Like a nuclear explosion, a revolution is a self-feeding process that uncontrollably draws on its own (human) resources not otherwise harnessed. Some revolutions yield obvious benefits, either immediately or within a decade, for example, the American Revolution (1776–1783). Others degenerate into terror, such as the French Revolution (1789–1796). Will and Ariel Durant, in their 11-volume *Story of Civilization* (1935–1975), cite about 5000 social upheavals of one degree or another.

Many of these upheavals were triggered by a book or tract: the equivalent of a critical point. Some were written in the form of novels, for example, Harriet Beecher Stowe's *Uncle Tom's Cabin* (1852), which preceded the American Civil War. Others were were diatribes, e.g., Thomas Paine's *Common Sense* (1776), at the time of the American Revolutionary War, and Friedrich Engels and Karl Marx's *Manifesto of the Communist Party* (1848), at the time of the concurrent revolutions attempted throughout much of Europe. In the 19th century, Henry David Thoreau's *Civil Disobedience* (1849) was ignored, but in this century his thesis bore enormous fruit in India's successful drive for independence from Great Britain and in this country by the civil rights drive led by Martin Luther King, Jr. The list also includes John Milton's *Areopagitica* in 1644 (on freedom of the press) and the profound scholarship of John Stuart Mill, especially his *On Liberty* (1859) and *Representative Government* (1861).

ECONOMICS

Economics has four levels or perspectives. The first level looks to the great *forces* underwriting the accumulation and expenditure of wealth, a perspective that this book calls *metaeconomics*.[17] From this perspective, money is only a medium of exchange useful to measure the relative magnitude of transactions. The second level—*macroeconomics*—focuses on a nation's economy, which increasingly takes international trade and finance into account. This level, and the remaining two levels, are *money* oriented. The third level—*corporate microeconomics*—focuses on business operations. Businesses range from a corporation that controls more assets than owned collectively by the poorest half of the world's countries, to a so-called mom-and-pop shop. The fourth level—*consumer-unit microeconomics*—is similar to business microeconomics but focuses on how individuals and family units cope economically. With these levels of perspective in mind, let us review the analytic contributions of five major economic thinkers: Adam Smith, David Ricardo, Friedrich List, Alfred Marshall, and John Maynard Keynes.[18]

Adam Smith's metaeconomic arguments in *The Wealth of Nations* (1776) described how various economic forces comprise an abstract system that depends on a kind of growth-oriented equilibrium. He also said that wealth was not money but rather the total productivity of a nation to meet its own needs. Moreover, the output of this productivity (whether reinvested as capital to produce more goods or directly consumed) possessed an inherent or theoretical value about which prices fluctuated. He also saw this phenomenon operating beyond effective control by man, saying it was guided by an "invisible hand." As such, his theory provides the main justification for capitalism and *laissez faire*.

Four decades later, David Ricardo published *The Principles of Political Economy and Taxation* (1817). In that book, he argued that production and growth weren't everything because the proceeds had to be distributed. If maldistributed, the benefits reached a point of diminishing returns. He also said that excessive taxation, rents, and protectionism had a negative effect on the economy. As a member of the British Parliament, however, he often witnessed his ideas overrun by political expediency.

The third scholar is Friedrich List. List understood the force of expediency and, after migrating to the United States from Germany, he became the father of political economy. His thesis was that laws of economics were neither universal nor immune to control by man. In *The National System of Political Economy* (1841), he argued that the task of a nation was not to worship value but to ensure the capacity to produce wealth. But unlike most of his contemporaries, he did *not* advocate any specific program of protectionism or incentives. Instead, he said that a government should act pragmatically and use whatever policies enhanced production, changing them when the circumstances demanded new policies.

The fourth economist in this group—Alfred Marshall—reasked the question: How does economics operate? In *Principles of Economics* (1890), he saw economics operating as a system with several dominant but competing or conflicting forces, seeking equilibrium, practicing what might now be called homeostasis (microeconomic bankruptcies) when it could not, and exhibiting catastrophe (depressions or runaway inflation) when that failed. He employed extensive diagrams but was less enthralled with pure mathematical explanations because too many factors influenced economics that did not lend themselves to quantification.

This brings us to the last of the five economists in this group—John Maynard Keynes, who was Alfred Marshall's most apt pupil and biographer. Keynes was an avid capitalist who sought to curtail the excesses of capitalism while spurring productivity and spending. In *The General Theory of Employment, Interest and Money* (1936), he argued that the *propensity* to consume was the key, and that consumer units would spend at least some of any additional income they earned, hence feed the economy. As such, governments must avoid recessions and depressions at all costs. To achieve that goal, he advocated a host of government policies to provide the necessary leverage, especially discretionary tax policy. Today, most free-world governments are Keynesian in their outlook, but these policies have not avoided recessions nor have they assured *steady* growth on any long-term basis.

One additional contribution rates mention, and that is John von Neumann's *Theory of Games and Economic Behavior* (1944). This book did not address economic foundations or national economics, but von Neumann did see economic competition as a variation on game theory, which arguably is an application of systems theory.

GEOPOLITICS

Geopolitics is political science combined with economics as viewed from a geographic perspective. It asks the penetrating question: Is geography destiny? Put another way: To what extent do the geographical attributes of a country determine its role in world affairs? In this context, geography includes relative position and size, configuration of coastlines and availability of harbors, natural resources, economic power, and military prowess derived from those attributes, though traditional geography remains the dominant factor.

That is to say, the world is primarily a physical system with its elements distributed in a more-or-less permanently defined arrangement or configuration. The geographic mass of each country is the source of its international clout as leveraged by relative position. For example, Great Britain, as an island nation, has a natural defense against attack and, by the same token, easy access to sea lanes. This enabled her to create a global empire. But her lack of geographic mass contributed to the decline of that empire. By contrast, the United States had all of the geographic advantages of Great Britain, plus considerable mass. This combination permitted her to become a superpower. Russia has the greatest land mass of any country but only limited access to the oceans. Accordingly, her former superpower status, based on land mass alone, has been significantly deflated.

The same lessons apply to the Roman Empire. Rome was able to create and control a massive empire for nearly 1000 years by incorporating vast territories into her system of rule. No other nation or culture remotely possessed the resources or infrastructure to take on that empire. Yet as documented so well by Edward Gibbon in his renowned *The Decline and Fall of the Roman Empire* (1776–1787), numerous smaller rebellions in the various provinces, combined with steadily increasing internal rot, served the same purpose.

Keep in mind that the distribution and shape of land is fixed, and each region has a predictable climate on average while natural resources do not fluctuate greatly in any one generation. Similarly, and with the very notable exception of the recent breakup of the Soviet Union, the boundaries of most nations have stabilized: United States, Canada, Mexico, virtually all Latin American countries, Western Europe (except the former Yugoslavia), Japan, China, Indonesia, India, Australia, New Zealand, the bulk of the Middle East, and two-thirds of the African nations. Thus, to rephrase the critical question of geopolitics: To what extent did geography contribute to this equilibrium or stabilized balance of power?

To answer that question, consider the arguments of two seminal works in geopolitics. The first was Alfred Thayer Mahan's *The Influence of Sea Power Upon History, 1660–1783*, written in 1890. The second was Sir Halford Mackinder's paper "The Geographical Pivot of History," delivered before the Royal Geographic

21. **The Roman Forum as the center of an international system. The Roman Empire declined and fell, yet in nearly 1000 years of existence it demonstrated clearly what could and could not be done in geopolitics. Fifteen hundred years of technological progress have not altered those lessons.**

Society in 1904. Mahan argued that seapower was the main determinant of national power *if* a nation had the geographic position, economic strength, and political infrastructure to exploit it. His book was an immediate success in Great Britain, which at the time was the world's foremost seapower and had used that power to build and sustain its empire. Additionally, Mahan's thesis awoke the United States to its own potential. Within 20 years, the United States rose from the world's sixth largest naval power to the second largest (and eventually to the largest).[19]

By contrast, Mackinder's paper recognized the enormous geographical mass possessed by Russia in the middle of the Eurasian continent. He predicted that she would become the world's dominant nation, echoing Tocqueville's observation in 1835, namely, that Russia and the United States were quietly growing into world superpowers. Yet within a century both Mahan's and Mackinder's theses deflated. The British empire no longer exists and her navy is literally rusting. And while Russia continues as the largest country in the world, it no longer exerts global hegemony, especially by way of political allies.

To restate the case, Admiral Mahan's book tacitly assumed the world was up for grabs. Within 50 years, several nations took up the challenge at the cost of 100 million lives. Germany became the foremost of the aggressors, followed closely by Japan. After the Axis powers were defeated, the former Soviet Union exerted itself for roughly 44 years without directly fighting any major wars.

The role of the United States followed a more curious path. In the same year as Mahan's argument, Frederick Jackson Turner observed that the U.S. frontier no longer existed; *manifest destiny* had run its course.[20] Thus, when it came to the Spanish–American war 8 years later, the United States took advantage of the victory to acquire additional territory: chiefly Puerto Rico, the Panama Canal Zone, the Philippines, and Hawaii.[21] Hawaii has since become an integral part of the United States, and Puerto Rico is a U.S. protectorate. On the other side of the coin, the Philippines has long since gained its independence, and U.S. control over the Canal Zone expires in the year 2000.

□ □ □

Admittedly, this tour of the social sciences seems like only a mass of facts and ideas with only vague relationships. Yet the operating fundamentals of systems are present. All sociological systems consist of elements—people and their institutions—and the interrelationships among them. The error has been to seek too much "system," ignoring the vast significance of human emotions and initiative. This washes out only in geopolitics, wherein the configuration of massive physical elements, and how those elements affect human behavior collectively, operates like a physical system. The other social sciences lack this anchor, and as such appear to be excessively fluid. Appear. We need to image geopolitics as a theoretical baseline for sociological systems. From that baseline, the other social sciences can be viewed as variance because of the greater opportunities to exercise human initiative.

PART III

Abstract Systems

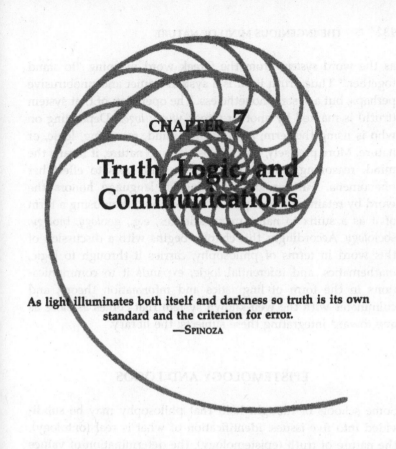

CHAPTER 7

Truth, Logic, and Communications

As light illuminates both itself and darkness so truth is its own
standard and the criterion for error.
—Spinoza

What is truth? Two thousand years ago, Pontius Pilate asked
Jesus that question, and philosophers have been asking it ever
since (and before too). Science devotes itself to finding truth
without bothering to define it except by pragmatic criteria. Can
such and such phenomenon be verified? Does thus and thus
theory predict new phenomena? Is the so-and-so concept gaining acceptance among experts? As one scientist put the case:
"Only God can define absolute truth, and I don't believe in
Him."

We begin our survey of applications of the thesis of this
book with the most abstract system of all—truth. The scholarly
word for the nature of truth—*epistemology*—shares the same root

as the word *system*, from the Greek word meaning "to stand together." Thus, truth itself is a system—quiet and unobtrusive perhaps, but a system nonetheless. The operation of that system (truth) is marked by another Greek word: *logos*. Depending on who is using the term, *logos* means: mind, reasoning, logic, or nature. More precisely, when *logos* refers to nature, it means the mind, reasoning, and/or logic that nature uses to effect her phenomena. Furthermore, the English language honors the word by retaining it as a pure transliteration, then using a form of it as a suffix to name many sciences, e.g., geo*logy*, bio*logy*, socio*logy*. Accordingly, this chapter begins with a discussion of this word in terms of philosophy, carries it through to logic, mathematics, and inferential logic, expands it to communications in the form of linguistics and information theory, and culminates with the well-known system that comes as close as any toward integrating these subjects: the library.

EPISTEMOLOGY AND LOGOS

Some schools of thought hold that philosophy may be subdivided into five issues: identification of what is real (ontology), the nature of truth (epistemology), the determination of values (axiology), the methodology for pursuing the first three issues (logic, reasoning, logos), and historiography (in the sense of how subsequent philosophers draw on and/or modify the work of their predecessors).[1] In terms of systems, the thesis of this book views elements as ontology, and the relationships and programming inherent in the configuration and interactions among those elements as epistemology.[2] *Pure* physical and abstract systems have no axiology; only sociological systems must contend with values, which in turn spawn the applied abstract systems of ethics, law, and theology. This chapter focuses on logos, while Chapters 5 and 6 gave a lick and a promise to historiography.

Elements are easy to understand; we focus here on epistemology (truth) and logos (the means of ascertaining truth). The two are inseparable because—as noted by Spinoza in the epigram that opens this chapter—the process of determining the nature of truth must itself be truthful. Truth may be universal or personal. The logic of metabolism is universal; your date of birth is personal. Elsewhere, one discovery can have global implications; another will have meaning only for its discoverer. Newton's work had global implications (to say the least); the discovery of a modest talent for playing a musical instrument seldom reaches beyond the front door. Lastly, truth may stand on its own or require the support of preceding derivations. A historical fact *per se* needs no support. It is ontological: something happened on such and such a date. Period. The "why" or "how" of that fact relates to many other facts (epistemology), but once established a fact stands as is, notwithstanding that another historian may question or deny it. By contrast, a mathematical theorem claims legitimacy only to the degree it flows unhindered from axioms (basic assumptions) and preceding theorems.

Yet all forms of truth have a common trait. That trait is *interior lines*. Interior lines refers to the unadulterated ease or simplicity with which elements in a system relate or are otherwise linked with one another. The Interstate Highway System perfectly illustrates interior lines because those highways link major urban areas with nearly direct routes that (in theory) require no stopping or detours. Similarly, the interior lines of truth are marked by the ability to go from one fact to the next—to link—without paradoxes, exquisite reasoning, or other forms of esoteric rationalization. This property is sometimes called *logical integrity*.

The colloquial expression for logical integrity is that such and such explanation has "the ring of truth" to it. The logical paths among facts and other elements of knowledge must *resonate* analogous to a fine musical composition, otherwise the attempted linkage becomes just so much "noise." Loose ends are tolerated *if* they suggest that another fact or element exists which remains to be discovered. Mendeleyev's periodic table of chemical elements was a classic case of this. For that matter, planets have been located because of gaps in a logical model. The planet Neptune was said to have been discovered "on the tip of a pen."[3]

Restated, truth depends on logical linkage. These linkages may reveal cause and effect, a correlation, a proportion, a sum, rules, or any other relationship. The specifics do not matter. The elements or facts must stand together to form a larger picture, which is the definition of a system, of which truth itself is a major case. However, this does not mean all observers will agree with that picture. In a satire of academe, Lewis Carroll had the Mock Turtle in Alice's Wonderland say: "I only took the regular course ... the different branches of Arithmetic—Ambition, Distraction, Uglification and Derision." In short, truth resists being picked apart. It does not convolute or entangle itself. Tweedledee—another character in Alice's Wonderland—explained logic in plain terms: "If it was so, it might be, and if it were so, it would be; but as it isn't, it ain't."

The opposite of truth is not ignorance. Ignorance is a state of mind before truth is discovered. Elsewhere, ignorance is insufficient awareness of sufficient facts or the inability to perceive relationships among known facts. Instead, the opposite of truth is falsehood. Falsehood either misstates facts or emphasizes a relationship that does not exist or which is negated by another relationship. When this situation is unintentional or unavoidable, the process is called trial and error (heuristics). A physician must often rule out various possibilities before arriving at an accurate diagnosis. This process identifies discontinuities in each alternative until only one is left that passes muster. If no satisfactory alternative remains, additional facts are sought. But when the falsehood is intentional, the action is fraud. Fraud ranges from playful blarney to the absurd "logic" employed by Hitler (in *Mein Kampf*) to justify the Holocaust. Bishop Fulton Sheen said that blarney was bologna sliced so thin it was delicious. No such levity attends the memory of Auschwitz. On the positive side, the process of ascertaining truth is the history of science and philosophy writ large. This process is also the basis for therapeutic analysis in psychiatry, psychoanalysis, and psychotherapy.

We think about facts and their potential relationships, from which we often propose theories. In science, as in life, most theories prove inadequate if not outright useless. Some theories (or ideas), however, gain tentative acceptance, and a few of those eventually earn global acknowledgment. For a while at least. The fact of the matter is that few scientific theories stand unchallenged for more than a century. Yet even when popular theories are debunked, many of their elements are salvaged and rewoven into a new logical cloth. A good scientist is therefore something of a rebel. He or she strives to find order amid logical chaos, even though the process may appear to overturn what is perceived as established order.

In sum, truth itself is a static abstract system while the methods to determine it constitute a dynamic system—logos. Euclid's *Elements* is a static system of truth. In contrast, playing according to Hoyle is a dynamic means of searching for a solution or answer. In a game, the search becomes strategy. In a courtroom, the search becomes a presentation of sufficient and convincing interior lines among equally sufficient facts. Truth is recognized because the configuration of relevant facts can be linked without strain. The logical integrity of the presentation— the interior lines—speaks for itself. We see this quite often by way of the question: What do the facts tell you?

Now let us see how well systems theory dovetails with the "Great Ideas" developed in the *Great Books of the Western World*. This set of more than 500 masterpieces written by 135 among the Western world's most noted thinkers is indexed by 102 "Great Ideas" delineated in the accompanying two-volume *Syntopicon*. The *Syntopicon* is further linked in excess of another 500 thinkers and several thousand of their works.

At first glance, the fit of the systems idea with the *Syntopicon* seems poor indeed; the word *system(s)* does *not* appear in the title of any books or papers included in either the first or second edition. For that matter, the word does not rate an appearance in the index of key terms that was added to the second edition of the *Syntopicon*. However, the word *is* found in the title to part III of arguably the most influential book in science ever written: *The System of the World* in Newton's *Principia*.

Moreover, the *Syntopicon* itself exemplifies a system, especially with respect to the interior lines of reference linkage. When

	Primary relationship		Secondary relationship
Systems in general	• Being • Metaphysics	• Nature • World	• Art • God
Factors	• Infinity	• Space	
Purpose	• Desire • Idea		• Form • Memory and imagination
Elements and attributes	• Element		• Man
Quantity and variance	• One and many • Quantity • Same and other • Universal & particular		• Mathematics
Initiative	• Chance • Opinion • Will		• Angel • Desire • Cause • Liberty • Courage† • Soul
Degrees of freedom	• Liberty • Slavery	• Tyranny	• One and many
Arrangement	• Art • Form	• Beauty	• Idea • Necessity and contingency
Dominant elements and focal points	• Relation		• Mechanics • Revolution
Operations and Effects	• Eternity • Time	• History	• Nature
Change and determinism	• Cause • Change	• Fate • Prophecy	• Choice • Custom and convention
Configuration mechanics	• Mechanics • Progress		• Evolution • Labor
Entropy and its reversal	• Immortality • Life and death		• Being • Eternity • Change • Infinity
Sustainability and resiliency	• Custom and convention • Habit		• Immortality
Equilibrium and homeostasis	• Opposition • Temperance		• Induction
Controllability and cybernetics	• Experience • Necessity & contingency		• Education • Sense
Power curves and catastrophe	• Prudence (avoiding catastrophe)		• Nature (abnormal phenomena) • Revolution • World (chaos versus order)

† One of the four great ideas not listed in the primary column

22. Systems and the Great Ideas. All but 4 of the 102 "Great Ideas" outlined in the *Syntopicon* of the *Great Books of the Western World* set can be anchored to generalizations on systems theory and chaos, or to categorical applications of that theory.

	Primary relationship		Secondary relationship	
Abstract Systems	• Truth	• Principle	• Form	
Means of seeking truth	• Dialectic • Hypothesis • Induction • Judgment	• Reasoning • Sense • Science	• Art • Beauty • Quality†	• Poetry† • Wisdom
Logical systems	• Mathematics	• Logic		
Information and linguistics	• Definition • Knowledge • Language	• Rhetoric • Sign and symbol		
Physical systems	• Matter	• Physics	• Metaphysics	
Inorganic	• Astronomy		• Chance • Change	• Progress • Relation
Organic	• Animal • Evolution	• Man • Medicine		
Sociological Systems	• Soul		• Angel	
The psyche and general psychology	• Memory and imagination • Pleasure and pain • Emotion	• Mind	• Desire • Habits	• Man • Sense
Development, education, dysfunctions	• Education • Happiness • Love		• Medicine (psychiatry) • Memory and imagination • Prudence	• Temperance
Group behavior	• Duty	• Family	• Signs & symbols	• State
Economics	• Labor	• Wealth	• Mathematics	• Relation
Nations and types of government	• Aristocracy • Democracy • Monarchy • Oligarchy	• Citizen • Government • State	• Freedom • History • Slavery • Tyranny	
Revolution, war, and world order	• Revolution • War and peace		• Custom and convention • Emotion	
Jurisprudence	• Constitution • Punishment	• Justice • Law	• Government • Opposition	• Duty • Judgment
Ethics	• Good and evil • Virtue and vice	• Sin	• Beauty • Courage† • Happiness • Honor†	• Quality† • Truth • Wisdom
Philosophy	• Philosophy	• Wisdom		
Theology	• Angel • God	• Religion • Theology		

22. *(Continued)*

you study the various summary essays and structured bibliographies, a sense of system gradually permeates your thinking. Furthermore, all but 4 of the 102 "Great Ideas" immediately relate to some aspect, element, or application of the theory developed in this book (and the 4 exceptions are highly subjective ideas). The accompanying two-page chart in Figure 22 illustrates these relationships. In the primary column, each

"Great Idea" is listed once under the systems aspect or application to which it most closely relates. In the secondary column, many of the "Great Ideas" are again listed (some more than once) if they also apply to other aspects of systems, albeit not as strongly as their primary relationship.

EUCLIDEAN PARADIGM OF DEDUCTIVE LOGIC

By far, Euclid's *Elements* (circa 300 B.C.) is the best-known and perhaps the most widely used model of deductive reasoning, and it is still used as the basis for teaching geometry today. Reportedly, it has outsold all other books in Western history except scripture, and Einstein said that he who was not transported by Euclid in youth could never hope to become a serious theoretician. Even Abraham Lincoln made a point of saying that he had "nearly mastered the [first] six books of Euclid."[4]

To be sure, the subject of geometry had a mundane start in life: re-marking the boundaries of farmland in the Nile valley after annual flooding. The simple way to lay out right angles was to make a triangle out of rope: three units on one side, four units on the second, and five units along the third. Generations later in Greece, Pythagoras generalized the concept by proving the sum of the squares of the sides of a right triangle equaled the square of its hypotenuse. Other mathematicians proved other theorems, but each selected his own assumptions (axioms). This meant that A might rely on any one of B's theorem to prove his own and vice versa. Euclid scanned this chaotic logic and imposed a singular system on it. First he established a primordial set of axioms. Then from these axioms, he re-derived all of the known theorems in sequence, perhaps creating a few new theorems to fill gaps.

As such, the system implied by the *Elements* is hierarchical. In hierarchical systems, the uppermost elements are dominant, and the principal interior lines descend from them. In Euclid's case, the axioms are the dominant elements, and all theorems logically descend from them in sequence. The logical integrity—the logical interior lines—are flawless.

The only thing missing is for these lines of logic to wrap back on themselves to prove their own axioms. That seems to be impossible (as Goedel may have indirectly shown in 1931).[5] Thus, all proofs in mathematics (and science by implication) remain provisional based on the stability of the axioms and assumptions from which the logic flows to explain the relationships among mathematical elements or scientific facts as the case may be. Because of this, any broad-based acceptance of a theory requires: (1) universal agreement on the axioms, (2) flawless derivations, and (3) plausible explanation of relevant facts. That is a tall order, but Euclid came very close. He hit on the crucial configuration (sequence) of these derivations, and eliminated all degrees of freedom among shifting theorems (each theorem is inextricably linked to all other theorems). It is true that Euclid's parallel-lines-never-meeting assumption continues to spawn some controversy, but this does not negate the value of the model.[6] Accordingly, Euclid's system remains unchanged and needs no further work to maintain its influence. Above all, Euclid was the first mathematician to demonstrate that logical systems are permutations and combinations of simple elements.

MATHEMATICS AS A SYSTEM

The whole of mathematics can be derived, in the same general format as Euclid's *Elements*, from a handful of axiomatic assumptions and notions. Bertrand Russell and Alfred North Whitehead demonstrated this with their *Principia Mathematica* (1910). This newer *Principia* did not earn the fame of Newton's original, but there were good reasons. First, Whitehead and Russell addressed mostly arcane theorems of interest only to other mathematicians. Second, their new *Principia* only confirmed what most mathematicians suspected. Third, mathematics rarely generates public controversy; hence, it seldom attracts publicity. In any event,

Euclid's *Elements* may be considered a subset of mathematics, or it may be considered the paradigm for mathematics.

The surprising thing is that most mathematics derives from only three basic notions (excluding recursion): (1) quantity or magnitude of elements (numbers, shapes, proportions), (2) addition/combining or its inverse subtraction/deletion, and (3) comparisons. Any two expressions can be compared. Either they will be equal or not. If not equal, one must be larger than the other. Mathematical operations consist of permutations and combinations of these notions. Multiplication is repeated addition. Division is repeated subtraction. A quotient of 8 results from dividing 24 by 3, which means the number of times that 3 can be subtracted from 24 until zero is reached. Integral calculus integrates addition and multiplication, while differential calculus integrates the inverse—subtraction and division. This simplicity does not mean every conjecture in mathematics is easy to prove. Some *may* be impossible to prove, for example, Goldbach's prime-number conjecture described in the first chapter.

INFERENTIAL LOGIC AND THE "EUREKA" EXPERIENCE

Inference means to find an underlying truth or generalization based on knowledge less than requisite to confirm it with absolute certainty. The typical difficulties of ascertaining all relevant information regarding the phenomenon or system at issue mandate this approach, which commonly takes the form of statistics. Statistics divides into two aspects. The first aspect is a form of accounting. So-and-so hit X number of home runs in such and such year. Variance within any set of related numbers is quantified by measures like standard deviation. The second aspect seeks approximate or probable truth when the exact picture: (1) is beyond reach, (2) is impractical or unnecessary to obtain, (3) remains desirable but prohibitively expensive, or (4) would result in unacceptable side effects.

For example, physics cannot pinpoint the exact location *and* momentum of any particle in an atom at any given instant in time. A manufacturer of explosives does not run a 100 percent quality-control check. Then, too, the next decennial census (year 2000) will rely more on statistical analysis than did past surveys, because of the exponentially increasing cost of on-site head counts. Lastly, exploratory operations on some forms of cancer, especially prostate cancer, have a track record of triggering more problems than they solve.

Statistical calculations are by nature shrouded with probabilities and approximations. When we know the facts completely, there is no need for statistics (except to catalog). When we don't know the facts completely, the recourse is to determine probability. Probabilities have two variants. The first variant is the probability that such and such event will occur, given established averages or distributions. Meteorologists and card players use this kind of probability. The second variant is an estimate of how close a statistical calculation comes to a presumed "true" measure. Theoreticians and researchers depend heavily on this variant.

How can proximity to truth be measured when the truth isn't known? The rationale is that few phenomena are investigated unless they give some hint of underlying relationship or cause and effect. Further, most such investigations are based on similar situations wherein the intent is to see if the calculations run parallel. Alternatively, the intent may be to obtain random samples to compare for consistency. Or functional relationships can be plotted on a graph to see the progressive effect of systematically changing the values of the independent variables. The technique is called *regression analysis*, and the resultant curve is called a *line of regression*. Similarly, samples taken of any population will distribute themselves around the true numbers in a predictable pattern. By reversing the calculations, various samples can be used to estimate the true answers with a high degree of accuracy.

The power of statistical analysis has carried over into newer disciplines, e.g., systems analysis and fuzzy logic, especially when analysis requires comparison of both quantitative and qualitative values, or the contributions of identifiable factors or

elements cannot be mathematically isolated. At times, the analysis may have to contend solely with qualitative and logical
factors that do not lend themselves to mathematics at all, for
example, the legal doctrine of circumstantial evidence. When
evidence lacks so-called smoking-gun certainty, a conviction can
still be obtained if circumstantial evidence points to the guilt of
the defendant beyond a reasonable doubt, including conviction
for murder (discussed in Chapter 19).

Inferential analysis is not infallible. The various techniques
such as statistics and fuzzy logic, can be, and frequently are
abused by: (1) pretense of more accuracy or predictability
than the data warrants, (2) biased selectivity of the data used
(3) misleading (often graphical) presentation of the findings, or
(4) ignoring of one or more major factors. Still, abuse does no
negate the inherent utility of statistics. In the absence of complete knowledge, the investigation of any system, as in the
investigation of any phenomenon in science, must rely on inferential logic in some form.

But can the deciphering of the underlying interior lines
relating various ideas or facts reach a kind of critical mass, so to
speak, and the mental counterpart to a chain reaction of understanding? Yes, and we call this the "Eureka" experience, an
experience typically accompanied by a slapping of the hand on
the forehead accompanied by an exclamation on the order of
"Oh, what fools we have been!" This is exactly what happened
when a number of well-known physicists were puzzling over the
behavior of an atomic nucleus in terms of—ironically—critical
mass and chain reactions. Niels Bohr had the starring role.[7]

This situation raises the question: How many pieces of the
puzzle, as it were, must be present before such insight is
obtained? The answer is a power curve. Up to a point, facts
accumulate without insight. Then, depending on many circumstances such as the experience of the observers, seemingly minor
increases of facts trigger the understanding. Lastly, the curve
reaches a point of diminishing returns because it is rarely, if ever
essential to accumulate 100 percent of the relevant facts before

sudden insight occurs. As such, the sudden insight itself constitutes an obvious if abstract critical mass and chain reaction.

LINGUISTICS

The logic and perhaps a few symbols of mathematics may be universal but not the words to express it. A mathematics article written in one language must be translated before it can be read by others who speak a different language. Fortunately, the universal nature of mathematics will simplify the translation. The same cannot be said for other sciences, and certainly not for less objective material. Virtually all translations are plagued by errors, a situation that is grist for anecdotes, one of which is shown in Figure 23.

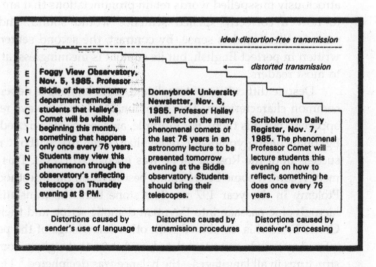

ideal distortion-free transmission

distorted transmission

EFFECTIVENESS		
Foggy View Observatory, Nov. 5, 1985. Professor Biddle of the astronomy department reminds all students that Halley's Comet will be visible beginning this month, something that happens only once every 76 years. Students may view this phenomenon through the observatory's reflecting telescope on Thursday evening at 8 PM.	Donnybrook University Newsletter, Nov. 6, 1985. Professor Halley will reflect on the many phenomenal comets of the last 76 years in an astronomy lecture to be presented tomorrow evening at the Biddle observatory. Students should bring their telescopes.	Scribbletown Daily Register, Nov. 7, 1985. The phenomenal Professor Comet will lecture students this evening on how to reflect, something he does once every 76 years.
Distortions caused by sender's use of language	Distortions caused by transmission procedures	Distortions caused by receiver's processing

23. **Entropy in communications.** Like most systems, the communication of information tends to disorder and misunderstanding. Linguistics and information theory strive to dampen that trend with myriad rules, techniques, and procedures.

All of this comes under the heading of *linguistics*. Linguists study languages as facets (elements) in a system of communication. Most thoughts and ideas can be expressed in any language, though not always with the same precision. One isolated tribal language had only four words for counting: one, two, four, and many. Still, all languages can express complete thoughts, commonly called *sentences*. Several thoughts can be linked by paragraphs or their equivalent. Conversely, the logic that flows through any sentence or paragraph can be parsed into its elements, analogous to derivatives in calculus. This means that if a thought remains clear, it can be communicated despite misspellings. Compare

Kin u reed n udderstn ds centnce?

with

Can you hear the sound of one hand clapping?

The meaning of the first interrogatory sentence is clear. The atrociously misspelled words retain pronunciations that are close to how a correctly spelled sentence would sound, and the sentence itself makes sense. By contrast, the second sentence is written in perfect English, but its content is meaningless, at least to most readers.

Despite difficulties of translation, most languages depend on common characteristics. This commonality is not always readily apparent, but it is there. For example, archaeologists puzzled over Egyptian hieroglyphics for more than a century without much success until the Rosetta Stone was uncovered (1799). This fragment recorded some benefactions bestowed on the priesthood by Ptolemy in the year 197 B.C. The stone bore the translation of hieroglyphic passage into Egyptian demotic script, and then into Greek. From this, combined with other knowledge of the period, a few key words were translated, and from that—given common structures in all languages—the balance was deciphered.[8] This led to a general translation of all hieroglyphic inscriptions—another neat case of a logical chain reaction.

But why do languages have so much structural commonality? Besides the limited number of sounds that the human larynx can utter, the million words or so in the world's most comprehensive language—English—can be subdivided into less than 1100 groups. One thousand of those groups were identified by Roget when he published his thesaurus in the 19th century. By the middle of this century, the list expanded until in the fifth international edition, the total reached 1073. Most of these groups are really pairs of antonyms, which reduces the total to less than 600. Moreover, the central idea of most of these paired groups is a combination or permutation of a few irreducible concepts. Then, too, all of the original groups were lumped into only six phyla, one of which was subdivided into three phyla in the third international edition. The fifth international edition recast these eight phyla into 15 categories. (We will visit these phyla and categories in Chapter 15.) The point here is that if truth is a system, and linguistics is a system for studying the expression of truth—and falsehood—then knowledge must also be a system. Note that the *Propaedia* of the *Encyclopaedia Britannica* outlines the whole of knowledge in one volume.

INFORMATION THEORY

Some aspects of information theory and linguistics are inextricably linked, especially the syntactical aspects. Both address effectiveness and efficiency in communications. Information theory differs by stressing the technical aspects, to include electronics, networks, optimal routing, database structure, accessibility, and cryptography. In other words, information theory either presumes that sender and receiver can emulate one another, or that their differences are irrelevant. It is not surprising, therefore, that the study of linguistics dates back to ancient times, whereas information theory came of age concurrent with communications and computer technology.

Nevertheless, the two disciplines interface on many points. Words and expressions in spoken languages are composed of permutations and combinations of a limited number of sounds (though the pitch or intensity of a sound, as well as punctuation, can be significant). When words are transmitted over circuits, or even just spoken, these nuances can easily wash out as a result of limitations in the technology. Mark Twain once noted to the London correspondent of the *New York Journal* that the report of his death was greatly exaggerated.[9]

Information theory goes far beyond the technical aspects of communications and networks. It is equally concerned with *accessibility* of information. Tables of contents and indexes in books are elementary ways to facilitate accessibility. When the 15th edition of the *Encyclopaedia Britannica* was first issued (1974), the publisher discontinued the index, perhaps assuming the one-volume *Propaedia* would suffice. So much for theory. When the complaints grew loud enough, the index was restored, this time in two volumes. Unfortunately, vast collections of knowledge and data do not lend themselves to simple indexes. For example, libraries are confronted with more than 100 published indexes and their kin. Elsewhere, computer science has spawned a subspecialty of database structure.

Information theory—in combination with psychology, sociology, and political science—also addresses how communication(s) sometimes trigger catastrophes. For example, within five years after publication of the novel *Uncle Tom's Cabin* (1852), five million copies were in print in 22 languages, a record that surpassed anything written to that time. Ten years later, Abraham Lincoln met the author Harriet Beecher Stowe, greeting her as "the little lady who wrote the book that started this big war."[10]

Lastly, information theory seeks to explain the granddaddy of all communications media. The chromosomes of the humble one-celled amoeba are said to contain 400 million bits of information, and those in a human cell have roughly five billion. This information is responsible for the orderly development of all organisms, including the intricate processes of cell differentiation, reproduction, heredity, and evolution. In short, the genes are, among other things, a computer that employs information theory better than we do.

LIBRARIES AS SYSTEMS

Libraries exemplify linguistics and information theory. For starters, libraries deal almost exclusively in information. The information is highly organized and usually indexed. In turn, most libraries are linked with many other libraries, if only by way of interlibrary loan service. As such, they constitute a classic system in their own right, primarily of the relational form. The elements of books and other recorded information are more important than the superintending system of the library. Furthermore, library systems have not changed much in the last 3000 years beyond the metamorphosis from papyri and scrolls to books and journals. Nor has much change been necessary until recent times. The idea was, and still is, to provide focal points and interior lines between knowledge and users.

However, the sheer volume of the material combined with the ever-increasing cost of buildings, books, and labor is forcing libraries to overhaul their methods, heralded by the name change from library science to information science. Libraries no longer oversee a manageable number of books and journals. They are inundated with an unmanageable flood of information that promises to intensify indefinitely. Were libraries to continue to operate on the old pattern, they would eventually arrive at that point where making information available in a timely fashion would be impossible. The entropy level would slowly rise to infinity.

The way out of this morass is to transform the printed material into electronic images combined with interactive electronic indexing. An image record records each page of written material as if it were a photograph. This preserves and redisplays information exactly in the form it was originally printed. Sometimes this process bypasses the hardcopy phase altogether, especially for journals, by publishing the text directly on compact disks or on centrally archived image files, with copies transmitted electronically to libraries. This offers numerous obvious advantages, such as eliminating the costs of handling, logging, storing, binding, and rebinding (as well as reducing space requirements).

□ □ □

In summary, nature's *logos* exemplifies the symmetry of all systems, and indeed, no physical or sociological system can be described or evaluated with resort to abstract terminology and the underlying logical and quantitative relationships among various terms. Actual systems, of course, may and usually do lack the perfections of abstract reasoning, but at least we can use such perfections—models, as it were—from which to form those descriptions and evaluations. The next step, then, is automation, because it makes us think long and hard about the nature of systems, and programming forces people to think with explicitly correct logic.

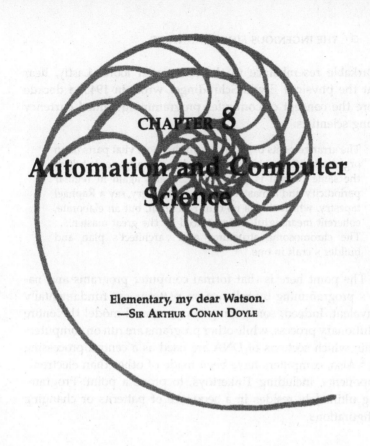

CHAPTER 8

Automation and Computer Science

Elementary, my dear Watson.
—Sir Arthur Conan Doyle

The purpose of this chapter, like the previous chapter and all subsequent chapters, is to apply the thesis of the book to a subcategory of applications, in this case automation. Computers and robotics subtended by automation are obvious physical systems, but the principles and rules by which they operate are abstract. More importantly, these principles underwrite evolution and genetics, as well as man-made machines.

As mentioned in the chapter on history, the automated loom had many of the features of modern computers except that its programming wove patterns of cloth rather than mathematical or logical patterns of data. Moreover, the *Colossus* computer used to break the German *enigma* code during World War II bore a remarkable resemblance to the automated loom. Lastly, hear what the physicist Erwin Schrödinger wrote in 1943, a decade before the concept of computer programming gained currency among scientists:

> The arrangements of the atoms in the most vital parts of an organism differ ... between an ordinary wallpaper in which the same pattern is repeated again and again in regular periodicity and a masterpiece of embroidery, say a Raphael tapestry, which shows no dull repetition, but an elaborate, coherent meaningful design traced by the great master. ... The chromosome structures are ... architect's plan and builder's craft in one.[1]

The point here is that formal computer programs and nature's programming by reliance on patterns are fundamentally equivalent. Indeed, some computer programs model the entire evolutionary process, while other programs are run on computers within which sections of DNA are used as a central processing unit.[2] Also, computers have been made of other than electronic components, including Tinkertoys, to prove a point: Programming ultimately resides in a sequence of patterns or changing configurations.

AUTOMATION MECHANICS

Automation consists of instructions stored by, or read by, a machine and then executed. When a person reads and carries out instructions exactly as stated, he or she acts like a programmed machine. The easiest way to envision a computer is to add memory to a calculator. Next, modify the calculator to record the key presses instead of reacting to them as commands. Lastly, modify the calculator to read its memory and carry out those commands *as a set*. This enables it to perform five of the six operations of a computer.[3]

Of these six operations, the first three are identical to the fundamental notions of mathematics: (1) to identify quantities, magnitudes, or shapes, (2) to add (or subtract) them, and (3) to compare them. The next two operations are self-explanatory: (4) to read and store data and instructions, i.e., input, and (5) to route the results of processing to a user, i.e., output (typically to a screen, a printer, or a database). The sixth operation is the ability to jump to another command out of sequence. Jumping is required whenever a condition can lead to two or more outcomes requiring different processing. Because programs are written sequentially, any restriction on jumping will cause all inquiries to be processed identically. For a payroll, every employee would receive the same pay.

Now let us build an imaginary computer by converting a familiar piano-roll player (a robot) into an information processor. The piano interprets holes in a scroll as instructions to sound notes corresponding to the position of the holes in any given row. To convert the piano into a computer, modify it by inserting sliding disks on the chords (like the "buttons" on an abacus). Then modify the levers so that instead of striking the chords, they shift the disks on the chords, thus creating a binary-digit abacus. Divide the abacus into two parts; one part to hold and carry out the current instruction, the other to record intermediate results within a sequence of calculations. Next, add linkage that will punch holes in the trailing end of the piano roll corresponding to the position of the disks on the abacus. As for jumping, further modify the player so that it will partially rewind the scroll to a previous row or jump ahead to a row out of sequence (as directed by the current instruction).

These modifications manifest the six computer operations. The machine recognizes symbols for quantities, accepts input, and displays output. It can add by taking an existing number in memory and combine it with a number (an *argument*) from the input scroll, using its abacus. It can also compare a number in memory with a number on the input scroll simply by comparing the pattern of holes with the alignment of the disks on the chords. Such comparisons, of course, are necessary if one condition (A is equal to B) requires such and such processing while a different condition (A is not equal to B) requires different processing or none at all. This is where the jump-type instruction comes in. The program must be physically written and stored as if it were a singular sequence of instructions, but you want different conditions to "jump" to different segments. In the case of complex branching, a computer breaks the logic down into a sequence of yes–no decisions.

The next task is to create a language that the machine can understand. The easiest solution is to draw imaginary lines down the length of the scroll, dividing each row into four fields. Use the left-column field to label rows in order to jump to them if necessary. Use the second-column field to record instructions. Use the rightmost two fields to record arguments the current instruction needs in order to execute. Within each of the four fields, let the rightmost position—when punched—symbolize the quantity 1. Let the first position to the left—when punched—represent the quantity 2. For 3, punch both holes. For 4, punch only the third position over, and so forth. Lastly, assign a separate number to each command.

Granted, this particular computer is Neanderthal, but it illustrates the basic process. First, automation is based on changes to configuration *and nothing else.* Second, more versatile computers consist of permutations and combinations of simple processes. To be sure, these computers use ingenious mechanical features and different materials to miniaturize and streamline the processing, yet the earliest electronic true computers were not much more sophisticated than the converted piano-roll apparatus described above.

Lastly, let us add a few features to this elementary computer: a keyboard, a printer, a random number generator, and some linkage that will translate words and decimal numbers into machine language. The random number generator need only be a simple device that reads a clock face and returns a value equal to the last digit of the number of seconds past the current minute, i.e., 34 seconds would generate the digit 4 (and 40 seconds would generate the value of 10). With a handful of instructions written into a simple program, the computer is ready to play automated blackjack. Figure 24 illustrates this program in detail.

SYSTEMS OLD AND NEW

Though the machine described in the preceding section would not do well in the marketplace, it does demonstrate that computers are at root mechanical devices for combining, sequencing, or separating data represented by bits that have only two states:

Logical instruction	Generic programming	Physical implementation
Generate a random number	a = RANDOM(10)	Position of sweep second on a clock determines the number. Record number by positioning designated disks in a binary-digit abacus.
Generate a second random number	b = RANDOM(10)	
Add the two numbers	Total = a + b	Compare disk positions of the two numbers in memory, do a binary add, record sum.
Print the sum	PRINT Total	Punch corresponding holes in roll to match value in abacus memory for variable *Total*.
Begin a loop for additional numbers	LOOP UNTIL Total > 20	Use a lever that trips when variable *Total* exceeds 20.
Prompt to generate another random number	PRINT "Another?"	Punch holes in roll equal to code numbers for each letter of the character string.
If yes, then: Fetch a random number, Add it to the current value of *Total*, and Print that value Else, exit the loop	IF INPUT = "Y" THEN Total = Total + RANDOM(10) PRINT = Total ELSE EXIT LOOP END IF	Compare key press pattern with the pattern for "Y." If they are the same, fetch a random number from the clock and add it to *Total*. If not the same, jump to row labeled as the end of loop.
If the value of the variable *Total* exceeds 21, print "Tilt"	If Total > 21 THEN PRINT "Tilt"	Compare pattern in abacus for *Total* with pattern for 21. If *Total* is larger than 21, punch holes equal to code numbers for T, i, l, and t.
Mark end of loop	END LOOP	Mark row on roll with label
End the program	END	Turn the machine off.

24. **Inner workings and hidden mechanisms. Computers project an image of complexity, but the innards consist only of banks of on–off switches and a number of inanely simple circuits. On this simplicity run the most complex programs imaginable.**

on or off, present or absent, yes or no, one or zero. But if the underlying concepts haven't changed much, the implementing technology has.

These improvements can be divided into four groups. The first group separates programs and data. Old systems typically bundled data and instructions in one package, called *batch processing*. This meant that answers had to wait until all necessary elements were assembled. But once data is recorded electronically in memory devices peripheral to the computer, any number of programs could be run against that data ad hoc, interactively and concurrently. The importance of this lies in data being more critical than analysis. Once a program is debugged, it can run an infinite number of times without revision, whereas data typically changes and thus requires constant maintenance and expense to keep it current.

The second group of improvements automates translation between English (or any written language) and machine language. Writing programs in machine language verges on masochism (actually it is worse), and the infuriating detail required is extremely prone to errors. This work load led to the development of assembler language, wherein words (called *mnemonics*) were used for commands instead of strings of zeros and ones. Assembler language must still be translated into those infernal digital strings, but at least that part of the chore was automated. Still, assembler language is tedious, so higher-order languages were developed that integrated various sets of assembler instructions into fewer and more readable procedural commands. A separate program (called a *compiler*) parses these commands into assembler instructions. FORTRAN (FORmula TRANSlation) and COBOL (COmmon BUSiness oriented LAnguage) were the first of these higher-order languages still in use. They reduced programming time by 80 percent (or more) while facilitating revision. Today, more sophisticated languages make the task of programming even easier.

The third group of improvements streamlines the physical mechanics. The functionality of the *Colossus* computer can today be fitted into a watch case and execute 10,000 times faster. Other improvements include processing of 16, 32, 64, or 128 bits of information simultaneously through literally millions of circuits. And perhaps the most important step was integrating the electronic circuitry into one small chip. That chip did not physically reverse entropy, but its low cost and high efficiency inaugurated a power-curve growth in computer usage that shows no sign of abating. Every small increment of further improvement evokes major increases to sales and profits.

The fourth group links computers into networks. Single computers are adequate for most scientific models but not for business. Business relies heavily on the telephone and other forms of communication. Once computers came of age, linking them by telephone networks became essential. All that was necessary was to add a piece of circuitry (called a *modem*) that re-formed blips of data into voice modulation and then back into blips. However, this technique could not accommodate the growing volume of traffic, so exclusive lines were created to transmit digital traffic directly. Today, the fastest of these networks can send the equivalent of an entire set of the *Encyclopaedia Britannica* in about 2 seconds.[4]

DATA REPRESENTATION AND ORGANIZATION

As described in previous sections, data is usually represented by strings of zeros and ones or their equivalent. Therefore, the only limitation to the amount of data a computer can store is the size of its memory. For all practical purposes, this size can be expanded to include every piece of information ever recorded. The difficulty lies in relating—linking—all of this data. Similar to the logical interior lines of theoretical truth, users must be able to link, hence access, data with neat, logical interior lines. Otherwise, information searches would evoke the needle-in-the-haystack syndrome. The two general forms for those lines are hierarchical and relational. Each form corresponds with its namesake in systems theory.

The criterion for the hierarchical form is that it be use extensively for a relatively singular purpose. The most widel known application is the airline reservations system (which : really a network of several similar systems). The first leve covers routings between cities. The second level lists all of th flights between each pair of cities. The third level records all c the seats on each flight (which depend on the aircraft used, th amount of intentional overbooking, and various fare categories The database structure and its accompanying program are orga nized to facilitate the hierarchical sequence.

The criterion for the relational form of database is tha different users require the same data in different arrangement Here the idea is to organize the data to facilitate access to an part desired. This means that users must prepare their ow access logic, but at least they know where the data is and how is arranged. Typically, a relational database groups strongl related data in separate files but minimizes the programme access-linkage within the file and between files.[5] This appears t work counter to the idea of interior lines, but reprogrammin access is much easier than continually modifying the intern structure of a database (which would mandate continuous rev sion of user-access programs anyway).

Efficiency of database systems or subsystems also depenc on two other considerations. The first is the quantity of dat over which a distinct power curve operates. If the databas contains too little data, it will largely go unused. Data can b added until a threshold or upswing point is reached, whereupc the power curve (of utility as a function of investment) turn sharply upward. But accumulating too much data leads * diminishing returns. The last dregs of information are seldo used yet entail high maintenance expense.

The rationale of this power curve is that most data releva to organizational purpose is logically related in some way, like jigsaw puzzle. When only a few pieces are in place, the picture difficult to envision. As more pieces are placed, the clarity of th picture grows rapidly and appears more or less complete wi only 85 to 90 percent of the pieces.

The second consideration is whether data is a prerequisite for action or is entered only after the fact. At one time, information systems relied on after-the-fact entry, and as a result the accuracy and timeliness was chancy. In time, this changed. As databases grew, the usage changed from mere recording of transactions to generating the necessary documents or equivalent (which updates the database concurrently), for example, airline reservations.

PROGRAMMING

A program is a set of instructions that tells a machine what to do in a language that it understands. In terms of control, programming has three variants. The first variant is a straightforward sequence of instructions that is executed in the order in which the instructions appear (although the same instructions may be executed repeatedly by reactivating the program as many times as necessary). The simplest examples include counting the number of records in a file and copying information from all records in a file.

The second variant introduces branching, whereby different conditions lead to different processing, especially when it is impractical to foresee the conditions in advance. The point to note is that branching adds a dimension to the logical paths of the program. That is, although the program is written in linear form, it does not execute that way. Instead, the execution resembles a traffic pattern or flowchart. This is why the ability of a computer to jump to instructions out of sequence is so important.

The third variant is three-dimensional processing, which takes two forms. The first form—hierarchical—divides the work load among several computers or among processing chips in the same computer. Some computers contain 9200 chips, resulting in an enormous increase in processing speed. To make use of these arrays (or linked computers), the program must ensure that no segment executes until all prerequisite processing is done.

The other form of three-dimensional processing—relational—is more intricate. In this form, separate computers operate independently with minimum linkage between them, and sometimes they rely only on a common clock. For example, consider a traffic control system that synchronizes the timing of stoplights in order to reduce queuing at intersections. If the signals at the intersections are controlled by a central program, the system would use the hierarchical form. In the absence of centralized control, each signal must operate independently, which requires the relational three-dimensional form. Still, some sort of clock is needed, and there are two options. One signal can maintain a clock and send that data to other units (without controlling them). Alternatively, each signal can have its own clock, calibrated to run with sufficient precision for the system's purpose. See Figure 25.

In practice, the one-dimensional variant is rare. Similarly, three-dimensional programming is also uncommon, and when it is used the hierarchical form predominates. This leaves the two-dimensional variant, and that is trouble enough. The problem is that most programs undergo repeated updating and enhancement, often at the hands of different programmers. After a while, the listing of the instructions begins to resemble spaghetti, hence the entropy of the program—at least in terms of readability—rises sharply. The logical interior lines, as represented by the sequence of instructions, is lost in a sea of distended comparison and "jump" commands directing the processing to go every which way.

Computer professionals seek to reduce this entropy by what is called *structured programming*. Structured programming organizes the commands into easily recognized blocks (analogous to paragraphs in literature), typically combined with indenting (exactly as done in written outlines). This layout facilitates human interpretation of what the program will do under various circumstances. The interior lines of the logic now extend to the programmer's perception. Modern programming languages make this structuring even easier.

This structured programming has a cousin called *modular programming*. Modularity is analogous to an organization chart.

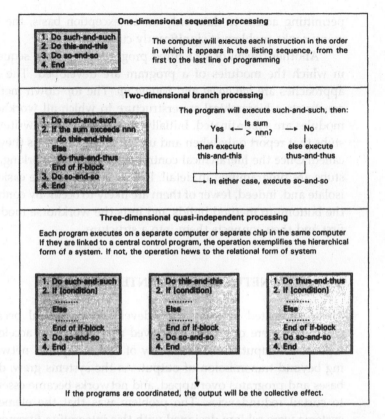

One-dimensional sequential processing

1. Do such-and-such
2. Do this-and-this
3. Do so-and-so
4. End

The computer will execute each instruction in the order in which it appears in the listing sequence, from the first to the last line of programming

Two-dimensional branch processing

1. Do such-and-such
2. If the sum exceeds nnn
   do this-and-this
   Else
   do thus-and-thus
   End of if-block
3. Do so-and-so
4. End

The program will execute such-and-such, then:

Is sum > nnn?

Yes ←— —→ No

then execute this-and-this   else execute thus-and-thus

→ in either case, execute so-and-so

Three-dimensional quasi-independent processing

Each program executes on a separate computer or separate chip in the same computer If they are linked to a central control program, the operation exemplifies the hierarchical form of a system. If not, the operation hews to the relational form of system

1. Do such-and-such
2. If [condition]
   ....
   Else
   ....
   End of if-block
3. Do so-and-so
4. End

1. Do this-and-this
2. If [condition]
   ....
   Else
   ....
   End of if-block
3. Do so-and-so
4. End

1. Do thus-and-thus
2. If [condition]
   ....
   Else
   ....
   End of if-block
3. Do so-and-so
4. End

If the programs are coordinated, the output will be the collective effect.

25. Programming in various dimensions. A program executing step by step is one-dimensional. Programs that branch based on conditions are two-dimensional. Programs that require two or more computers independently and concurrently are three-dimensional.

The idea is to subdivide processing requirements into compact modules, which can be called in any sequence. Additionally, most programming languages prevent these modular units from updating data unless programmers take pains to permit it. Otherwise, too many degrees of freedom would operate—not from the program but from the programmers. Large projects with hundreds of subprograms typically employ many programmers working in comparative isolation on different modules. By permitting access to data only on an exception basis, the too-many-cooks problem is significantly curtailed.

Another cousin of structured programming is the sequence in which the modules of a program are developed. The two approaches are *top down* and *bottom up*. The top-down method constructs a hierarchical superstructure to which all workhorse modules are subordinated. Initially, these modules are written as stubs that report only when and under what conditions they are called. Once the hierarchical control mechanics are working, the stubs are programmed in detail. In this way errors are easier to isolate and, indeed, fewer of them are likely to occur. By contrast, the bottom-up method builds and tests the workhorse modules first and then connects them into a structure.

NETWORKS AND INTEGRATIONS

Most automated systems were developed piecemeal because computers were originally perceived as machines to accelerate repetitive computational tasks, few of which required networking beyond transmission of output. As the systems grew, databases and programs overlapped, and networks became essential to curtail redundancies. The problem was that the elemental systems were seldom designed with this integration foremost in mind. The solution lay in translation or what the trade calls *emulation*. Emulation reconfigures information sent into a form the receiving machine can understand or, alternatively, causes its host to act as if it were compatible with the receiving machine. Unfortunately, two systems can have so many differences that the cost of the emulation is higher than a new overarching system.

Integrating separate computer systems can take several forms.[6] The easiest form to implement is linking two existing systems. This is the relational form, which requires sharing of data or software (or possibly some items of hardware) on an as-needed basis; hence, these linkups tend to be primarily communications or network oriented. The second form attempts to create a distinct hierarchy in the linkage in which the existing systems come under the control of a central unit (or dominant regional units). These integrations are the most difficult to implement because the overarching control usually works at cross-purposes with the intent if not the design of the constituent units. This problem is similar to the ongoing difficulties that arise between the objectives of government and the wishes and propensities of individual citizens or in families between parents and children.

The third form is integral. This means the member systems are no longer needed on a stand-alone basis to meet local requirements; hence, those requirements can be met equally well if not better with a fully integrated system. However, the presumed improvement to efficiency may gloss over details to the point where the project fails. For a mechanical parallel, consider home workshop integrated power tools. These machines combine five or more power tools in one unit, e.g., a table saw, lathe, drill press, grinder, and so forth (at perhaps 70 percent of the total cost of the tools bought separately). The problem is that the apparatus must be reconfigured for each application, and compromises are often made in the design to achieve this versatility. When the owner has limited space, does not need maximum precision, and doesn't mind the rigmarole, the integrated machine solves many problems. In other situations, that all-in-one machine may be a source of bother more than of satisfaction.

Two other points. First, the problems of integration can be reduced by relying on standardized hardware and operating systems for different systems. For obvious reasons, this policy reduces the need for emulations when linking or joining the units together. Nature mirrors this technique by cloning more-or-less identical chromosomal structures for every cell generated in any given organism, starting with the fertilized egg.

Second, a computer system can experience catastrophe for the same reason that a critical mass of fissionable material sets off a chain reaction. One breakdown leads to more than one new breakdown. In the case of computers, the explosion is replaced by the entire system going dead. Perhaps the best-known case of this occurred on January 16, 1990. On that date approximately half of the long-distance telephone network within the United States shut down for about 12 hours, losing tens of millions of dollars in profits in the process.

This telephone network consists of hundreds of nodes or switching points. Traffic is normally routed along the least-crowded routes. When the traffic between New York and Washington is overloaded, additional traffic might be routed by way of Chicago or, for that matter, any other node or sequence of nodes. Because of a minor error in one line of programming that operated in each of the nodes, the first time that one of the nodes reached a certain capacity, it diverted *all* of its traffic to other routes, not just the excess. This led to a chain reaction shutdown because the remaining nodes each had to bear exponentially increasing loads they were programmed to reroute.

CYBERNETICS AND DECISION SUPPORT PROGRAMS

Norbert Wiener's *Cybernetics: Or Control and Communication in the Animal and the Machine* (1948) marked scientific recognition of the relationship of man-made automation with nature's variation. As a result, many analysts presumed that machines could be made to mimic human intelligence—artificial intelligence. This is not what Wiener was driving at, a point that he made abundantly clear in a later book: *God and Golem, Inc.* (1964). Instead, Wiener focused on the mechanics whereby a biological organism was "programmed" to learn from its experience. He accomplished this by making comparisons of man-made machines with the physiological mechanics of living organisms. There was no intelligence to it (except implicitly in the design). It was a matter of programming. That is to say, if man-made machines are to "learn," they must be programmed explicitly to do so.

This led to what we now call *decision support programs.* These programs mimic human processing of complex data and alternatives. For example, decisions to buy and sell various investments involve complex mathematical computations. A program/system for this purpose can be developed to suggest or recommend that certain transactions be made, or it can be programmed to initiate those transactions directly. The latter is commonly used by pension funds and other large holders of stocks and bonds. These programs were largely responsible for triggering the chain reaction that resulted in the massive one-day drop in the Dow Jones average on October 19, 1987. Corrective actions have been taken to prevent this from happening again, by automatically shutting down the stock market when prices fall below a specified threshold differential—the first "war" by proxy using opposing computers.

The ability of machines to perform these feats is based on the fact that most logical thinking consists of permutations and combinations of simple comparisons, additions, and so forth, in tandem with access to various arrays of data. But can computers be used where the facts are incomplete? Yes, at least to the extent that relationships among facts are perceptible and the "gaps" can be "filled" with computation of statistical probabilities.

This concept leads to two general models for decision support programs. When all of the facts and relationships are known but too complex for manual processing, the system uses a rule-based model. By contrast, when the facts are not known or the relationships are not clear, a pattern-based model is used. The rules-based model is sometimes called cookbook decision making, an expression that evolved from decision support programs designed for medical diagnosis (which is often called cookbook medicine). However, many of those diagnostic systems are more accurate than doctors and are nearly unbeatable when used by a competent physician.[7] The first of these programs, named *Mycin*, was developed in the 1970s by the medical college at Stanford University as an aid for training doctors. Today, hundreds of similar programs exist, and at least one forensic expert predicts that the day will come when physicians will be sued for malpractice for *not* using them.[8]

The pattern-based model was originally developed for analysis of images, for example, in paper-currency change machines to identify counterfeit bills. Advanced applications go beyond imagery and are sometimes called Sherlock Holmes models. The famed detective was forever deducing conclusions from seemingly insignificant facts if the pattern or configuration of those facts bore similarity to previous cases. See Figure 26.

But are decision support programs cybernetic? Do they modify their own programming based on their record of previous processing? Most of them do not; a few do. Among those few are chess-playing systems that "evaluate" their opponent's strategy and then modify their "moves" accordingly. How can a machine do this unless it was programmed to do so in the first place? Answer: It can't. Still, it is sometimes easier to write a program that can modify other programs as necessary rather than to modify all of those programs in advance, like the nib of pen that conforms differently to the pressure and physical writing style of each owner. This situation is especially practical when: (1) the range of possible events is broad if not infinite, (2) only a fraction of the events will likely occur, and (3) that fraction is not easily predicted. Not many situations meet these criteria. The major exception is the task of creating a universe with all of its solar systems and forms of life, all of which had to evolve over billions of years without further intervention.

ARTIFICIAL INTELLIGENCE

The most profound question in computer science is: Can a computer replicate human thought processes? The name given to this study is *artificial intelligence*. So far, no one has been able to replicate human initiative, and the only way to give a machine the appearance of emotions and feelings is to robotically mimic physiological manifestations. Elsewhere, a program may give the appearance of creating fascinating output, but that output is always the result of the program and can be predicted.

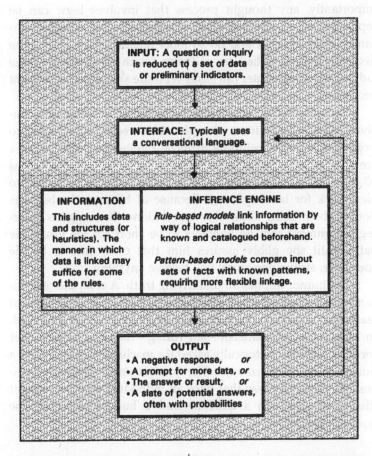

26. Decision support models. Computers that mimic human thought processes rely on logical relationships among data. When the relationships are well known, the model uses rules-based programming. Otherwise, it must resort to comparing patterns of relationships.

On the other hand, science is increasingly convinced that the brain is constructed along the lines of a computer. Zones in the brain are set aside for memory, and most memory is permanent (storage) memory. At any given moment the psyche is conscious of only a tiny fragment of that information. More importantly, any thought process that involves logic can be replicated on a computer, and the computer will execute thousands of times faster and do so more reliably. Thus, unless the decision making relies on initiative—with or without value judgments—a computer can easily replicate any objective thought processes.

Understandably, many persons recoil at the idea of a machine replicating human thought. This recoil often leads to pejorative remarks belittling artificial intelligence. Still, recall a scene from the recent film *Sleepless in Seattle*. A precocious 10-year-old girl computer-hacks a plane ticket from Seattle to New York for her boyfriend. Because of his age, the boy has doubts about being allowed to travel alone. She suggests entering a comment that he happens to be short for his years. He balks until she glibly assures him that once data is in the computer, airline personnel will believe anything.

The humor implies an underlying truth. A computer system can reach a point where it appears to take on a life of its own, at least with respect to the sociological system in which it is used. In 1950, the mathematician Alan Turing—the chief programmer for the *Colossus*—rhetorically asked if a machine could replicate a human thought. His answer: The question itself was inappropriate. His reasoning was that once an individual could no longer distinguish between a machine and a human response, the question was moot.[9]

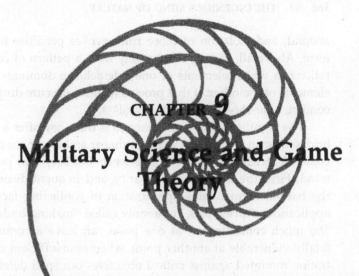

CHAPTER 9

Military Science and Game Theory

All great captains have done great things only by conforming to the rules and natural principles of war.... whatever may have been the audacity of their enterprises and the extent of their success, they have succeeded only by conformity to rules and principles.
—NAPOLEON, MAXIM CXII

I beat the Russians every time, but that gets me nowhere.
—NAPOLEON, Moscow, 1812

Some years ago in Baltimore, a bridge tournament had to share a facility with a kennel show. Halfway through, the host issued a bulletin: Although the barking and yelping were deplorable, the dogs were bearing with it very well. Some games, then, could be said to be the moral equivalent of war. Or perhaps the immoral equivalent. Games are also part of a spectrum from card and board games, through sports, to lethal conflict. The Duke of Wellington said that the battle of Waterloo had been won on the playing fields of Eton.[1]

The common feature within this spectrum is that players vie to win. The contest unfolds on a field of some kind: table, stadium, or continent. Tactics and strategy are employed. Rules abound, and violation of those rules evokes penalties in some form. Above all, victory results only from a pattern of configurations in which elements of one side achieve dominance over elements of the other. If that process remains obscure during the conflict, after-the-fact analysis reveals it.

The significance of such analyses is that they offer a logical bridge to span the gap between nonlinear and linear science—in colloquial terms, to turn art into science. But keep two points in mind. First, the obvious nonlinearity, and frequent disorder, of the battlefield mandates optimization of conflicting factors or application of principles, commonly called "making trade-offs." Too much concentration at one point can leave a commander fatally vulnerable at another point, whereas insufficient concentration mounted against critical objectives can spell defeat. Second, even minor changes in circumstances can mandate a new set of trade-offs. In other words, military science cannot be reduced to linear, formulaic thinking, but a more comprehensive logical approach might be useful—another application of fuzzy logic.

PRINCIPLES OF WAR

It is easy to understand why a commander lost a battle or a war after the fact. The military profession calls this twenty-twenty hindsight. The ideal is to preclude defeat by foresight. A few gifted commanders have done so consistently with little formal training (in military art): Alexander, Caesar, and Napoleon (prior to 1812). Most other commanders must struggle with abstract principles. The military profession formalized nine such principles in the 1930s, and since that time historians have used them with unmitigated success—in hindsight.[2] Commanders in the field have not. The implication is that the formal principles of war must be valid but too complex to employ in the confusion of battle, or too numerous to eke out a working balance (analytical equilibrium). Nevertheless, from a systems viewpoint these nine principles can be alloyed into four guidelines, and these guidelines can be stated in terms that apply equally to games, sports, and the battlefield.[3]

The first guideline is that *a commander or player should have a crystal-clear idea of what he intends to achieve and then reduce that goal to a sequence or set of crucial (and preferably concrete) objectives.* This is the same thing as "purpose" essential for all systems, and the military calls it the principle of the objective.

The second guideline is that *resources should be sufficiently concentrated to prevail in the contest for each crucial objective, allocating the fewest possible resources to all other purposes.* This merely combines the principles of mass and economy of force, and, in this application, *mass* is nearly synonymous with *concentration.* Concentration can be achieved by mass (clout), or an especially effective arrangement (leverage), or any combination of clout and leverage. And it follows that if the bulk of resources are concentrated on crucial objectives, the balance must be rationed miserly among lesser objectives.

The third guideline suggests that *the player or commander must maintain sufficient initiative to maneuver his resources in order to achieve the necessary concentrations, ideally by keeping the opponent in the dark as to intentions but at least sufficient to reduce vulnerability to counterattack.* This guideline combines no less than four principles: offensive, maneuver, security, and surprise. Offensive means to retain initiative, not to attack *per se.* Sometimes this means attacking, and sometimes it means lying in wait until the opportune moment. Initiative is therefore useful only if you can maneuver resources while preventing the opponent from degrading that initiative. The ultimate way to do this is to employ surprise. Unfortunately, surprise is difficult to achieve in practice; hence, you usually settle for keeping the opponent unsure of your intent as long as possible.

The fourth guideline is that *the player or commander should control and configure his resources with as much simplicity as possible, especially with lines of authority radiating from one individual at every level of subordinate control.* This measure combines the principles of simplicity and unity of command. The intent is to reduce chaos by reducing degrees of freedom. Hierarchical lines of authority reduce the potential for conflicting initiatives, while simplicity reduces the opportunities for chaos to evolve—the unwanted increase of entropy.

One last point. The ideal is to select objectives that return the highest dividends for the least investment in resources, and this can often be achieved by focusing on the opponent's weak "points" or, that failing, on those "points" from which his strength radiates. The colloquial expression is "going for the jugular." The systems equivalent is nullifying the opponent's dominant elements or at least displacing them from locations (focal points) where they could otherwise be effective. Granted, these neat focal points do not always exist, for example, the strength of the opponent's armed forces may reside in its sheer mass. But sometimes they do exist, as in the Battle of Midway in early June 1942. Though heavily outgunned, Rear Admiral Raymond Spruance's carrier task force sunk all four Japanese carriers, which caused the still numerically superior invading fleet to beat a hasty retreat and never again venture that close to U.S. territory in force.[4]

GAMES, SPORTS, AND WAR

The four guidelines developed in the previous section apply equally to games, sports, and war, though in some cases the options are limited. The common goal is to occupy, or at least reach, a designated area or zone (directly or by proxy), or to eliminate the opponent's ability to hinder *your* initiative. Figure 27 is a matrix that compares the guidelines with games, sports, and war. In military science, eliminating the opponent's ability to fight is called *absolute war*. This may or may not require occupation of various areas and, if so, the occupation may be transitory. Anything less than absolute war is *limited war*.[5] Success in limited war is problematic because the opponent retains the ability to continue fighting.

In sports, absolute war has no equivalent except in boxing when an opponent is knocked out (or when an opposing team is

	Games	Sports	War
Define the goal in crystal-clear terms, then reduce that goal to one or more crucial objectives that are essential to attain it.	In most games and sports, the goal is to attain a better score than the opponent or to reach a designated zone first. The objectives usually reduce to a sequence of plays. Games typically alternate moves, while sports rely on interactive plays closer to war.		Military force is rarely effective on abstract objectives. It works on physical objectives (which should lead to the abstract goal).
Concentrate resources to prevail against crucial objectives, allocating the lowest resources possible to other purposes.	Concentration is achieved by gaining more assets than opponent has or by rendering his assets ineffective.	Concentration is so effective in sports that it can destroy the spirit of the contest, e.g., the flying wedge. Rules often weigh against it.	Victory needs superior force at the critical (or crucial) time and place. This force may use either clout or leverage, but it must dominate.
Maintain the initiative to maneuver resources (to attain necessary concentrations) while keeping the opponent unsure of your intent.	Rules impose severe restrictions on maneuver (at least compared to the conduct of war), objectives tend to be transitory, and surprise is difficult to achieve between skilled players. Still, this principle governs the action of play and determines if the contest will be exciting.		This principle is the crux of battle because it needs constant vigilance under pressure. Others can be ensured beforehand.
Control and configure resources with as much simplicity as possible, especially with unity of command at every level.	Most games have simple rules, while unity of command is typically inherent by way of a single player on each side.	Control must prevail in team sports, especially when the play is fast-paced and continuous. Else, the opponent will likely pick you apart.	By definition, any lack of simplicity increases degrees of freedom. Thus initiative hastens disorder. In war, this is called *fog* or *friction*.

27. Principles of war condensed. The nine principles of war are too numerous for practical application in advance. They can be alloyed into four guidelines that apply to games, sports, and war alike.

too exhausted to vie effectively). Most other sports stress reaching a specified zone, for example, the goal (or net) in hockey, lacrosse, and soccer; the end zone in football; home plate in baseball; and the finish line in a race. Furthermore, except for non-relay races, most sports require that the player carry a token (typically a ball) into the zone or at least kick, throw, or slam dunk it in. Tennis and similar games vary this by using a fluid zone, i.e., placing the token out of effective reach of the opponent. The general exception includes the exhibition-type sports: gymnastics, ice skating, and so forth, wherein the pattern of dynamic configurations becomes the criterion for success.

In games and sports, the goal is specified by the rules. In war, the goal or purpose may be explicit, as in World War II (unconditional surrender of the Axis powers), or it may be nebulous, as in the Vietnam conflict (contain communism). Or the goal may change radically with a shift in the political wind, e.g., the recent U.S. involvement in Somalia. As for concentrating resources against specific objectives, this is commonly seen in the batting order in baseball, the sacking of a quarterback in football, and a ruthless sequence of punches in boxing. The third guideline—maintaining initiative to maneuver resources to achieve essential concentrations—is self-evident in all forms of games, sports, and war. And the concepts of unity of command and simplicity are inherent in games and sports by way of the coach but are often violated in warfare. The violation is especially prevalent in combined (allied) commands in which member countries retain the right to reject or modify orders.

LEVELS OF PERSPECTIVE

Bridge is not won with a single trick; football games are rarely decided by a single play; and few wars are resolved by a single battle. But can the outcome of a conflict result from anything more than the sum of its moves, plays, or battles? Why was Napoleon forced to retreat from Moscow after an unblemished string of tactical victories? The answer requires a twofold approach. First, grasp the five levels of perspective that operate in war.[6] Second, understand the geometry and density factors that operate differentially at each level.

The first level of perspective is the personal one. This is especially significant in games because the outcome depends entirely on the players. In most cases, this outcome reduces to the player's skill, which directly translates into the higher levels of tactics and strategy. In sports, the personal level of perspective usually takes the form of courage and determination to succeed in spite of any shortcomings. Alternatively, an entire team may be so determined to win that they become invincible against a superior opponent, e.g., the U.S. hockey team during the final game at the 1980 Winter Olympiad. In war, the personal level appears in two forms: extraordinary bravery on the part of an individual and charismatic leadership ability on the part of a commander. At the end of World War II, Audie Murphy (the war's most decorated soldier) was as well known as General MacArthur.[7]

The second level of perspective is *tactics*. In games and sports, tactics usually refer to specific plays or moves. The dynamics of conflict are partitioned into discrete tactical units: plays, downs, serves, and so forth. A few sports are based on more-or-less continuous play—for example, soccer, basketball, and lacrosse—yet even in these contests fouls, time-outs, and lulls effectively divide the play into segments. In war, tactics usually apply only to intense battles lasting anywhere from an hour to a few days. And several million times a day, members of the animal kingdom engage in mortal combat: the cheetah running down the impala, the mongoose snapping the cobra's neck, one moose killing or severely wounding another to gain dominance in the clan. It's all a matter of tactics.

The third level of perspective packages a series or sequence of tactical encounters and is called an *operation* (formerly a *campaign*). Securing the beachhead at Normandy after D-Day was a major operation, and it took 8 weeks before the Allied forces broke out from their narrow holdings. In baseball, an inning is the equivalent of an operation. In football, operations typically comprise the sequence of plays during which one team has possession of the ball. In bridge, it is a hand. In animal combat, operations are rarely seen. The supposedly comical roadrunner (*Geococcyx californianus*) is one of the few exceptions. This bird usually prevails in its weekly fight with a rattlesnake by *not* fighting directly. Instead, it teases the snake, dodging each strike until the snake is exhausted. At that point, the feathered creature steps in, kills his prey, and feasts. In military terms, the operation consists of a series of tactical encounters analogous to how Russia defeated Napoleon.

The fourth level of perspective is the game or contest as an entity. For war, this is usually called a theater of operations, though a war may be fought in two or more theaters concurrently as did the United States during World War II.

The fifth level of perspective is political in nature—the aspirations riding on the outcome of the conflict. A Saturday-night poker game is not the stuff of national interest; the Olympics can be. Recall the scene from *Chariots of Fire* where Eric Liddell refused to compete in an event because it was scheduled on the Sabbath. The Prince of Wales tried to persuade Liddell that the prestige of Great Britain was at stake but to no avail.

Interestingly, the outcome of a major conflict can hinge on the attitude opposing commanders hold toward one another, especially when bravery and charisma are factors. For example, Wellington could have prevailed over Napoleon at Waterloo sooner than he did (and spared many lives). The opportunity arose when a British marksman had Napoleon in his sights and asked Wellington for permission to fire. Wellington responded: "No. Generals commanding armies have better things to do than to shoot one another."[8]

THE GEOMETRY AND DENSITY OF CONFLICT

Games, sports, and war all demonstrate the crucial role of density (hence geometry) as the level of perspective rises. In *chess*, opponents start with 16 pieces *each* on a board with 64 squares. The density is 32 pieces to 64 squares. This 1:2 ratio is analogous to a military operation and provides ample opportunity for the strategy (and maneuver) for which chess players are renown. Now, cut the number of pieces in half, and lop off half the board *on each edge*. This leaves 8 pieces per player, for a total of 16 pieces on 16 squares (4 × 4). This 1:1 ratio leaves no room for maneuver, forcing players into a brawl. Any chess game played under such rules would end quickly, as do tactical encounters on the battlefield.

Lastly, return to the original configuration, then quadruple the number of pieces *and* the number of squares per side. This results in 128 pieces dueling on 1024 squares, a ratio of 1:8. Under this configuration, players would be at a loss on how to employ brilliant strategy. Instead, they would rely on gradual attrition. This is what happens in most wars. Evenly matched sides can bat it out for years until both sides are exhausted, even if one side retains a slight edge, for example, the Western front during World War I. Or one side can employ semiguerilla tactics until the opponent is worn down to the point of vulnerability. Napoleon encountered this in Russia, as do rattlesnakes when they take on a roadrunner.

This overall situation may be graphed, as shown in Figure 28. Let us call this graph the power curve of strategic opportunity. As discussed above with the game of chess, most tactics are canned and rely on clout, whereas operations offer the maximum opportunity to use leverage as a means of overcoming insufficient clout. At the theater level, this opportunity shrinks; hence, the outcome again depends more on clout than leverage.

The implication of this graph is that once a war degenerates into a contest of strength (sheer mass), strategic operations will not have much effect on the outcome of that war. World War II offers several excellent examples to support this view. The Allied *Operation Marketgarden* attempted to envelop (by parachute) German forces in the vicinity of Nijmegen on the Rhine River, but most of the Allied troops were surrounded and captured. On the German side, the so-called *Battle of the Bulge* also failed. In this operation, German forces concentrated most of their remaining force at one "point" (in the Ardennes forest), then broke through Allied lines in a surprise attack intended to reach to the Atlantic. The Allied lines buckled, but eventually the German momentum ran out. They were surrounded and defeated piece-meal, though significant numbers managed to retreat behind the main German line.

There *are* exceptions to this geometry-and-density business. The first exception occurs when the theater of war is constricted and therefore the area of operations more or less coincides with the entire theater of war (sometimes called a theater of operations). The best example in this century was Korea. There, MacArthur's operational envelopment at Inchon severed the North Korean line of communications and forced the North Koreans to release their grip on the Pusan perimeter in the

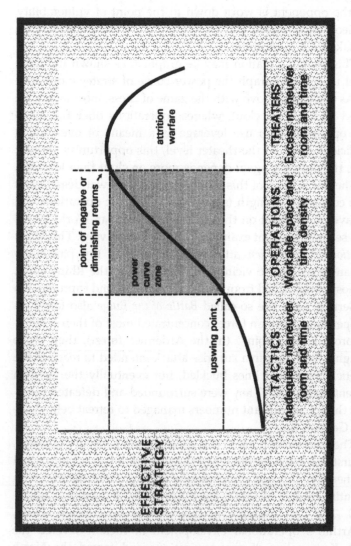

28. **Strategy Power curve. The intensity of battle leaves little time for strategy, while the conduct of war in a theater offers too much time and space for strategy to be effective. The power curve of opportunity resides between them in operations and campaigns.**

southeast corner of the peninsula. Within a matter of weeks they were chased out of South Korea, thus concluding the war as initially staged.

The U.N. forces then decided to extend the war into North Korea itself, at which point China entered the war.[9] She did so with sufficient force to evoke a rubber-sheet version of World War I trench warfare. During the remaining two and one-half years of fighting, the southernmost battleline was never more than 100 miles from the northernmost line, and after six months the line alternated in a zone about 40 north–south miles in width.

The second exception occurs when the defender has unusually effective interior lines. Interior lines mean that he can reconfigure his resources at much greater speed and flexibility than can his attacker. When combined with the inherent resistive strength of fortifications or their equivalent in obstacles, these interior lines can be used to nullify the attacker's strategic maneuver by not providing anything to maneuver about. Instead, massive force must be employed and then not always with success. U.N. forces (primarily U.S.) held out against the 12-to-1 force superiority on the part of the North Koreans along the Pusan perimeter until MacArthur outmaneuvered them.

The third exception occurs when the attacker possesses such overwhelming force that strategically brilliant operations are unnecessary for the offensive and futile on the part of the defense. This situation is common when developed nations invade aboriginal lands with full intent to stay. The outcome of three centuries of the U.S. war against American Indians was never in doubt, nor was the subjugation of the aboriginals in Australia and especially the Maoris in New Zealand by the British.[10]

CULMINATING POINTS AND OVERSTRETCH

Military history offers hundreds of cases whereby an attacker overextended his resources and was consequently routed if not beaten—the culminating point. In harsher terms, any offensive past a culminating point is forward motion on the road to perdition. Clausewitz put the case this way:

> Once the mind is set on a certain course toward its goal . . . it may easily happen that arguments which would compel one man to stop and justify another in acting, will not easily be fully appreciated. Meanwhile the action continues, and in the sweep of motion one crosses . . . the line of culmination, without knowing it. . . . This demonstrates without inconsistency how an attacker can overshoot the point at which, if he stopped and assumed the defensive, there would still be a chance of success. . . . An attacker may otherwise take on more than he can manage and, as it were, get into debt.[11]

Army Field Manual FM 100-5 *Operations* restates this concept

> Unless it is strategically decisive, every offensive operation will sooner or later reach a point where the strength of the attack no longer significantly exceeds that of the defender, and beyond which continued offensive operations therefore risk overextension, counterattack, and defeat. In operational theory, this is called the culminating point. The art of the attack at all levels is to achieve decisive objectives *before* the culminating point is reached.[12]

The consequence of passing a culminating point in battle is quite different than doing so in a theater of war. When a battle is thrown back because of overextension, the loss will seldom upset the overall force ratio or change the outcome of the war. When the entire war effort crosses that point, the outcome may change radically. Confusion arises when a battle is associated with a turning point in war. For example, the battles of Stalingrad in eastern Europe, El Alamein in the Mediterranean, and Coral Sea in the Pacific are said to have been the strategic turning points of World War II. These battles did mark the farthest tide of the Axis powers but, like Napoleon, Hitler lost the war the moment he attacked the Soviet Union. As Russia did on Napoleon's invasion in 1812, the Soviet Union "traded space for time" until the German forces became overextended, then destroyed those forces ruthlessly.

29. Moscow's invisible fortress. Napoleon's retreat from Moscow was the most dramatic instance in history of crossing a culminating point. Ninety-five percent of his forces were killed, captured, or severely wounded before he made his escape.

This is the nature of most dictators. Initially, they *do* reverse entropy, typically by creating order out of chaotic conditions, yet in most cases that order is or becomes destructive. Not long after, Sir Winston Churchill recognized Hitler's rise to power as absolute evil. Such evil was less apparent but just as real when 2400 years earlier Pericles led his nation into the fatal Peloponnesian war against Sparta. Thucydides recorded Pericles as saying to the Athenians that their country had a right to their services "in sustaining the glories of her position." Pericles then intoned that such glories were "a common source of pride" and that as such citizens "cannot decline the burdens of empire and still expect to share its honors." Then he admitted that what Athens held was "to speak somewhat plainly, a tyranny" and that to take it "perhaps was wrong, but to let it go unsafe."[13] How, then, can one grow indefinitely without crossing a culminating point? Only in stature.

TECHNOLOGY AND THE BEHEMOTH SYNDROME

Military dogma stresses that while tactics undergo continuous revision, the principles of war are perennial. It would seem, then, that tactics are a function of technology, while strategy remains the province of configuration in a zone of operations. Furthermore, as the principles of war remain constant, changes in tactics cannot go too far astray and, if so, they must cycle about some line of regression that represents the principles.[14]

The basis for that cycle is this: The firepower of lightweight weapons may be concentrated against a heavier but less mobile target. In time, these mobile weapons tend to become enhanced, augmented, or otherwise overdeveloped until *they* become behemoths and thus fall prey to newer, cheaper, more dispersed weapons. In turn, these new wonders sooner or later develop into yet another generation of behemoths.

Examples abound. Fortresses eventually fell prey to siege machines and artillery. Cavalry and phalanxes of infantry fell to hordes of arrows descending on them from the famous English longbow and later from the machine gun. The battleship yielded to the airplane. Actually, the Japanese almost did us a favor in sinking battleships at Pearl Harbor; the loss forced the United States to give higher priority to aircraft carriers. Today, the airplane—and the tank for that matter—are losing ground to anti-whatever missiles. And eventually missiles will yield to energy-beam weapons.

GAME THEORY AND PARAMETERS

Game theory is a branch of mathematics that concentrates on the quantitative and relational logic involved in making choices in various conflicts or at least among alternatives that pose conflicting factors. As such, game theory subdivides into three categories: (1) the absence of dominant conflict, such as solitaire in which there are only choices; (2) direct conflict, where the success of one party is offset by an equivalent loss of the other (often called a *zero-sum game*); and (3) alternatives that involve significant trade-off decisions.

Game-theory models use two approaches. The first approach is called the *extensive model*, which catalogs all possible alternatives at each juncture point, then weighs the advantages and disadvantages of each insofar as they can be reduced to numbers or logic. This model is similar to a rules-based decision support program. The second approach—often called the *normalized option*—develops generalizations on different types of choices players may face, similar to a pattern-based decision support system. Obviously, the normalized approach must be used when the choices are multitudinous, and it is often the more practical course of action when the options are less numerous.

This leads to the doctrine of *calculated risk*. In most conflicts, success demands the taking of at least some risks. A calculated risk occurs when the chance of succeeding is measurably greater than failing. In a set of conflicts, therefore, the continuous application of calculated risk means that some engagements will be lost but more—the majority—will be won. Applied to investment, the practice is called diversification and dollar averaging, which ultimately is an application of statistical determinism.

One last point. The problem with too rigid an adherence to game theory is that minor changes in factors can greatly change the strategy required to prevail in a conflict. For example, the dynamics of football are largely related to parameters such as the geometry of the field, the four-downs/10-yards rule, the rule against pass interference, the relative positioning of the infield markers, and so forth. Change the four-downs/10-yards rule to three downs or 20 yards, and the strategy of the game will change radically. In effect, this is what happens in war. The parameters are never quite the same. Hence, while the principles of war remain constant, their application varies considerably.

The overriding point to keep in mind is this: In the absence of neat linearity and therefore the presence of potential intensely disruptive disorder, only sheer mass or ingenious patterns will suffice to avoid degeneration or defeat.

□ □ □

In the preface to this book, I mentioned that studies in military science provided seminal insight for the development of the physiogenetic thesis. Furthermore, this chapter noted that Napoleon held that such studies comprised "the science of sciences." As this book goes into production, it is apparent that the military profession has become the first to fully grasp and apply chaos, complexity theory, and related fields outside the halls of academe. *Marine Corps Doctrinal Publication 6: Command and Control* (1996) removes all doubt.

The thrust of that revised doctrine clearly recognizes that control of all elements on the battlefield is all but impossible. Instead, the commander should focus of the unfolding pattern of those elements, then nudge various elements into better position whenever the pattern begins to lose efficacy. It is only a matter of brief time before the other services follow suit, and, indeed, they all now comment on this new perspective in their various journals.

PART IV
Physical Systems

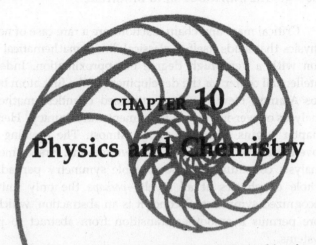

CHAPTER 10

Physics and Chemistry

> The whole burden of physics seems to consist in this—from the phenomena of motions to investigate the forces of nature, and then from these forces to demonstrate the other phenomena.
> —ISSAC NEWTON

The old *Candid Camera* television series included an episode whereby passersby were invited to sit at a dais-type table to sample pies stacked up by the dozens. Unbeknownst to the volunteers, the two men sitting at the ends of the table were actors. One actor walked over to the other and methodically smashed his face with a cream pie. The supposed victim just as methodically retaliated. It was not long before the scene degenerated into a hilarious melee. Obviously, this scenario was a sociological counterpart to the phenomenon of a chain reaction in physics, and the critical mass, as it were, occurred when the volunteers slackened their inhibitions against assault and battery.

Critical mass and chain reactions are a rare case of nonlinear physics that lends itself to classical, linear-mathematical evaluation with a very high degree of approximation. Indeed, the intellectual center for the development of the first atom bombs—Los Alamos, New Mexico—witnessed countless mathematical analyses of ever-increasing refinement and accuracy.[1] Hence, this chapter begins with this topic. Almost. The opening section covers dimensional analysis, for two reasons. First, dimensional analysis constitutes an indisputable symmetry pervading the whole of physics at all levels—*perhaps* the only universally recognized symmetry. Second, it is an abstraction, which therefore permits a smoother transition from abstract to physical systems.

DIMENSIONAL ANALYSIS

Physics and chemistry have many subdivisions, and sometimes the only linkage seems to be energy transformation and the equations used to measure it. Slim as that linkage may be, all physical equations are based on dimensional analysis. Not only must equations balance quantitatively, they must do so in the physical units the quantities represent. These units are permutations of *mass* (matter), *length* (space), and *time*. The implication is that if mass exists, it must exist somewhere in space. If more than one element of mass exists—and of course, they do—such elements perforce have an arrangement or configuration in space. And if they move—no debate there either—the configurations will change with the passage of time. By definition, such changes are patterns.

The virtue of dimensional analysis is that it instantly identifies fallacious equations in the physical sciences. For example, we can say that a velocity of 60 miles per hour equals a velocity of 88 feet per second, because there are two implied conversion factors of [1 hour/3600 seconds] and [5280 feet/mile]. Furthermore, both sides of the equation measure velocity (length per unit of time). But we *cannot* say that a velocity of 15 miles per hour equals an acceleration with the same number. The "dimensions" differ. Velocity is length per unit of time. Acceleration is velocity (length per unit of time) per unit of time, i.e., the rate of increase of velocity, not velocity *per se*.

In specific terms, the principal measurements in physics are velocity, acceleration (or deceleration), mass, momentum, force, energy, and power. The first two are *kinematic*, which means that no mass is involved. The others are *kinetic*, which means that mass *is* part of the equation. Another way of looking at this is that kinematics is kinetics in which points of mass are reduced to abstract geometric points. When a driver is tagged for speeding, the ticket does not normally cite the mass or weight of the car.[2]

Of the kinematic measures, *velocity* is the easiest to understand. It means the units of length traveled per unit of time, e.g., 88 feet per second. Dimensionally, this is expressed as:

$$\text{length/unit-of-time, } or$$

$$L/T$$

Acceleration is the measure of an increase in velocity. If a car is moving with a velocity of 40 miles per hour (mph), and a second later it is moving at 42 mph, the car is accelerating at the rate of 2 mph for each second of time that passes (or what is the same, in 1 second its velocity increases by 2 mph). At that rate, the velocity will reach 60 mph in another 9 seconds. Dimensionally, this is expressed as:

$$\text{(length/unit-of-time) per unit-of-time, } or$$

$$(L/T)/T, or$$

$$L/T^2$$

Now let us add mass to the consideration. The product of mass and velocity is called *momentum*, and the dimensions are:

$$\text{mass} \times \text{(length/unit-of-time), } or$$

$$ML/T$$

A heavier car will do more damage than a lighter car on impact when both have the same velocity. When mass is accelerated, the dimensions are mass times acceleration:

mass × (length/unit-of-time)/unit-of-time, *or*

$$M(L/T)/T, \textit{ or}$$

$$ML/T^2$$

This is called *force*, and it leads to one of the most basic equations in physics:

Force is the product of mass and acceleration, *or*

force = mass × acceleration, *or*

$$f = ma$$

By the law of inertia, an object in motion will continue in motion at the same velocity (speed plus direction) unless and until acted on by another object, which may be friction and air resistance. To accelerate an object, force must be applied, and, when applied, the acceleration will be in proportion to the force and inversely proportional to the mass. It takes more force to accelerate a heavier mass at the same rate. If an object encounters resistance, it will decelerate. To maintain current velocity (in the absence of a vacuum), it must be "pushed" with sufficient force to counter the force of friction. Except in the void of outer space, it takes force just to maintain velocity.

Energy is defined only in terms of effect, e.g., the capacity to do work (potential energy) or doing work (kinetic energy).[3] Work usually consists of, or results in, changing the configuration of a system in some form. Mathematically, energy can be said to measure force applied over length; hence, the dimensional analysis is:

$$M \times (L/T) \times (L/T), \textit{ or alternatively}$$

$$M \times ((L/T)/T) \times L$$

Both expressions reduce to ML^2/T^2.

The last item in this discussion is *power*. Power is defined as the time rate of doing work, i.e., the sustained capacity to do work, both of which mean the capacity or ability to expend energy. The dimensional units of power are:

$$M \times (L^2/T^2)/T, \text{ or}$$

$$ML^2/T^3$$

Power should not be confused with concentration. It is possible to focus beams of energy onto one spot and thereby concentrate power, but the power expended would be the same without the concentration. Efficient arrangements or configurations make the available energy more effective—a sharp knife obviously cuts more easily than a dull knife.

The underlying issue here is that if all of these measures are linked mathematically and dimensionally, are they not also linked ontologically—derivatives of the dynamics of mass (or matter) moving in space? Whenever energy is exchanged (kinetics), there also is momentum, force, and power.[4] The second proposed analysis in Appendix A addresses this issue.

CRITICAL MASS AND CHAIN REACTIONS

Nuclear chain reactions exemplify catastrophe. When a radioactive atom decays, it typically splits into two atoms and releases the leftover mass as: (1) subatomic particulate matter: helium nuclei (alpha particles) and electrons (beta particles) and (2) photons, usually in the form of gamma radiation. The adjectives are the first three letters of the Greek alphabet, intentionally chosen as a kind of dignified ABC.[5] No element heavier than lead is stable, and many isotopes of lighter elements are also radioactive. They all decay by an inverse exponential function known as *half-life*. Half-life means that in *n* units of time, one half of the atoms will decay. In another *n* units of time, half of what remains will decay, and so on until the last atom splits. This process is a function of mathematics. As the atoms decay, fewer and fewer of them remain to emit the wherewithal (neutrons) required to split the remaining atoms and fewer of their kin remain as targets.

But we do *not* call radioactivity a *chain* reaction because the number of reactions per "set," so to speak, gradually reduces to zero. A chain reaction occurs when each "set" of reactions is larger than the previous "set." This is accomplished by concentrating highly radioactive material so that this greater-than-unity ratio is achieved, analogous to the fact that riots can ensue among thousands of people jammed into a plaza but are not likely with the same number of people dispersed over the countryside. Restated, the key is to ensure that particles released by the decay of one atom immediately trigger, on average, more than one new split. If the ratio is 1:2, one split will lead to more than a half-trillion splits by the 50th "set." The cumulative total exceeds one trillion; the elapsed time, a few microseconds.

What are the factors that lead to a ratio greater than unity? The most common factor is the activity level of the radioactive element in use. The shorter the half-life, the less mass it takes to sustain a chain reaction. More neutrons released by decay will be available per unit of time. It is entirely possible to take a specified mass of one element that is barely subcritical, replace a plug of it with a more radioactive substance, and thereby achieve critical mass. However, if the half-life is too long, no amount of additional substance will achieve *criticality* (the "point" at which critical mass occurs). Six billion tons of uranium-238, *unaided*, will *not* sustain a chain reaction; less than 100 pounds of uranium-235 will.[6]

The second factor of critical mass is the length of the paths (interior lines) between the radioactive atoms in the mass. If a barely subcritical mass is compressed, distances among the atoms will decrease, thus increasing the time-density of reactions. It is true that the ensuing chain reaction will quickly reverse the compression by blowing up the mass, but in those few microseconds a sufficient number of atoms will split to sustain the reaction. This is especially true if the reaction begins with the release of a package of free neutrons and the material is tamped with a cover that reflects escaping neutrons back into the fissionable material. The technique is known as *implosion* and was used for the test at Alamogordo (Trinity, July 16, 1945) and for the bomb dropped on Nagasaki. The bomb dropped on Hiroshima was a gun type in which one subcritical mass was literally shot into the hollow of another subcritical mass.[7] But because the interior lines of the implosion technique are tighter, it required much less total material than the projectile methodology: between 10 and 15 pounds of plutonium, about the size of an orange.[8]

The third factor is the lack (or presence) of obstacles along the interior lines. In any given amount of radioactive material, impurities will delay a chain-reaction output, hence mandate higher mass. In peaceful nuclear power, this technique is used to control the chain reaction so that it operates just above critical mass. The first atomic pile (at the University of Chicago) used graphite rods (inserted into graphite blocks) to enhance further the releasing of neutrons, and cadmium rods (also inserted into the graphite blocks) to absorb them, hence control the chain reaction. Figure 30 pictures the various methods to attain critical mass.

This raises the issue: Even if the chain-reaction ratio just barely exceeds unity, would not the reactor pile *eventually* melt down if not explode? The answer is yes, and this is exactly what happened at Three Mile Island and later at Chernobyl. The potential disaster is prevented by reducing the ratio below unity when the chain reaction goes too far, a classic case of a homeostatic control of equilibrium run amok. An analogous situation occurs with a common kitchen oven with a faulty thermostat. In a crunch, a cook can turn the heat down when the oven gets too hot, then turn it back up when it falls too low—yo-yo baking.

So much for fission. Fusion is the inverse of fission. In fusion, heavy isotopes of hydrogen merge to form helium, emitting any excess as various subatomic particles and radiant energy. As more energy per reaction is released in fusion than fission, thermonuclear weapons have stronger yields by one or two orders of magnitude. And because it takes tremendous

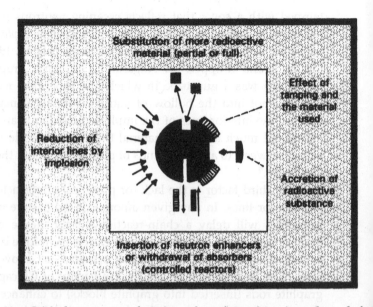

30. Critical mass as a function of configuration. A nuclear chain reaction requires a threshold mass, but that threshold varies with: (1) the element used, (2) arrangement, (3) tamping, (4) the presence or absence of inhibitors, and (5) the dynamics of accumulation. These factors suggest crucial configuration is a more accurate expression than critical mass.

pressure to initiate a fusion reaction, a fission trigger is required. It is also possible, in theory, to control a fusion "reactor," and this has been done in the laboratory for fractions of a second at very high cost.[9] Hopefully, the cost will shrink to the point where fusion power becomes commercially practical. The great benefit is that this process could yield an essentially infinite supply of energy *without* producing radioactive waste.

Meanwhile, we must settle for nature's fusion reactor called the sun, the chain reaction of which operates more or less at a net ratio of unity. Otherwise, it would quickly burn out or, alternatively, explode. The latter takes a larger mass than our sun has. However, other suns—stars—have the requisite mass and therefore do explode into humongous fireballs and then collapse into so-called black holes, a compact mass so dense that nothing can escape its field of gravity, not even Houdini. This collapse accelerates all fusion and fission, even of stable elements, toward Ag_{47}—the point of equilibrium for the nuclear packing-fraction curve.[10] Thus, while the grand finale of a supernova–black hole process combines a bang and a whimper, it also radiates the twinkling of a silver lining.

ENERGY EXCHANGE

In most physical interactions, energy is exchanged as a variation of a zero-sum game; the cumulative loss of the losing units equals the gain of the receiving units, though in virtually all cases, some of the gain is unusable and even dangerous (excessive friction) and some of it is lost to bystander units (like sparks). We cannot see or even visualize energy, except in mathematical terms, but we can often see and/or feel, and always measure the *effects* of its kinetic exchange: gears turn faster (mechanical energy), elements heat up (temperature, or thermodynamic energy), current flows (electrical energy), and so forth. That is to say, the loss or gain of energy in a unit, element, or system changes the properties or attributes thereof in some form or to some degree.

These transforms are ubiquitous, and many of them are found in the so-called power train of a nonnuclear-powered ship. The prevalent naval architecture requires no less than seven energy transformations. First, fuel oil is burned. Second, this energy transforms water into superheated steam (at 600 degrees Fahrenheit). Third, the pressure of this steam rotates a turbine. Fourth, the extension of the turbine's axle becomes the armature of an electrical generator. Fifth, the generated electricity powers a motor. Sixth, the motor's armature extends into a shaft at the end of which is a propeller. Seventh, the propeller substitutes for oars, pushing sea water away from the ship, which because of action and reaction drives the ship forward. Or, when the rotation, or pitch, is reversed, the propeller chews into the water in order to move the ship astern. This particular system illustrates four common forms of energy exchange. Burning fuel oil releases chemical energy. Heating water to steam is thermodynamic energy. Rotating the turbine, armature, and rotor exemplifies mechanical energy. And the generator, of course, produces electrical energy.

The symmetry here is that most of these transformations involve changes in position and motion or momentum. The energy obtained by burning fuel oil results in net lower energy levels of the electrons of the contributing molecules (which typically break down to simpler chemical elements). The superheating of water/steam increases the energy content of the electrons (and increases the circumference of their orbits about their respective nuclei). The actions of the mechanical transformations are obvious, and electrical energy obtains by "pushing" electrons down a wire (of which, more in the next chapter).

Interestingly, some energy transfers/transforms occur only in the presence of a catalyst—analogous to an agent—especially in molecular formation. The catalyst itself does not change, but without it the action will not happen or will proceed too slowly or only incompletely. In physical systems, catalysts operate like a reflector or refractor of sorts. They serve to concentrate or at least direct the effects of various interactions, the same way a magnifying glass can concentrate rays from the sun to initiate a burning process.

STATES OF MATTER

States of matter are a function of the energy level of orbiting electrons. As these electrons pick up speed, they become more ornery, and less willing to associate closely with their peers. As such, low levels of electron energy result in solid-state configurations. Near the lowest possible level of energy (the temperature of which is called *absolute zero*), atoms suddenly become good conductors of electricity. When the temperature approaches absolute zero to within a few billionths of a degree, atoms tend to form into humongous molecules that can best be described as goo.[11] In the opposite direction, increased temperature/energy levels render a substance more pliable (less brittle) but leave it as a solid.

At a certain threshold level of energy, atoms and molecules will no longer tolerate regimented structure, changing from a solid into a fluid. A fluid may be liquid or gaseous depending on atmospheric pressure. The lower the pressure, the sooner the liquid changes to a gas (in terms of further energy/temperature increase). In the absence of pressure, solids change immediately into a gas on rising past its "freezing" point. At that, some molecules change from a solid to a gas in spite of normal atmospheric pressure, for example, carbon dioxide (so-called dry ice). Note that liquids under pressure in an environment where gravity is artificially negated—space shuttles—remain liquid but break up into "floating" globs in which the atoms/molecules are held together by surface tension.

The plasma state—not to be confused with blood plasma—occurs at very high temperatures. In this state, electrons gain so much energy that the electromagnetic field/force operating within an atom can no longer hold them in orbit. The nucleus remains intact, but in the absence of electrons the substance changes some of its fundamental properties, most notably that nonconductors become good conductors (the same phenomenon that occurs near absolute zero). Incidentally, the plasma state is not all that rare in nature. For example, lightning sends an electrical current down through the atoms of the atmosphere, momentarily raising their temperature to the plasmic state—nature's arc welder.

The upshot of this is that changes in states of matter can be arbitrarily portrayed as a cycloid—the same as the shape of the chambered nautilus shell—as shown in Figure 31. To put the case another way, matter exhibits discrete intervals between states because electrons occupy distinct rings. The position of these rings with respect to the nucleus shifts outward or inward corresponding to different energy levels (temperature). As electrons pick up energy, the increase acts analogously to centrifugal force and thus jumps out to a new orbit.

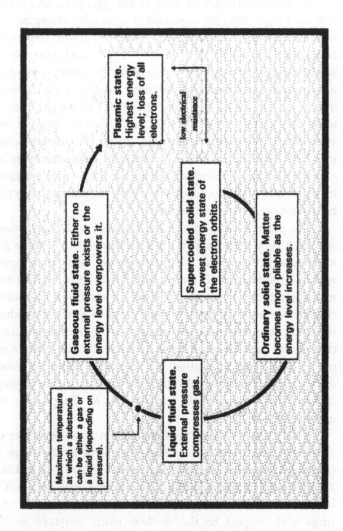

31. **States of matter.** States of matter can be portrayed as a cycloid. This cycloid represents variations in internal arrangement, to include coagulation of atoms into macromolecules as the temperature approaches absolute zero within a few billionths of

Within the image:

Plasmic state. Highest energy level; loss of all electrons.

low electrical resistance

Supercooled solid state. Lowest energy state of the electron orbits.

Gaseous fluid state. Either no external pressure exists or the energy level overpowers it.

Ordinary solid state. Matter becomes more pliable as the energy level increases.

Liquid fluid state. External pressure compresses gas.

Maximum temperature at which a substance can be either a gas or a liquid (depending on pressure).

Moreover, electrons gain energy by units called quanta, not continuously, or what in mathematics is called a *step function*. This was the phenomenon that confronted Max Planck at the turn of the century. By the same token, an atom exhibits a finite number of equilibrium states; only a small percentage of possible configurations work well. The theoretical configurations that lie between these states of dynamic equilibrium cannot adequately corral the electrons, but fortunately the imbalances quickly force the configuration into a more stable arrangement. Moreover, if it were not for these discrete operating configurations, electrons would fly off in all directions and by a chain reaction pull the earth apart at the seams. But we digress.

All of this leads to the question: Can a fluid in the form of liquid change to a gaseous form before it reaches the normal threshold temperature for the event (a function of atmospheric pressure)? That is, can the atoms increase their energy level without changing temperature? Answer: sort of, namely, the hydraulic jump. When water cascading down a chute reaches a certain threshold velocity, it will suddenly jump to roughly twice its height and reduce its velocity by half (thus maintaining the same flow rate). The jumping is a matter of the internal hydro-dynamic force vector being sufficient to momentarily overcome gravity. Those hydrodynamics act that way because the liquid picks up momentum to the point where its molecules will *attempt* to fly apart in gaseous form. Atmospheric pressure suppresses this mutiny by forcing the flow into a slower configuration. Incidentally, this phenomenon is quite common, and you can sometimes see it in an ordinary street gutter during a rainstorm.

PERIODICITY AMONG ELEMENTS

Mendeleyev identified periodicity among chemical elements by writing down the properties of each known element (63 at the time) on separate cards and then rearranging these cards on a table in columns of similar properties and in order of increasing molecular weight. Never underestimate the value of low-tech solutions. With the subsequent discovery of the inert elements, the present-day periodic table contains seven tiers, and each tier (except the last) terminates with an inert element. The first tier contains only two elements. The next two tiers have eight elements each, and the two after that have 18. The sixth has 32, and the seventh is incomplete, consisting of six naturally occurring, radioactive elements and the rest artificially created. The progression of tiers is based on the limited workable configurations of electrons, although it's not quite that simple. Some elements have more isotopes than others, and some of the isotopes of stable elements are radioactive.

Nevertheless, certain truisms prevail. All nuclei (except the first isotope of hydrogen) consist of neutrons and protons. These particles are bound together by enormous force (the strong nuclear binding force) to the point where it is exceptionally difficult to disengage them. Each element in the periodic progression adds one proton and one corresponding electron, while the number of neutrons increases at a progressively higher rate.[12] All lighter elements have at least one isotope that contains an equal number of protons and neutrons. Not until the 31st element (gallium) do neutrons outnumber protons in every known isotope.

Next, protons are said to have a positive charge; electrons, a negative charge; and neutrons, none. The neutrons—perhaps in conjunction with other particles—obviously play a role in holding the protons together because like charges repel. (The ordinary hydrogen atom with one proton does not need a neutron.) Furthermore, the ratio of neutrons to protons increases as the molecular weight increases, which probably has something to do with spherical geometry, analogous to the hexagonal- and pentagonal-shaped pieces that make up a soccer ball. If the pieces were kept the same size but the diameter of the ball greatly increased, the pieces would no longer fit. Either more of one kind, or a different pair of shapes, would be necessary.

Neutrons are slightly heavier than protons, and a free neutron decays into a proton and an electron. Meanwhile, the electron, with a mass of roughly 1/2000th of a proton, carries an equivalent but opposing charge. This difference in mass permits it to move in orbit while the nucleus more or less stays in place. Furthermore, electrons determine how elements combine or otherwise behave, and the determination is made largely by the electrons in the outermost shell. Atoms seek equilibrium. Inert elements achieve that equilibrium in their own right, though they can contribute to a compound under limited circumstances. In all other elements, the unbalanced electrons of one atom will seek equilibrium with the unbalanced electrons of another atom (formerly called *valence*).

Briefly, an element with one bachelor electron will tend to combine with an atom that lacks only one electron to achieve stability. Or an element that needs two electrons will combine with either an atom possessing two excess electrons or with two atoms that offer one each. This is an oversimplification, but the complexities are permutations and combinations of this trend. Furthermore, the elements in any one column of the periodic table differ from one another significantly despite their similarities. Finally, when the nucleus grows too large, it becomes inherently unstable probably for the same reason that a small construct made out of Tinkertoys is fairly stable, whereas a much larger one is wobbly and easily falls apart.

GEOMETRY OF THE PERIODIC TABLE

The pyramidal structure of the periodic table (2, 8, 8, 18, 18, 32, 17+) is, in part, misleading. The 18-element tiers comprise two periodic repetitions that lack an inert element between them. Instead, three metals appear in the center. The same is true of the 32-element tier, which arguably is a third 18-element tier with an additional 14 elements shoehorned in at a single point. Thus, with the logical rearrangement as shown in Figure 32 (a minor variation of the so-called *short-form table*), the periodic table can be transduced into a neat helix. You can see this by making a photocopy of that figure and rolling it into a cylinder. As the helix descends into each succeeding tier, the properties of the

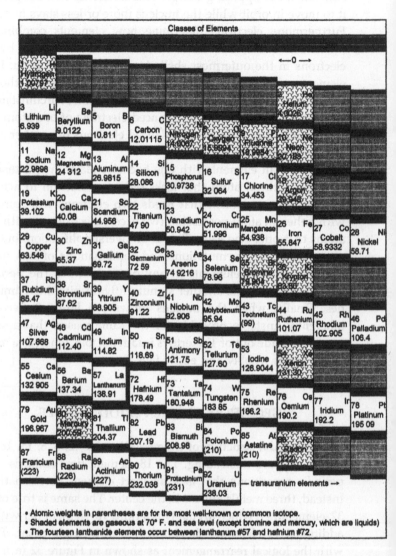

32. The periodic table of elements as a helix. Splitting the periodic-table rows with more than eight elements into two rows reveals the helical periodicity of elemental configurations. Wrap a photocopy of this chart into a cylinder to see the helix.

elements in each column shift. For example, the properties of silicon (class IV) veer to the right by demonstrating some properties associated with class-V elements. In the reverse direction, carbon (the first of the class-IV elements) exhibits more class-III properties than does silicon. The substitution of three metals for an inert gas at the end of the fourth, sixth, and eighth tiers can be explained by this shift to the right as the progression proceeds down the helix. The progression overshoots the internal balance mark, i.e., the configuration at that point lacks the necessary equilibrium for inertness.

The last point to note is the special position of the carbon atom and its four marriageable electrons. We know that carbon has the ability to form the most compounds, far in excess of a million.[13] It can also crystallize on a planar surface as graphite (which is therefore slippery) or three-dimensionally as a diamond, sometimes with perfect tetrahedral structure throughout. In addition, carbon forms itself into a 60-element geodesic-type molecule called *fullerite* (named in honor of Buckminster Fuller). This extreme versatility lends itself to complex organic constructs, not unlike the basic "wheels" in a Tinkertoy set. It is no coincidence that biochemists use a variation of this toy to build extensive molecular models, including DNA.

GENETIC PHYSICS

The concept of reducing biology, or at least a large slice of it, to physics is not exactly new. Alfred J. Lotka published *Elements of Physical Biology* (later retitled *Elements of Mathematical Biology*) in 1924, and Nicholas Rashevsky wrote *Mathematical Biophysics: Physico-Mathematical Foundations of Biology,* last updated in 1960. Still, the thrust of both was to document the orderliness of physiological processes. Much earlier, Gregor Mendel documented a certain degree of numerical orderliness in heredity. But the underlying mechanics of the extraordinary phenomenon of genetics, and evolution for the matter, escaped many of them, though Rashevsky was clearly aware of biological "weaving."[14]

Look at the side-by-side photographs in Figure 33. At first glance, the juxtapositioning of a nuclear-explosion mushroom cloud with the stately, dignified giant sequoia tree seems ludicrous. The explosion is an uncontrolled catastrophe, the force of which can equal a million tons (or more) of ordinary chemical explosive. The chain reaction starts with a handful of fissions and fusions, then proceeds by geometric progression to quintillions of them within an infinitesimal fraction of a second. By contrast, the biological chain reaction—binary fissions—takes roughly 2500 years to progress from a seed weighing about 1/5800th of an ounce to a behemoth that tips the scales at up to 2,700,000 pounds—the estimated weight of the General Sherman tree shown in Figure 33. And what is the difference? Time. The organic chain reaction is highly controlled and, except in the first stages of development, does *not* proceed with geometric progression.

33. Variations on chain reactions. A nuclear explosion is a well-known case of a chain reaction and among the most destructive. But the same fundamental process also applies to the creation and development of organisms and, ironically, the giant sequoia tree increases its weight from seed to mature tree by a ratio of up to 250 billion.

Also, keep in mind that catastrophe can be either constructive or destructive in its effects. This reversal of entropy can unfold with suddenness and obviously radical change, or it can proceed with *apparent* order. Apparent. The efficacy of a seed weighing 1/5800th of an ounce multiplying its heft by a factor of 250,000,000,000 (250 billion) is radical indeed.

Sociological systems offer a parallel here, in that some of the most radical military and political leaders emanate such an aura of quiet, bona fide dignity and methodicalness that they can turn a part of the world upside down, so to speak, and still be regarded as dour bureaucrats. The classic case was Major General Adolphus Washington Greely (1844–1935), thrice-wounded veteran of the Civil War, first U.S. Army private to rise to general office rank, Arctic explorer, cofounder of the National Geographic Society, university-level professor of geography with only a high-school education, confidant of more than ten U.S. presidents, and mentor of the radical General William "Billy" Mitchell.[15] On Greely's 91st birthday, Congress awarded him the Medal of Honor "For a lifetime of splendid public service." Not a special gold medallion, *the* Medal of Honor.[16] Yet he was so dour and unobtrusive that he is all but forgotten today. Granted, this case distracts from physics, but the intent is to illustrate how we fail to recognize catastrophe and radical growth when it comes to pass in a dignified guise.

□ □ □

Now we need to look at engineering, which is the technological face of physics and chemistry—the methodology by which we change and combine configurations of the material world to meet our needs.

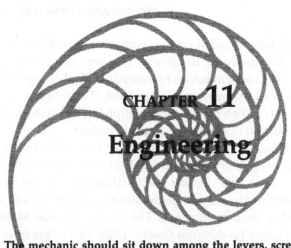

CHAPTER 11

Engineering

The mechanic should sit down among the levers, screws, wedges,
and wheels, considering them as the exhibition of his thoughts, in
which a new arrangement transmits a new idea to the world.
—ROBERT FULTON

The word *engine* **derives from** Latin words meaning "talent"
and "to beget." Thus, *engineering* means talented creation of
something presumably useful and, as such, is another way of
expressing the design of a system. Furthermore, Robert Fulton
suggested that mechanics (as engineers were sometimes called)
can apply their skill to both physical and abstract systems. In
practice, however, the term is limited to physical systems,
encountering derision when used elsewhere, for example, so-
called social engineering.

Until recent times engineering applied only to inanimate
systems. The extension to biology arose with "smart" prostheses
linked to nerve tissue, earning the moniker *biomedical engineering*.

In turn, biomedical engineering led to *genetic engineering* and to manipulation of many forms of biological tissue. It is but one small step, then, to extend the concept of engineering to the entire genetic process—how nature engineers the autoconstruction of an organism from a fertilized egg. This chapter focuses on inanimate engineering; the next three chapters, on organic engineering.

CIVIL ENGINEERING

From the earliest times of recorded history, man has built structures for his memorial. The best known of these ancient structures are the pyramids in Egypt. Their design is massive, stable, and enduring. But it is one thing to erect a cairn of squared-off rocks; quite another to build the cathedrals of Europe that appeared about 2000 years later. During that period, builders learned to arrange stone more efficiently. This is civil engineering: to configure the least amount of material into a "permanent" structure that meets one or more human needs, ideally with a finesse that also qualifies it as a work of art.

Although most civil engineering projects do not move perceptibly, they must resist imposed loads and forces without collapsing or weakening to the point of reduced capacity. The key is to transfer loads, before they can do damage, to a much larger system that easily absorbs these loads without flinching. The earth meets that criterion handily, but there are times when the earth itself must first be engineered, called soils engineering or stabilization. The most famous case where stabilization was taken for granted is the campanile of the cathedral in Pisa.

Forces have only two general effects on building materials and the earth's surface. The first effect is *compression* (pressure), which results in disintegration or crumbling unless the material has the ability to withstand it, or is designed to compress, like a spring. The second effect is *tension* (pulling), which results in the substance being torn apart unless it is strong enough to resist the expansion. The rotational variant of tension is called *torque*.

The clearest way to see how this works is with a concrete beam. Concrete resists enormous compression, but it tears apart easily even when reinforced with embedded steel rods. Hence, concrete is an excellent building material for foundations but not beams. As shown in Figure 34, when a load is placed on a concrete beam, it causes the beam to sag. This sag compresses the upper half of the beam while expanding the lower half in compensation. Given the poor tensile strength of concrete, the beam shatters or cracks.

The ultimate causes of this weakness are geometry and density. In the upper half, the pressure forces the same amount of material into slightly less space. In turn, the beam morphs into an arc, which means the outer half tends to expand. The way to

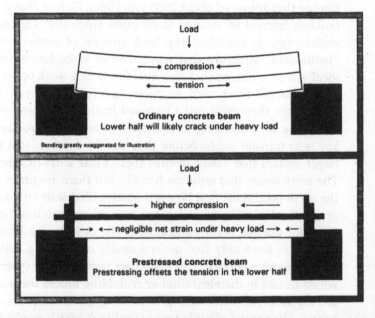

34. The push and pull of engineering. Engineering can modify unfavorable patterns of compression and tension. Sometimes a simple reconfiguration can eliminate a major problem, for example, prestressing concrete to give it the strength of a steel beam.

remedy this problem is to put lateral stress on the beam *before* it is loaded. This is done by installing steel plates at the abutments and threading nuts at each end of a rod connecting the plates. When a load is applied, the lower half of the beam will try to expand, but the prestressing force will counter the would-be expansion; hence, the concrete will not crack. The upper half, of course, will be compressed even more by the prestressing, but concrete's high compressive strength will handily resist crumbling. This prestressing technique is a good example of changing a configuration without changing its arrangement. The elements are placed under pressure without perceptibly changing their relative location.

Longer spans require at least a major change in configuration if not material. As beams "grow" longer, they must also "grow" thicker in order to sustain the same load. Their increasing deadweight is analogous to the futility of loading an aircraft with too much fuel. Hence, beams are impractical much beyond 100 feet in length, mandating a new configuration such as the common truss or arch configuration.[1] And beyond roughly 2000 feet, the only efficient design is the suspension bridge. Like the straight beam, the exponentially increasing deadweight is counterproductive and prohibitively expensive.

At that, the suspension configuration has practical limit in the 4200- to 4600-foot range. Otherwise, the towers would become too tall and massive in proportion to the roadway. Had the Golden Gate Bridge at the entrance to San Francisco Bay been built with a clear span from shore to shore (about 5400 feet), the structure would have expanded from a graceful work of art into a monstrosity.[2] Incidentally, the design of the suspension-bridge section of the nearby Bay Bridge had to cope with a 9000-foot span. The solution was to construct a concrete "island" at the midpoint, then build *two* suspension bridges.

One last point here: If systems are programmed—be it in the form of written instructions or by virtue of configuration—where is the program for a static structure? It is in the flexing and bending (equilibrium mechanics) of the materials that transfer loads without rupturing or weakening those materials. Recall the discussion on the *Titanic* in Chapter 2. It sunk because the welding of the watertight-compartment bulkheads (walls) lacked the resiliency to absorb the shock waves reverberating from the collision with the iceberg.

MECHANICAL ENGINEERING

Mechanical engineering includes most physical systems whereby the dynamics are critical to the operation. Some machines and plants are stationary, for example, an oil refinery. Some move or at least are transportable (though not for the purpose of conveying passengers or materials), for example, a road grader or a food mixer. Others *are* used for conveyance. Some machines are designed to do work while others transform and route energy to other machines that perform work. Yet irrespective of the intent and design, all machines depend on the configuration of their elements or, more accurately, on the pattern of changes in the configuration. This is the meaning of dynamics.

Many machines are driven by a motor or equivalent, but the most ubiquitous applications are hand tools: hammers, screwdrivers, pliers or snips, and wedges. The configurations of these simple machines always rely on some combination of clout and leverage. Leverage is usually called mechanical advantage, which is easily seen in a pulley system or a car jack. Lifting 2500 pounds *directly* is beyond the capacity of any human being, but it is easy when mechanical advantage is employed. Lifting 2500 pounds 1 foot is equivalent to lifting 250 pounds 10 feet, or 50 pounds 50 feet. Furthermore, the 50-foot lifting can be done by cyclic operation, as with a car jack lever.

Screwdrivers apply leverage by way of handles that are wider than the blade. For a demonstration of this, try using a screwdriver by grasping it along the shaft rather than the handle. The leverage of pliers and snips is demonstrated by trying to use them in reverse. The same applies to a hammer by way of the arc of its swing, but this force is compounded by the clout of its head (aided by gravity for downstrokes).

Note, too, the limited operations that machines can perform: compress, add, subtract, turn, or rearrange (or reconfigure). Roads and dams reconfigure the landscape. The planing of lumber removes material. Generators and motors turn. Garbage compactors compress. Crowbars are used to pull material apart. If this sounds like mathematics, it is. Kinetics is nothing more than kinematics plus mass, and kinematics is applied mathematics.

The permutations and combinations may be infinite, but the underlying concepts remain the same. It is little wonder, then, that virtually all new inventions draw on previous technologies to one degree or another. Some commercial washing machines are still user-programmed for different types of loads by way of plastic punch cards that are similar to the cards used by the first automated looms and by early computers.

In some applications, energy must be transferred over a long distance or under circumstances that render pulleys and the like impractical or unreliable, say, a car lift in a commercial garage. In most cases, a pipeline or variant supplies the need. Aqueducts and canals have been used for moving liquids since ancient times. Today an enclosed pipe is more common, which can also move solid material in the form of a slurry, particularly coal.

Pipeline equivalents are also used to transfer energy directly, in which case the technique is called *hydraulics*. The design of brake lines in cars long ago changed from mechanical linkage to hydraulic lines, though emergency brakes still rely on wire. Hydraulics constitute an extremely efficient means of transmitting force or energy. A pressure of 100 pounds per square inch (psi) at any one point often results in a *usable* pressure of 99 psi at all other points in the enclosed pipe. Mechanical advantage works by depressing a piston a greater distance in order to lift a heavier mass by a lesser distance, e.g., a hydraulic car jack.

Electrical transmission is a variation of hydraulics, whereby electrons are sent bounding down a wire or cable. Whereas a cable may appear to be a solid to us, to an electron it is 99.99 percent hollow. The problem is that the electrons are "locked" in orbit. To the extent these electrons resist being dislodged, the

material is a nonconductor. Conversely, to the extent electron
dynamics facilitate exchange, the material is a conductor. The
electrons successively knock other electrons from one atom or
molecule to the next, much like a series of billiard balls placed at
intervals in a channel. In this case, the pressure is called *voltage*
the interference, *resistance*; and the resulting flow, *current*.

CONVEYANCE ENGINEERING

While machinery assumes millions of different forms, conveyances tend to merge into a handful of paradigms. To be sure
these paradigms differ markedly among categories—highway
rail, sea, and so forth—but within any category the variations
tend to be more a matter of style than function, at least once the
development phase is over. The apparent reason for this singu
larity is that the requirements are more intense and partially
contradictory; hence, only a few configurations have the flexibil
ity to strike the necessary balance. The most widely used
conveyance power plant in the world—the internal combustion
engine—has withstood all attempts at replacement, at least fo.
fossil fuels.

For land transportation, wheeled vehicles dominate. The only
significant competition is the tracked vehicle, and it too is driven
by wheels in the form of gears meshed to racetrack-shaped tracks
Moreover, most wheeled vehicles—bicycles and motored kin be
ing the principal exception—have two or more axles with two
wheels (or sets of wheels) per axle. The reason is the extreme
efficiency of the wheel combined with the need for vertical
stability. Furthermore, virtually all land vehicles depend on gear
or equivalents for mechanical advantage, either to increase torque
for heavy loads (or initial acceleration) or, the opposite, to increase
speed when power requirements are minimal.

Interestingly, the common automobile evolved into a single
genus that varies less than *Homo sapiens*. As among humans
some cars are larger, or in better condition, or are stronger than
others. Yet today, arguably the only significant differences are

the adornments and styling, i.e., the body shell, which alternates between bulbous and boxy.

The train is also a wheeled vehicle, trading operator-controlled steering for tracks. However, the use of trains is declining (except for heavy bulk freight), supplanted by wheeled vehicles and, increasingly, air transport. The reason is that when power plants were first added to conveyances, they were too heavy and required too much maintenance for wagons. Initially, only ships could bear them. As the design of engines grew somewhat more compact, they were applied to trains, which were a variation on a string of towed canal barges. In time, the engine shrunk to the point where it could be accommodated by wagons and buggies; hence the advent of the automobile and, with it, virtually unlimited mobility.

As for ships, the earliest powered models were paddle wheelers—a common waterwheel operation in reverse. Eventually, these wheels were reduced to propellers and moved aft perpendicular to the wake. The wood hulls also gave way to steel, and slowly—some would say sadly—the sails disappeared. From that point forward, most ships have been templated from a common paradigm that varies primarily by what is conveyed: freight, bulk cargo, liquids, or passengers. Only the aircraft carrier and the submarine vary significantly in configuration. The carrier substitutes for a small island while the submarine resists water pressure from all directions.

This leaves the airplane, which is the most fascinating of all conveyance designs. The chief problem is that powered flight *begins* at a point of diminishing returns. On the earth's surface, the weight of a conveyance engine and fuel load inherently transfers to land or water. In the air, a glider barely resists the pull of gravity. Add an engine and the total weight will send it crashing to the ground. The Wright brothers succeeded only by enlarging the wings of a glider, thus providing more lift for the plane.[3] At that, it took several years of experimentation before the first successful flights were made, and another five years to develop controls sufficient for the pilot to maneuver the aircraft.[4] Still, the airplane was largely ignored until about 1912, when it suddenly caught the public imagination, culminating in the famed dogfights over Europe in World War I.

A few early prophets—among them Generals Giulio Douhet and William "Billy" Mitchell—saw the potential of the airplane as an instrument of war, far ahead of parallel commercial insights.[5] However, general disarmament after the war curtailed this impetus while commercial development sought an ideal configuration. This configuration was achieved with the Douglas DC-3 model in late 1935, which became perhaps the most notable upswing point in the history of transportation. From that time forward, the configuration of most new aircraft evolved from it. Of more than 10,000 DC-3s produced, several hundred remain in commercial use, including one used for tourist flights over San Francisco Bay.

Man, of course, learned to walk and run before he built machines in which to ride, but the same is not true of flight. The idea of man-powered flight elicited only derision until the middle of this century because early attempts gave the *Three Stooges* competition for slapstick. Yet in the end ingenuity won

35. The almost indestructible DC-3 aircraft. Few machines endear themselves to mankind. The DC-3 is an exception. Its near-perfect configuration made it the most forgiving aircraft in aviation history, and several hundred are still in commercial use.

out. The task, like that faced by the Wright brothers, was to create a glider that would sustain the weight of pedal-to-propeller machinery *plus* the pilot while staying aloft at slow speeds. Planes do not float in air like dirigibles. Instead, they are lifted by differential air pressure above and below the wing. As velocity decreases, so too does this differential, eventually reaching stall speed. At that speed, gravity becomes the dominant force. Solution: "inflate" the configuration, i.e., lower the density, and use extremely lightweight materials. This increases the total lift capacity and reduces the weight that must be lifted. The *Gossamer* series are the best known of these machines, and one of them retraced the mythical flight of Icarus, this time successfully.

ROBOTIC ENGINEERING

Robotics can be arbitrarily divided into three perspectives: (1) morphology, (2) degree of programming, and (3) status as a system versus a subsystem. *Morphology* is the common perception of a robot as a mechanical clone of man or at least his arms and hands. In practice, the vast majority of commercial robotic systems are designed to substitute for human *labor*, not human *form*. As such, the configuration may or may not resemble anything human. The human body may be a marvel of efficiency, but it is prone to break down with some tasks. Mechanical cranes, for example, are infinitely more reliable than the human backbone for lifting, and cranes do not particularly resemble human form. On the other hand, robotic welders on automated assembly lines bear striking resemblance to arm-and-hand movements.

The second perspective on robots focuses on *programming*. All machines rely on programming if only by mechanical design, e.g., the linkage of pistons, shafts, and axles in an automobile. In practice, however, most robotic systems employ formal, *modifiable* instructions and have done so ever since Jacquard added this feature to automated looms around the year 1800. The third perspective compares stand-alone systems with control mechanisms for another machine or system. The automated loom is essentially a stand-alone system. Examples of robotic *controls* include programmed thermostats and computers found in all newer automobiles. These automobile computers modify engine "behavior" based on input from that behavior itself (analogous to a speed governor), notably in the form of pollution controls. The majority of stand-alone robotic systems are networks in some form. The best known are automated assembly lines, light-rail passenger (or materials-handling) systems, and power distribution nets. These networks have analogous counterparts in anatomy and physiology: respectively, the gastrointestinal tract, the circulatory system, and the nervous system.

The automated assembly line is Frederick Taylor's time-motion technique carried to its logical extreme. If humans can be reduced to de facto automatons, why not substitute de jure apparatus? Place a chassis at the beginning of a rail line and rely on various machines to attach components in sequence. This automation process is cheaper than human labor and, in most instances, more reliable and faster. For example, the automobiles manufactured by the former East German Trabant plant were so poor that what was formerly West Germany had to replace the plant with an automated model. This new plant employed only 10 percent of the former labor force (which admittedly raises social if not ethical issues).

The automated rail system is a variation of an assembly line in which the attachments are temporary. This may sound silly, but it is exactly what happens with automated mail carts guided along invisible electronic tracks. The carts beep at each station to signal clerks to drop off and to pick up mail. The Pentagon in Washington has used robotic carts for more than a decade with considerable success. Elsewhere, entire subway systems have been automated in which the only employees seen by passengers are security guards.

Energy, too, is usually routed by robotic systems, automatically redistributing power as required to meet local peak loads and drawing it from other sources when necessary, or selling it when there is slack. These networks are largely silent and invisible. However, they are subject to (pseudo) catastrophic breakdown, as occurred in New York on November 9, 1965. In *Connections*, James Burke—referring to a tiny relay mechanism—said that precisely at 5:16:11, the failure of a simple electronic relay "set in motion a sequence of events that would lead, within twelve minutes, to chaos." Burke then described the magnitude of that disorder, namely, that roughly 80,000 square miles of an extraordinarily industrialized and highly populated region came "to a virtual standstill."[6] The details are unimportant except to say that this is just one more application of a chain reaction—the so-called butterfly effect that initiated the science of chaos. Burke called it the trigger effect.

ENERGY ENGINEERING

Energy engineering can be divided into three approaches: (1) efficient utilization of nonrenewable sources, primarily fossil fuels; (2) utilization of renewable sources, especially solar, wind, and tidal power; and (3) searches for new sources of energy. Efficient utilization of nonrenewable sources takes two forms: active, such as designing more efficient machines, and passive, meaning to reduce or eliminate the need for energy (e.g., better household insulation and working at home via computer networks).

A variation on efficient utilization—and this applies to all forms of energy—is to store it, then use it only as necessary. This approach may seem childishly obvious, but keep in mind that kinetic energy does not store well. It does not exist in the same sense as matter because it is a function of motion in some form, and the only way to "store" motion is to corral it until needed. The simplest example of corralling energy is the flywheel, but the flywheel is too cumbersome for large-scale operations, for example, hydroelectric power where the flow of water cannot always be correlated with electrical-energy requirements. To solve this problem, one electric power generation plant uses excess power during periods of low usage to pump water to a reservoir maintained at a higher level. During high-usage periods, that stored water is released down the spillway to augment routine power generation.

Lastly, one other source of potential energy remains untapped: common magnetism. To be sure, electrical power is based entirely on this phenomenon, but that is *induced* magnetism. Electromagnetism is induced by transforming mechanical rotation into electromagnetic current, which is then used to power motors, and so on. Now if so-called permanent magnets could be substituted for electromagnets, the need for an external power source is eliminated. The problem is that permanent magnets have a steady linear pull in one direction, whereas electromagnetic fields can be shaped in the round (windings) and thus operate with alternating polarity. Question: Can existing lineal force be transformed into angular force (torque) without an external power source?

As illustrated in Figure 36, no less a physicist than Sir Isaac Newton thought the feat was possible.[7] In an electromagnetic motor, magnetized fields of opposite polarity turn an armature. As the configuration changes to the point where the magnetic fields have the same polarity, the circuitry changes so that opposite polarity is brought to bear. This is not possible with a permanent-magnet motor, but the angular momentum of the armature could be used to physically shift one of the magnets out of the way momentarily until the next magnet came in range. The key is to shift the stator magnet before the rotor magnet gets too close. Otherwise the arrangement would use more power shifting the magnets than can be obtained from the natural magnetic attraction.

The disadvantage to this concept is that the power takeoff will consume at least 90 percent of the generated torque. Also, permanent magnets must be periodically remagnetized, which can be done by "stroking" them with another magnet. As such, generation of a kilowatt of power might require a machine weighing at least a ton. Still, the price is right.

One last point. Do not underestimate the power of magnetism. It is roughly 100,000,000,000,000,000,000,000,000,000,000,000,000

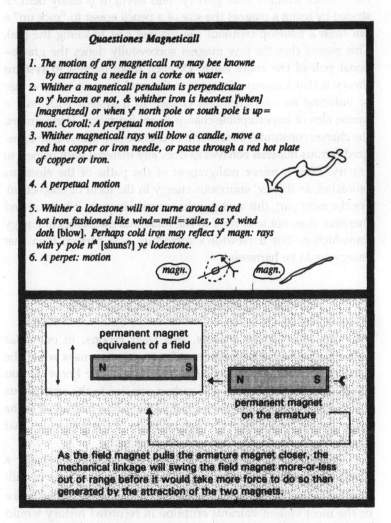

Quaestiones Magneticall

1. *The motion of any magneticall ray may bee knowne by attracting a needle in a corke on water.*

2. *Whither a magneticall pendulum is perpendicular to y' horizon or not, & whither iron is heaviest [when] [magnetized] or when y' north pole or south pole is up= most. Coroll: A perpetual motion*

3. *Whither magneticall rays will blow a candle, move a red hot copper or iron needle, or passe through a red hot plate of copper or iron.*

4. *A perpetual motion*

5. *Whither a lodestone will not turne around a red hot iron fashioned like wind=mill=sailes, as y' wind doth [blow]. Perhaps cold iron may reflect y' magn: rays with y' pole n^th [shuns?] ye lodestone.*

6. *A perpet: motion*

magn.   magn.

permanent magnet
equivalent of a field

N   S

N   S

permanent magnet
on the armature

As the field magnet pulls the armature magnet closer, the mechanical linkage will swing the field magnet more-or-less out of range before it would take more force to do so than generated by the attraction of the two magnets.

36. Magnetic engines. Although Newton pondered magnetic perpetual-motion machines (using that term explicitly), he meant the conversion of existing linear force to a rotational variant, at least until such machines wore out. That limited process is theoretically possible by way of the configuration dynamics pictured in the lower frame.

(10^{38}) times stronger than gravity. This strength is easily demonstrated by using a magnet the size of a pencil eraser to "pick up" a pin from a tabletop (without the magnet first touching the pin). This means that the tiny magnet successfully defies the gravitational pull of the entire earth. Of even more interest to systems theory is that a common magnet is the world's most extreme case of inducing an overwhelming change in a system with only a minor, almost imperceptible change in configuration. Furthermore, the change constitutes a rare form of catastrophe. How? Unmagnetized ferrous material behaves as does any material with respect to gravity. Then a minor realignment of the paths of the electrons unleashes, as it were, enormous energy in the form of magnetism. For the most part, this energy goes to waste because it is linear and therefore does not lend itself to the rotational mechanics of energy transduction. But if Newton's conjecture was correct, then that energy could be harnessed.

EVOLUTION OF THE SOLAR SYSTEM

Why do eight of the planets, and their moons, in our solar system consist of inhospitable rocks or gaseous mass while the remaining planet—earth—thrives with life as we know it? You might as well ask why a ton of carbon remains in the amorphous form of coal while a few micrograms of it spring into life. It is the same answer: the different configurations of the molecules that carbon can form under different circumstances.

However, the creation of a solar system, and especially a livable planet, is a violent process that takes hundreds of millions of years. By way of comparison, the strongest earthquake or the most violent volcanic eruption in recorded history would be a sniffle. Nevertheless, the specifics remain controversial, and each proffered explanation must be traced back to the big bang of cosmology. The only thing we can state clearly is that nature relied on a physiogenetic process of some kind. One configuration led to the next, and so forth, and each new configuration was a function of both the previous arrangement and the attributes of the elements. If so, this process leads to some interesting speculations.

For example, look at a globe with Tahiti (150° west longitude, 17° south latitude) closest to your eye. From this perspective, most of the visible hemisphere is water, which means the bulk of the land mass resides on the other side. If the earth did not rotate about its own axis during the early years, the original land mass may have been formed by an outward thrust impelled by centrifugal force. Later, other forces pushed the plates apart (originally called *continental drift*, now *plate tectonics*). Most of the earth's land mass can be refitted as if the continents were pieces of a jigsaw puzzle.

WEATHER AND ECOLOGY

Nature herself is an able practitioner of engineering. We see this clearly in the spiderweb, the dolphin's sonar, the ants' colony, the chambered nautilus shell, the electric eel (at 650 volts), and the hummingbird's hovering flight that puts a helicopter to shame. But these unusual phenomena tend to overshadow a much, much larger inorganic system, namely, ecology with its distinct subsystem of weather. The popular notion of weather is that it is erratic, fickle, unpredictable, and in general, as Mark Twain noted, something about which to complain without results. Interestingly, the former maximum-security prison on Alcatraz Island in San Francisco Bay permitted the "residents" one visitor per month, but restricted conversation to family matters, pending legal cases, and the weather.

Contrary to popular opinion, weather *is* a system. Despite its apparent irregularities and local catastrophes in the form of hurricanes and the like (which meteorologists say arise from a chain reaction of circumstances), it operates with an overall steady pattern that permits agriculture. Yes, droughts occur, but without weather most species would face extinction from a lack of food, and only a small fraction of *Homo sapiens* could survive via greenhouses. To put the case another way, weather exhibits a

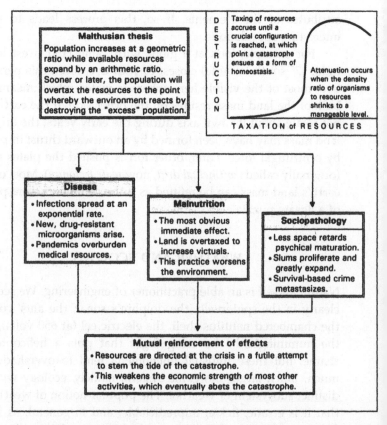

Malthusian thesis

Population increases at a geometric ratio while available resources expand by an arithmetic ratio. Sooner or later, the population will overtax the resources to the point whereby the environment reacts by destroying the "excess" population.

D E S T R U C T I O N

Taxing of resources accrues until a crucial configuration is reached, at which point a catastrophe ensues as a form of homeostasis.

Attenuation occurs when the density ratio of organisms to resources shrinks to a manageable level.

TAXATION of RESOURCES

Disease

- Infections spread at an exponential rate.
- New, drug-resistant microorganisms arise.
- Pandemics overburden medical resources.

Malnutrition

- The most obvious immediate effect.
- Land is overtaxed to increase victuals.
- This practice worsens the environment.

Sociopathology

- Less space retards psychical maturation.
- Slums proliferate and greatly expand.
- Survival-based crime metastasizes.

Mutual reinforcement of effects

- Resources are directed at the crisis in a futile attempt to stem the tide of the catastrophe.
- This weakens the economic strength of most other activities, which eventually abets the catastrophe.

37. **Malthusian catastrophe.** The Malthusian thesis states that when the carrying capacity of an environment, or the earth as a whole, is overtaxed by excessive population, it will react violently to rid itself of the excess.

highly dynamic form of equilibrium, and its disruptions lack the force to destroy all life on earth. The earth has had only one evolution of the species.

On the other hand, it is possible that mankind could trigger such destruction. For a locally confined example of this, consider the case when the former Soviet Union diverted water from the Sea of Aral for irrigation. The receding of the lake affected the weather adversely, which destroyed the increased crop production without refilling the lake.[8] Elsewhere, we have encroaching deserts in Africa, the depletion of rain forests in much of South America (especially in the Amazon River basin), the adverse affects stemming from global warming, the greenhouse effect, acid rain, dwindling natural resources, and fruitcakes.

This is what ecology is all about. The earth possesses enormous resiliency, but when we—or any species for that matter—tax the earth's resources beyond its carrying capacity, nature reacts with a vengeance, leading to mass starvation and/or epidemics. Thomas Malthus saw this clearly in 1798, when he wrote *On the Principle of Population*. That essay set off a storm of protest that endured for two centuries, but today scientists and social scientists see overpopulation as a major, if not the greatest, threat to human survival.[9] We may have already passed the culminating point of ecological resiliency, even to sustain the *present* world population. We will revisit this issue in the chapter on sociological systems operating on an international basis, because mankind has choices on this matter. In the interim, we will turn to organic engineering.

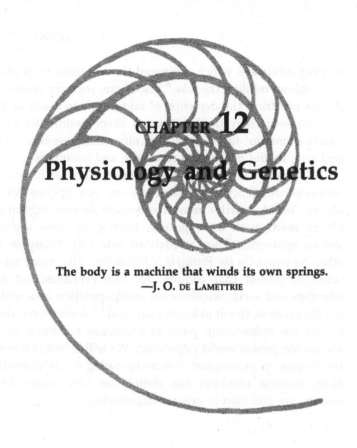

CHAPTER 12
Physiology and Genetics

The body is a machine that winds its own springs.
—J. O. DE LAMETTRIE

We come now to the crux of the book—the application of the physiogenetic model to organic phenomena. The focus here is on zoological organisms on the grounds that if the model explains the physiology and genetics of animals, it can handle the comparatively simpler world of botanical life. But before going down that path, it seems appropriate to insert a personal note to describe how this concept came about. It happened on Christmas night of 1973. I had come down with a miserable, influenza-like cold, but was just curious enough to find out something more about its cause, a virus. Looking that term up in the *Britannica*, saw a micrograph of a T-4 bacteriophage, and all hell broke loose in my mind.

222

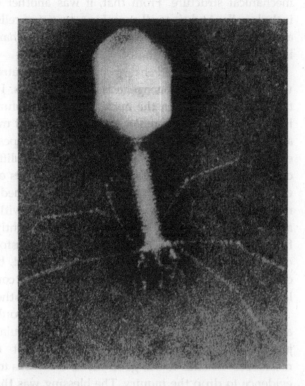

38. Biological machinery. The idea that biological organisms adhere to different designs and rules than man-made machinery is easily dispelled by looking at an electron micrograph of a T-4 bacteriophage [here magnified 200,000× (reproduced at 65%)]. It resembles a robotic spider and operates similarly to the lunar landing module.

I had been working on some problems in physics, developing by that point a vague notion of what eventually became the physiogenetic thesis of this book. But the idea of extending that notion to organic systems had not dawned on me. On the other hand, I had considerable experience assembling nuclear weapons. The upshot: I jumped to the conclusion that if a subcellular microorganism as sophisticated as a virus seemed to operate mechanically, other microorganisms would also, despite lack of

mechanical structure. From that, it was another short step
include cells. And since all organisms consist of cells, I might ju
as well throw organisms into the bargain—a chain reaction
instant insanity.

Worse, I concluded that oncogenesis—the cause of cancer
was a kind of critical mass in the nucleic acids. I remember
that critical mass in nuclear material was a function of t
isotope used, and presumed that chromosomal molecules, li
all molecules, are formed with different isotopes of requisi
elements. These isotopes operate somewhat differentially,
course, but the so-called law of averages washes out the diffe
ences in *almost* all cases. Yet, the same law implied that on ve
rare occasions, the isotopes could accumulate with an unfavo
able, twisting bias. This might render DNA slightly more for
ful, which in turn could trigger neoplastic transformation. T
odds against this happening were infinitesimal, but given t
countless trillions of cells in the body, the odds could sooner
later exert themselves in one cell somewhere in the anatomy.

It is easy to see how erroneous—*idiotic* would be a mo
accurate term—this line of thinking was; molecular biology h
jurisdiction. Yet the superintending concept of biologi
mechanics—thanks to our good friend T-4—was too strongly
evidence to drop the inquiry. The blessing was that I had ve
little knowledge about molecular biology, and struggli
through James Watson's *Molecular Biology of the Gene* was a cho
at least for me. But I *may* have been asking the right questio
and therefore was able to put the bits and pieces of evidence in
a coherent picture without previously formed misconceptions

But first, a major caveat. This chapter, and the next two, a
not reportorial summaries of biological sciences. Rather, they a
an application of the physiogenetic thesis to that knowledge.
such, the discussion relies on terms pivotal to systems, physi
and chaos, but which are not yet coin of the realm in biology.
turn, this approach gives the *appearance* of oversimplifying if
trivializing a century's work. Appearance, not intent. We kn
that complex phenomena are reducible to permutations a
combinations of simpler actions and transactions. According

we also sense that if there is a singular dispatcher controlling related but greatly diverse phenomena, then that dispatcher must rely on a plain symmetry. This does not mean we can reduce the biological sciences to a grade-school-level course, but even the most sophisticated mathematician begins his learning with simple arithmetic.

At any rate, let us have at it. *All* organisms above single-cell species consist of two or more cells, exclusive of dead tissue like hair and transitory material. Furthermore, *all* cells reproduce and multiply by the same process called *binary fission* regardless of any differences in complexity and molecular biology. This symmetry appears in millions of species, and so there must be an underlying set of common principles. True enough, ordinary cellular fission, asexual reproduction, the conception stage of sexual reproduction, and parthenogenetic reproduction differ in very important details, but the same overall process prevails.

The mechanical operation of the cell nucleus during binary fission is called *mitosis,* with the single exception of *meiosis.* Meiosis (*meowsis* in *Felus catus*) cuts the complement of chromosomes in half—right down the middle, so to speak—so that the male sperm cell and female egg cell can combine to create a new set of (diploid) chromosomes. However, there is no meiosis in parthenogenesis. The mother's eggs have a complete set of chromosomes, and this form of reproduction occurs as high up as reptiles (genus *Cnemidophorus*—whiptail lizards) on the evolutionary "scale" of development. In terms of biological complexity, that's between 90 and 95 percent of the way along the path from *protozoan* to *Homo sapiens.*

ORGANIC PHYSIOGENESIS

What follows in this section goes beyond conjecture, but it is *not* recognized principle. Rather, it is part of a structured theory that appears consistent with all physical phenomena, at least at or above the level of an atom. The key is *cyclical reversal of force.* In dynamic systems, certain elements and/or forces will dominate,

because of their magnitude or their occupying a favorab
position (and sometimes both). In billiards, the cue ball is th
dominant force, at least initially, because it is the only one
which momentum may be directly imparted by a player.

In an automobile driven uphill, the dominant forces are th
chemical reactions in the cylinders, which drive pistons, whic
via connecting rods turn a crankshaft, the torque of which
transmitted to the wheels, which overcome the force of gravit
that otherwise prevents the automobile from "climbing" the hi
But when that automobile crests a hill, the driver permits th
forces to reverse. He or she lets gravity become the domina:
force. Of course, gravity would quickly accelerate the autom
bile to a dangerous speed so that force must, in part, b
countered. The driver can apply mechanical brakes, but it
easier, and safer, to use the engine as a brake. The power tra
continues to operate as before, but this time the path of forc
energy transductions is from the wheels *back* to the pistons, th
friction of which offsets the pull of gravity. Interestingly, th
reversal can also be seen (and heard) in a performance
Rachmaninoff's *Second Piano Concerto*. For most of the perfo
mance, the pianist follows the lead of the conductor, but durir
the dramatic passages the conductor follows the lead of th
artist. Even Toscanini yielded. Sometimes.

The same reversal of force seems to occur in organ
"engines"—the chromosomes within the cell nucleus. Durir
most of their lifetime, the chromosomes are contained by th
activities of surrounding molecules, like a compressed sprir
used in car suspension subsystems. When a car is put on a li
the wheels "drop" with respect to the chassis and the sprin;
expand. In organic mechanics, a cell's life span depends on ho
long the mechanics can hold out before locally aging or breakir
down. When this happens, the pressure on DNA is lessene
with the consequence that they expand, at least sufficiently
initiate new chemical activities that literally rend the cell in
two parts, albeit with a finesse that keeps the ripping und
control. The cascading molecular mechanics prevent the c

from "renting its garment"; instead, they cut it with a precision that a seamstress would envy.

The mechanism for keeping chromosomes compressed is probably steric hindrance resistance. *Steric* refers to the arrangement or configuration of atoms (and molecules) in space, like the meshing of teeth in opposing gears. However, the cellular reversal of force can fail to cycle properly or, by cumulative mutation, the chromosomes can exert more force than normal. The latter results in the uncontrolled growth seen in malignant tumors. In the meantime, consider Figure 39, which portrays the physiogenetic model as it operates in biology. Also consider the p53 gene, the operation or misoperation of which is now strongly implicated in the failure of the body's immunology to contain many cancers. Illustrations unmistakably depicting steric hindrance have even made it to the pages of *Newsweek*.[1]

THE FIVE FUNCTIONS

Cells are bureaucracies of sorts. Most have similar shapes and gross internal structure, starting with their configuration as a walled fortress, thanks to membranes. Second, with the notable exceptions of eggs and the axon extension of neurons, the size of most cells from smallest to largest range only over a single order of magnitude. However, the morphology (shape) of cells can differ markedly, especially blood cells and neurons. Third, virtually all cells contain genetic material, and this material is the dominant feature of the cell nucleus. Fourth, most cells have the same five functions. The balance of this section summarizes how these five functions operate within a cell, as shown in Figure 40. The remaining sections of the chapter extend the discussion to how each function works in an encompassing organism.

The first function is *logistics*. Every living cell must accept (and sometimes find) nutrients and other material to sustain itself apart from any other function. Each cell must also expel the waste products consequent on converting this material (either

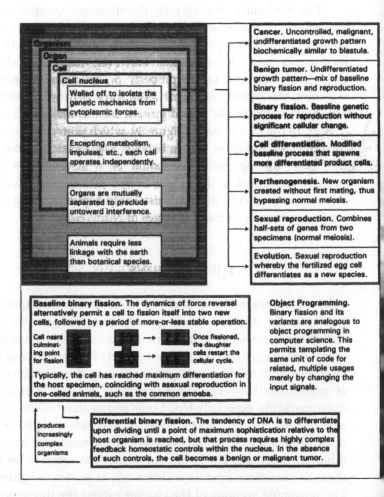

Organism

Organ

Cell

Cell nucleus

Walled off to isolate the genetic mechanics from cytoplasmic forces.

Excepting metabolism, impulses, etc., each cell operates independently.

Organs are mutually separated to preclude untoward interference.

Animals require less linkage with the earth than botanical species.

Cancer. Uncontrolled, malignant, undifferentiated growth pattern biochemically similar to blastula.

Benign tumor. Undifferentiated growth pattern—mix of baseline binary fission and reproduction.

Binary fission. Baseline genetic process for reproduction without significant cellular change.

Cell differentiation. Modified baseline process that spawns more differentiated product cells.

Parthenogenesis. New organism created without first mating, thus bypassing normal meiosis.

Sexual reproduction. Combines half-sets of genes from two specimens (normal meiosis).

Evolution. Sexual reproduction whereby the fertilized egg cell differentiates as a new species.

Baseline binary fission. The dynamics of force reversal alternatively permit a cell to fission itself into two new cells, followed by a period of more-or-less stable operation.

Cell nears culminating point for fission

Once fissioned, the daughter cells restart the cellular cycle.

Typically, the cell has reached maximum differentiation for the host specimen, coinciding with asexual reproduction in one-celled animals, such as the common amoeba.

Object Programming. Binary fission and its variants are analogous to object programming in computer science. This permits templating the same unit of code for related, multiple usages merely by changing the input signals.

produces increasingly complex organisms

Differential binary fission. The tendency of DNA is to differentiate upon dividing until a point of maximum sophistication relative to the host organism is reached, but that process requires highly complex feedback homeostatic controls within the nucleus. In the absence of such controls, the cell becomes a benign or malignant tumor.

39. Organic physiogenetic model. The simplicity of physiogenesi applies fully to organic phenomena, but the intricacies, combine with the distinct relationships between environment and organisms lead to a much more complex model.

Concept of Force Reversal
Organic force reversal is a highly structured, systemic variation on equilibrium, whereby entire subsystems or distinct parts of subsystems practice a kind of mutual give-and-take. At times, this give-and-take is distinctly cyclical, while at other times it is ad hoc, depending on the stresses and trauma encountered by the organism.

DOMINANCE OF 1st PART
DOMINANCE OF 2nd PART

Function	Cell		Organs/Organism	
Logistics	Although metabolism is an essentially fixed process, reversal occurs in the slightly alternating emphasis between nutrient intake and waste removal.		Distinct, cyclical patterns of force reversal are especially evident in the systolic/diastolic operation of the heart, as well as in respiration.	LUNGS / HEART
Repro-duction	Cellular mitosis alternates with a stable period of cellular operation. And mitosis is equivalent to asexual reproduction in single-cell organisms.	MITOSIS / STABILIZED	The opportunity to reproduce is limited to brief periods of time; otherwise more frequent reproduction would destroy the organism.	FERTILE / INFERTILE
Control/Loco-motion	Controlled locomotion is not normally available at this level as the mechanics are too sophisticated for a cell. Instead, directed movement is achieved via a medium.	BLOODSTREAM	Voluntary initiation of motion is subject to feedback from muscle tissue. That feedback can itself redirect movement (autonomous reaction).	VOLUNTARY / AUTONOMIC / MUSCLE TISSUE
Equili-brium/Homeo-stasis	Membrane walls keep the tremendous internal cellular forces in check, both within the nucleus and the cytoplasm "ring" (except during mitosis).		This pervasive feature is best demonstrated in the sympathetic and para-sympathetic systems that regulate an organism's physiological processes.	⊕ GLANDS ⊖ / COMPRESS / RELEASE / OPERATIONS
External	Mitosis and initial cellular growth overwhelms the reactive force of adjacent cells, at least until that growth overextends itself and attenuation sets in.		Specimens exert only a passive, short-lived dominance over the earth, but if a species overtaxes the environment, a Mal-thusian reaction follows.	LIFE/GROWTH / taxing dominance / passive dominance / DEATH

40. Equilibrium and force reversal in physiology. Nature regiments the ordinary concept of equilibrium into structured patterns whereby subsystems within an organism alternate dominance cyclically, to include all five basic functions at all levels.

directly or by a sequence of energy transforms) to its own sustenance. To achieve these ends, single-cell organisms trade directly with the environment, but the vast majority of cells in all but the simplest multicellular organisms depend on logistical subsystems within the host for delivery and trash pickup. They get their breakfast in bed, and the host also takes care of the bedpan.

Cell membranes are very good (but by no means perfect) at permitting the necessary material to pass through (in both directions) while blocking matter that should not pass through. Once the material maneuvers into a cell, inbound and outbound lorries (endocytotic and exocytotic vesicles) transport it internally in a role analogous to blood cells. The transformation of the material to release energy (transduction) is achieved by miniature internal organs (organelles), in this case mitochondria.

The second function is *reproduction*. All physical systems eventually decay or rot despite any ability to temporarily reverse entropy. Cells are no exception, though tenure varies greatly. In *Homo sapiens*, white blood cells last 48 hours, whereas brain cells can endure for 60 years. Still, the grim reaper waits for them all. Most cells escape this fate by fissioning (dividing) into two new cells. The familiar four-stage process of mitosis (prophase, metaphase, anaphase, and telophase) is a subset of an encompassing cycle of cellular life.

The difficult question is how multiple chromosomes synchronize their activities. Three explanations appear feasible. First, an organelle (or one of the chromosomes) could serve as a timer, broadcasting signals when the time is ready. Alternatively, all of the chromosomes could have sufficiently accurate clocks to "wake up" at nearly the same instant. If those techniques are too precarious, the mechanics could rely on symbiosis. Under this alternative, the first chromosome to unwind emits a shock wave—like a pebble dropped into a still pond—that nudges its cousins into action. To be sure, the succeeding mitoses will also emit shock waves, but by that time the cellular dynamics would be active enough to ignore the additional shock.

The third function includes *control* and *locomotion*. Most single-cell organisms tumble about rather than swim purposefully despite any cilia enhancing the process. By contrast, most cells in a multicellular organism are stationary with respect to other cells. The principal exceptions are the blood cells, although these transporters are more barge than tugboat. On the other hand, the vast majority of zoological organisms are mobile and self-propelled, and the bulk of their control mechanics—the nervous system and muscles—concentrate on this function. Within a cell, movement is continual and ultimately regulated by instructions contained in DNA. This is frequently illustrated by drawings of DNA unzipping, communicating with RNA, and so forth—the principal focus of molecular biology.

The fourth function covers *equilibrium* and *homeostasis*. Dynamic systems have two options. Either they must control and sustain their configuration and forces, or they must face destruction. This containment process is effected by homeostasis. For the most part, cellular homeostasis is aided immensely by the inner and outer membranes, which constrain operations. Said the poet Robert Frost: "Good fences make good neighbors." The cell also reacts in different ways to various input material and to stimuli that do not enter the cell but nevertheless transmit momentum (impulses) in some form across the membrane. However, in multicellular organisms, cellular homeostasis is *comparatively* rudimentary, else there would be little need for the complex neural network and the immunoresponse subsystems. Millions of substances, microorganisms, and radiations can invalidate cellular operations. To expect a single cell to provide its own defense against that massive array of opponents would be the equivalent of Pitcairn Island prevailing in a war against all of the major powers allied against it.

The fifth function is implied, namely, *external purpose*. A few cells have no known purpose, e.g., cells making up the appendix "off" the cecum (in *Homo sapiens*). But most cells have one or more purposes beyond their own sustenance. Single-cell organisms serve a wide variety of functions as members of the environment—destructive, constructive, or both. Multicellular organisms also trade with the environment as part of ecology, again sometimes destructive but more often constructive. (Man's greatest organic enemies are subcellular/cellular microorganisms.) Within an organism most cells contribute to the environment only as an integral part of their host. A notable exception is the chlorophyllous process, which in converting carbon dioxide into carbohydrates has the side effect of releasing oxygen back into the atmosphere.

To state the case another way, most cells exist to support the host organism, not the environment, and, in this context, can be subdivided into four zoological types: muscle cells, nerve cells, connecting or supporting cells (of which blood cells are the most well known), and epithelial cells, which control movement of various substances through skin, membranes, and other walls.[2] Some members from all four types also process material or realign their innards as a result of being called into action. Botanical organisms do not normally require separate muscle and nerve cells; the diurnal opening and closing of some plant material occurs by way of cyclical expansions and contractions of cells.

In addition, many cells must coordinate two (or more) of the five functions and, similarly, all organisms must coordinate all of *their* functions. Most of this coordination is passive by way of a design that keeps various functions and organs from interfering with one another. The design doesn't always work as intended, of course, and failure sometimes leads to an early death. Still, within a cell we see a clear and well-defined structure. For any specific cell in a multicellular organism, the need to locomote and to defend against the hordes of enemies is minimized. Intramural control is then effected by the actions of the nucleus while the heavy-duty work of energy transduction (both for internal needs and as a contributor to the host) generally takes place in the cytoplasm, which lies outside the nucleus.

ORGANIC LOGISTICS

The majority of work performed by most organisms concentrates on satisfying internal logistical requirements (e.g., eating, breathing), and much of the balance is expended indirectly to support these operations. In *Homo sapiens*, this logistical machinery takes up 80 to 90 percent of the torso (much less in the limbs) as shown in Figure 41. Furthermore, the majority of external work for most

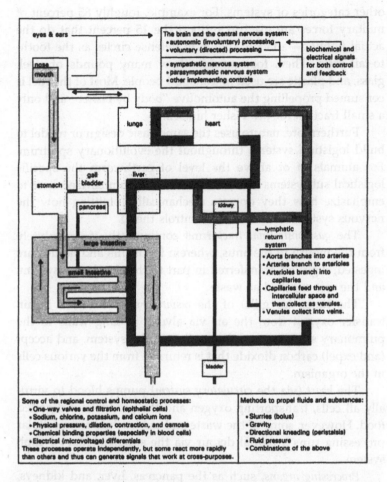

41. Mechanical schematic of organic logistics. Most of an organism's physiology operates to sustain itself, thus closely resembling a schematic for a mechanical system.

people consists of eating, drinking fluids, going to the store to obtain these products (or growing them), cooking or otherwise preparing them, working at formal labor to earn the money to buy them, and getting rid of the waste.

On the other hand, this overweening proportionality is common in other physical systems and in the physical aspects of other categories of systems. For example, roughly 85 percent of military forces are used to support the 15 percent that do the actual fighting—a factor known in defense circles as the tooth-to-tail ratio. Then, too, consider how many pounds of steel, glass, and plastic are used to transport people. Most of the fuel is consumed propelling the automotive "body by Fisher" and only a small fraction for Mr. Fisher himself.

Furthermore, nature uses the same basic design or model to build logistical systems throughout the evolutionary spectrum. For animals at or above the level of earthworm the specific logistical subsystems are well known, and summarized here to emphasize how they operate mechanically (omitting how the nervous system autonomically controls them).

The *gastrointestinal tract* runs continuously if convolutely from the mouth to the anus, wherein foodstuffs and liquids are ingested, digested, transferred in part to the circulatory system, and the rest expelled as waste.

The lungs (or gills) of the *respiratory system* suck in air, transfer oxygen from the air via alveoli in lung walls to the pulmonary subsystem of the cardiovascular system, and accept (and expel) carbon dioxide that is returned from the various cells in the organism.

The *heart* (via the *circulatory system*) pumps blood to virtually all cells, transporting oxygen and the products of digested food. However, some of the waste and carrier fluid from cellular processing runs a partial detour via the supplementary lymph system.

Processing organs, such as the pancreas, liver, and kidneys, perform yeoman work. The pancreas produces enzymes to catalyze the digestive process, bicarbonates to neutralize stomach acids, and insulin (and glucagon) to regulate glucose levels in the bloodstream. The liver is a multifunction organ tasked to produce various bloodstream proteins, coagulation factors, globin, and cholesterol (among other material). The kidneys filter the processed waste from cells, expelling the cell-processed waste and excess water as urine.

We are more conscious of these logistical subsystem processes than all of the other physiological processes combined, and they are also the most prone to malfunction. More than half of all deaths are directly attributed to disorders such as heart attacks (myocardial infarctions) and various blockages (embolisms, thromboses, scleroses) including stroke. Additionally, cancers and other disorders can lay waste to these subsystems.

One more point. Birds and mammals, which represent the upper five percent or so of evolutionary development, are nearly the only organisms that maintain more-or-less steady-state temperatures regardless of the environmental temperature.[3] This central heating and cooling is achieved largely by overtaxing the same organs and processes as "cold-blooded" forebears. Understandably, the greater load imposed on these organs subjects them to more extensive and/or more frequent disease, explaining why the human disorder rate is much higher than experienced by lower-order vertebrates, such as fish, amphibians, and reptiles.

REPRODUCTION AND CELL DIFFERENTIATION

Although sexual reproduction of a multicellular organism appears to differ radically from routine asexual cell division, the mechanics are similar. As mentioned, the only real difference is that two organisms supply half-sets of genes rather than one organism providing both sets (meiosis). At that, even this difference is bypassed in species that reproduce parthenogenetically.[4] But nature depends on meiosis to the point that we now call it the central dogma of genetics. This dogma states that while chromosomes (DNA) may transmit information to RNA and other molecules—typically to effect a construction or modification of yet other molecules—the reverse is not possible except what may occur by random mutation, or, occasionally, either by so-called jumping genes that swap places or by the action of retroviruses.[5] Apparently, even the tremendous forces unleashed during binary fission are insufficient to bypass this dogma.

The reason is to ensure a divergent range of species. The mutational exceptions to the dogma are insufficient to account for the vast evolutionary process, even over billions of years. By mixing genes over that same period, hundreds of millions of mutations were generated, the cumulative effect of which is known variation or, what is the same, the development of new species. As a result, nature creates a variety within a species, the alternative being for all people within each gender to look the same. Except for size, one gnat appears to be the same as every other gnat. The same is not true for higher-order animals, and certainly not humans. Lastly, the dogma ensures sustainment of at least some species. Laws against incest have a sound biological basis. Continuous, same-source reproduction produces too many stunted offspring, especially among higher orders. The historian Will Durant posited that the entire civilization of ancient Egypt declined early from a sociological chain reaction emanating from a preponderance of marriages between brothers and sisters.[6]

From a mechanical viewpoint, nature relies primarily on one sex to do most of the work for reproduction, and by definition this is the female. The exceptions are well known. For example, in the sea horse (*Hippocampus erectus*) the male carries the fertilized egg to term. Elsewhere, one species of fish (*Monopterus albus*) transsexualizes itself from female to male after bearing its young, and unfertilized eggs of some organisms can artificially be induced to reproduce *without* sperm. And most earthworms are hermaphrodites.

Nature also spawns many deformed specimens within a healthy species, though the criterion for deformity is in part subjective. For example, persons who grow exceptionally tall or short could be considered to be deformed, but as long as their morphology retains a reasonable proportionality to the norm, the outcome is considered deformed only in extreme cases, e.g., dwarfism and giantism. The hard question is not how mutations occur (readily explained by differing configurations among genes) but how a fertilized egg (zygote) and its ensuing blastula differentiate within a few cycles of chain-reaction cell division.

This process is remarkable because in the absence of detrimental environmental factors or genetic defects, organisms turn out to be physiologically normal the vast majority of the time.

Part of the answer lies in the evolutionary process itself. As discussed in more detail in the chapter on evolution, nature did not create one species after the next in sequential progression. Instead, she created a handful of cellular and simple multicellular prototype organisms, from which various species spun out into specialized subprototypes, which then went on to generate the array of species—a very long-term chain reaction called *cladistics*.

So if the organic physiogenetic model is valid, the gestation of an organism must follow a similar chain-reaction process. The development of an embryo adheres to a neatly defined sequence of ever-more-detailed templates. We also know that with the exception of stray mutations, the chromosomes in every cell within an organism are identical. It follows that cell differentiation must result from selective use of the genes that constitute DNA. Earlier blastocoel (binary) fission might use a grosser level of gene activation than do later mitoses. Alternatively, certain combinations of genes could operate with less precision than later combinations. With either option, think of this as a hierarchical sequence of skeleton keys. The keys from each succeeding tier are able to open fewer and fewer locks until the final tier whereby each key opens only one lock.

This leaves the question: How does each cell in the early blastula know which key to use? If the physiogenetic model is correct, three explanations are feasible. The first option is that the cells could differentiate immediately but imperceptibly. This means that the configuration of dominant forces in the cell changes *initially* shy of a threshold value that activates change, analogous to the so-called all-or-nothing phenomenon of signal transmission in some neurons.

Under the second option, blastocoels closest to the wall of the uterus might exert, or cause to be exerted, the decisive influence. Under the third option, the process could be controlled by symbiosis. In this symbiosis, the first significant differentiation might occur in any of the cells, which would send a signal or impulse to the other cells instructing them to differentiate in different ways. This third option is consistent with the fact that some cell divisions are triggered by external chemical signals. Yet any such symbiosis must take an overarching aspect of fetal development into account, namely, that the fetus must be supported by way of the umbilical cord and its own developing circulatory system (which operates differently before birth because the fetus does not breathe). This means that the fetus must develop in a certain orientation with respect to that cord. Otherwise the modified logistical system essential to support gestation would fail. In two words (for a fetus): cord=life.

CONTROL AND LOCOMOTION

Under the general physiogenetic thesis, the three ontological roots of physical systems are matter, length, and time. As such, the only other ontological property is the arrangement or configuration of matter or elements in space. These arrangements can only be changed by motion, and work can only be effected by an exchange of momentum or energy among elements. It is no surprise, therefore, that the control a zoological organism exerts over itself is associated with motion. All vertebrate animals, and most lower-order species, are capable of propelling themselves. One species—the cheetah—can accelerate to a speed of 70 mph in three seconds.

Organic motion assumes one of three forms. The most common form is the locomotion of the organism as an entity or at least some distinct part of it, for example, lifting an arm. The second form is the pattern of motions used to operate internal organs, of which the circulatory system is dominant. The third form occurs within cells and organs as they change configuration to fulfill various needs. This third form ranges from expansion of the uterus to accommodate a growing embryo, through rearrangement within neurons and perhaps synapses, to recording memories analogous to storing data on computer disks.

Control of most biological motion or movement is autonomous, which means that control is exercised by homeostatic or cybernetic mechanisms that do not require conscious decisions. However, some higher-order organisms may be conscious of what occurs afterward, for example, when you touch a hot stove. Even when conscious decisions are made, much of the work is effected by acquired programming. For example, if you are thirsty, you unconsciously (almost) activate an organically stored, formal program of instructions that waltzes you through the motions to quench that thirst. That program may consist of more than 20 distinct segments or actions, each of which may require hundreds of minute physiological steps (what computer science calls *utilities*).

The central nervous system (including the cerebrum source) is arranged to accommodate all of this, complete with separate channels to effect the yin and the yang of various functions—the sympathetic and parasympathetic subsystems. A signal transmitted along one channel tells an organ to compress; a different signal sent along the parallel channel tells it to relax. At the head of this system resides the cerebrum. It is a very quiet machine, and some resident cells linger for 60 years (especially cells that store outdated theories). There are no pain sensors in the brain, and only infrequently do brain (primarily glial) cells *initiate* a malignant growth, though many cancers originating elsewhere reach the brain by metastasis.

Question: If the psyche is something apart from the physical brain, how could a minor chemical imbalance or an almost imperceptible deformity lead to insanity? If you cut a critical circuit in a computer chip containing millions of circuits, every program run on the host computer will execute bizarrely, if at all. Yet the logic of the program remains intact apart from the hardware. Interestingly, the book and subsequent film *Awakenings* (1992) portrayed how the psychiatrist Oliver Sachs temporarily "awakened" a number of institutionalized patients from catatonic stupor by pharmaceuticals. Presumably, this altered brain circuitry sufficiently to allow their psyches to again manifest themselves.

EQUILIBRIUM AND HOMEOSTASIS

The design of an organism is nearly sufficient to coordinate biological requirements *passively* when that organism remains in a static condition and the environment remains equally cooperative, like a new car kept in a hermetically sealed garage filled with helium. The closest that an organism comes to that condition is when it is frozen (cryogenics). Indeed, some organisms have been put into this cryogenic state and subsequently brought back to life.[7] Under normal environmental conditions, organisms are faced with no less than seven classes of dynamic requirements, many of which conflict.

First, in an 80-year human life span, the heart beats about two and one-half billion times and cumulatively pumps millions of gallons of blood. The digestive system processes roughly a million pounds of food while the body generates countless trillions of new cells.

Second, the body must accommodate thousands of activities that generate severe stress. Each one of these activities taxes various organs in different ways. Third, the body (in birds and mammals) must maintain a relatively constant temperature despite environmental temperatures that range from 70° below zero to 120° above. The body must also cope with different atmospheric pressures at various elevations.

Fourth, the body must destroy, neutralize, or otherwise ward off hundreds of thousands of invading viruses, rickettsias, parasites, and poisons. Furthermore, many if not most cancers are destroyed before they are able to expand beyond a few cells.[8] In some instances, the organism programs itself to defeat these invaders more effectively in subsequent battles (acquired immunity). Fifth, the body must recover from the occasional traumatic injuries resulting from falls, crashes, burns, wounds, and other mishaps. Accidents are the leading cause of death among younger people, but the vast majority of victims survive, more often than not without permanent damage or handicap.[9]

Sixth, each individual body must have the potential to cope with any of several thousand well-documented disorders that afflict man. Sometimes this pathology proves fatal because it overwhelms any defense the body can mount. More often, it does not, mandating that the body struggle and then cope with the aftermath. If it were otherwise, nobody would reach their first birthday, and the species would rapidly face extinction.

Seventh, the body must maintain accurate timing and cyclical mechanisms. Subsequent to birth, the body continues its growth program with changing proportions and, at puberty, sexual characteristics are activated or made more pronounced. Additionally, the body depends on numerous cycles in order to function, of which the circadian rhythm and menstrual cycles are the best known.

These tasks make for a strenuous, lifelong workout, yet the operations are kept separate as far as possible and always consist of permutations and combinations of simpler processes. Still, intricacies remain and when disorders occur, they can be difficult to diagnose and sometimes exceptionally difficult to repair without inflicting more damage (iatrogenesis). Worse, this homeostatic balance is so finely tuned that even a minor deficiency can turn the system against itself and eventually destroy the host organism by an autoimmune process, such as *systemic lupus erythematosus* (SLE) which attacks many subsystems, taking special aim at connective tissue. Figure 42 presents a rudimentary idea of the hierarchy of this biological homeostasis.

EXTERNAL PURPOSE

Almost every organism has its place in the environment—some destructive, some constructive, and some both. The proportion is partly a matter of subjective judgment, but without a doubt

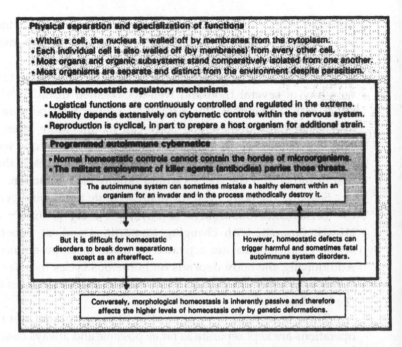

Physical separation and specialization of functions
- Within a cell, the nucleus is walled off by membranes from the cytoplasm.
- Each individual cell is also walled off (by membranes) from every other cell.
- Most organs and organic subsystems stand comparatively isolated from one another.
- Most organisms are separate and distinct from the environment despite parasitism.

Routine homeostatic regulatory mechanisms
- Logistical functions are continuously controlled and regulated in the extreme.
- Mobility depends extensively on cybernetic controls within the nervous system.
- Reproduction is cyclical, in part to prepare a host organism for additional strain.

Programmed autoimmune cybernetics
- Normal homeostatic controls cannot contain the hordes of microorganisms.
- The militant employment of killer agents (antibodies) parries those threats.

The autoimmune system can sometimes mistake a healthy element within an organism for an invader and in the process methodically destroy it.

But it is difficult for homeostatic disorders to break down separations except as an aftereffect.

However, homeostatic defects can trigger harmful and sometimes fatal autoimmune system disorders.

Conversely, morphological homeostasis is inherently passive and therefore affects the higher levels of homeostasis only by genetic deformations.

42. Conceptual schematic of organic homeostasis. Nature implements organic homeostasis in three distinct stages or levels: (1) the simplicity of physical separation and specialized functions, (2) routine homeostatic control mechanisms, and (3) programmed, cybernetic autoimmune subsystems.

some organisms play a greater role than others. Furthermore, it has been estimated that at least 95 percent of all species that ever existed are extinct, and the world seems to get along quite well without dinosaurs. Moreover, even under the most favorable conditions, nature continues to eliminate a few hundred species every year (and perhaps creates a few new ones).

The fact of the matter is that most species exist only as integral elements in a huge ecological system. Nature doesn't give a damn about who lives or dies, and more or less condones their eating one another *ad infinitum*—like bank acquisitions—including a few otherwise beloved chimpanzees who kill and devour newborns within their own clan. However, when too much or too little of anything pushes the balance past the ecological culminating point, nature reacts with Malthusian violence. This does not gainsay the belief that every living thing may be sacred to one degree or another, but nature takes a different viewpoint. The earth is an organism, and it will do what it has to do to get rid of excess species or excess overtaxing of its crust despite any sacredness of life other than its own collective mien. Nature is patient. And ruthless.

Now it is said that the survival *instinct* is inherent in all organisms, but that depends on the definition of instinct. Instinct can mean the collective, passive effect of various biological processes within an organism. We survive, in part, by the simple acts of breathing, eating, and sleeping. Or instinct can mean that an organism possesses an explicit, overarching system that autonomically ensures survival. Much below *Homo sapiens*, the closest nature comes to this is homeostasis. Or the organism may *consciously* seek to preserve itself, and this is quite evident in man, and to a lesser degree in other mammals and birds.

We note repeatedly that man is the only species that survives by way of modifying the environment to his taste rather than reacting to it by biting or moving. Indeed, the whole practice of medicine depends on this capability. We may feel a great kinship with chimpanzees, and, indeed, chimps can make a few elementary tools and even do a little gardening and farming. We may have been far more destructive than constructive in our dominance over the environment, and often paid the price. And we may in time grow even more destructive, yet all of those points beg the issue. For better or worse, for love or hate, for scholarship or stupidity, nature grants jurisdiction of the earth to mankind.

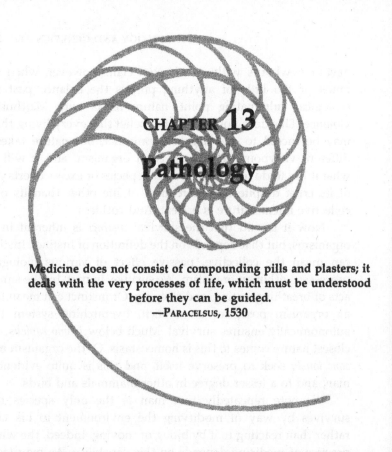

CHAPTER 13

Pathology

Medicine does not consist of compounding pills and plasters; it
deals with the very processes of life, which must be understood
before they can be guided.
—PARACELSUS, 1530

Scanning any authoritative encyclopedia of medicine suggests
that the human body has about as much stability as an egg
tossed out on a major highway at rush hour. More than 2000
disorders are known to afflict human physiology, although less
than 50 account for the overwhelming percentage of deaths and
hospitalizations.[1] On the other hand, research and practice over
the past century have nearly doubled average life expectancy. So
if conventional practice has improved life and life expectancy as
much as it has, there must be a reason. That reason is called the
physical theory of medicine, namely, that most medical disorders stem from the body's physical system gone awry. Consequently, these disorders are best remedied by equally physical

means. And if that is so, it would seem that there is a symmetry underlying physical pathology.

That symmetry was clearly seen by Gustav Eckstein, who died at the age of 93 conducting grand rounds at a teaching hospital. Dr. Eckstein lamented that the profession of medicine continues to see various diseases only as various sets of symptoms, ignoring "the corrective dynamics—what the body is striving to do." Alluding to the ghosts of Claude Bernard and Walter Cannon, he advocated "the not necessarily easy task of searching ceaselessly for the *rebalancing* factors in an imbalanced body and head. The pattern of imbalancings and rebalancings is the sickness." He concluded that "disease is a process aiming at recovery, partially destructive, its intent *equilibrative*, slanted toward health, but, missing that, accepting the downslope toward death."[2] To be sure, the practicing physician can seldom afford the time to ruminate on this grand overview; the exigencies of specific pathologies demand here-and-now treatment. But medical academe does have that time.

In this chapter, we attempt to condense the bulk of known pathology into a handful of mechanical categories as an application of the organic physiogenetic model, keeping Eckstein's mechanically oriented perspective in mind. However, we must also note dysfunctions themselves are not so much applications of the symmetry but deviations from how the encompassing organic system *should* work. Indeed, the argument can be made that much if not most of our understanding on the details of physiology arose from dealing with pathology. In order for the practice of medicine to cure disease, it must first understand how the system or subsystem operates normally and, from that, determine what went wrong.

Note also that many diagnostic tools now adapt the instrumentation of physics: nuclear resonance scanning, radionuclide scanning, positron emission tomography (PET scans), and diagnostic ultrasound. Moreover, computers are used routinely for the analysis of lab tests, especially blood samples. Lastly, computer models increasingly mimic everything from the operation of various pharmaceuticals on cells to mapping out the entire genetic process.[3] In short, indicators increasingly point to physics as the ultimate basis of medicine.

THE MECHANICS OF DISORDER

Every element and operation of every physical system—man-made machinery or organism—is subject to breakage, wearing out, malfunction, or failure. Anything that can go wrong, will go wrong, at least some of the time. In addition, there is no way to guarantee that any machine will meet the criteria under which it was designed. As the human body is the most complex machine known to man, it is not surprising that, collectively at least, it is so often and so diversely in need of repair.

By way of analogy, some cars can be driven for several hundred thousand miles with few visits to the garage while others have trouble attaining a fifth that mileage and, at that, spend half the time in the shop. Similarly, some people live to old age with few visits to a doctor or hospital while others find medical assistance is a way of life and then die young.

For both automobiles and organisms, the causes of disorders and death are: (1) design and manufacture/gestation, (2) the intrusion of debilitating environmental factors, (3) wear and tear/aging, (4) faulty operation of mechanical subsystems/organs, and (5) systemic oxidation and other chemical buildup of destructive deposits/infections and tumors. For organic structures, faulty design equates with a genetic flaw while improper manufacture shows up as a congenital defect. Negative environmental factors range from malnutrition to viruses. The biological parallel to wear and tear is obvious. Next, compare the bursting of a fuel line with the rupture of a blood vessel. As for faulty processing, electrical malfunctions can be as difficult to diagnose as a quirk in the biological system that regulates heartbeat. This leaves deposits, which in a car include carbon buildup in a fuel injection system. In the body, the equivalent is atherosclerotic plaque in the aorta and calcium deposits in an aortic valve. Another deposit—rust—can turn malignant and literally eat an automobile as it metastasizes, not unlike the way a cancer consumes its host. A stronger parallel occurs in nuclear reactors, wherein a chain-reaction oxidation can quickly destroy the entire pile.[4]

The analogy goes further than this. A design defect in an automobile can lead to excessive wear and tear, which leads to faulty processing, which accelerates the progress of disorders emanating from environmental factors or hastens deposit buildup. Similarly, medical disorders can arise from many sources. A congenital defect can trigger an autoimmune response, which lays waste to a minor organ, which creates an embolism that eventually blocks a critical vessel culminating in death. All well and good, says the critic, but the physiology of *Homo sapiens* bears little resemblance to an automobile, at least not in terms of complexity. No argument. But increased complexity arises solely from permutations and combinations of simple underlying processes.

The analogy does break down in matters of proportionality. Automobiles of all kinds in most locations experience a strikingly similar proportionality among disorders, with notable exceptions occurring in temperature and climate extremes. Sand, not rust, is a major problem in desert regions. The same is not true for the physiological machine. In the poorest countries, environmental factors are the primary cause of death. In highly developed countries, by contrast, wear, tear, and cancers are the operating factors in perhaps 75 percent of deaths. Proportionality also varies markedly with age and even social mores. For example, Mormons—who eschew alcohol, cigarettes, and caffeine—experience a cancer rate only one-third to two-thirds that of the general population.[5]

The goal of medicine, as in automotive mechanics, is to identify the causes of various disorders or at least clearly diagnose them, and then do something about them. The ideal is to restore the body to a state as if the disorder never existed. For this ideal, nature herself sets the standard. For example, the body not only fully recovers from the vast majority of colds, it improves itself by activating its immune-system programming to defeat that particular virus the next time around. That failing, medicine tries to restore the body to its previous status as closely as possible. When that approach fails, and death appears inevitable, the patient can be made more comfortable by way of palliative care.

The auxiliary goal of medicine is to *prevent* various disorders from taking root or, that failing, to detect them early in order to minimize both damage and treatment. The two major exceptions are: (1) psychiatry, which often strives to improve the patient's life beyond anything he or she has previously known, and (2) plastic surgery, which more often than not caters to vanity, though it may be essential to restore appearance (and some functions) after injury or to correct serious genetic and congenital deformities.

The next five sections take a closer look at the five thematic sources of medical disorders respectively, but keep several points in mind. First, a disorder may arise from a singular and easily identified cause, for example, ingesting a toxic substance. At the other extreme, a disorder may result from a sequence of events drawing on all five sources. Between these extremes, a disorder may stem from several concurrent sources, for example, diabetes mellitus type II (adult onset) which typically arises from a combination of genetic disposition, a partial organic dysfunction, poor diet, and being overweight. See Figure 43.

Second, medicine uses various terms to describe disorders, to include *disease, dysfunction, deformation, deficiency, syndrome.* At the next level of detail, an extraneous growth may be a tumor, cyst, plaque, sclerosis, thrombosis, embolism, or clot. Yet from a physiogenetic perspective, they are all disorders. The biological machine did not turn out the way it should have, or it isn't operating up to snuff, or something is missing or distorted.

Third, some disorders are fatal instantly while others are all but harmless, and this may vary with the individual. For example, penicillin has saved countless millions of lives, but it can and does kill a few people by way of a fatal anaphylactic reaction.

Fourth, the medical profession *of necessity* is not systems oriented. Instead, most specialties concentrate on a specific tract,

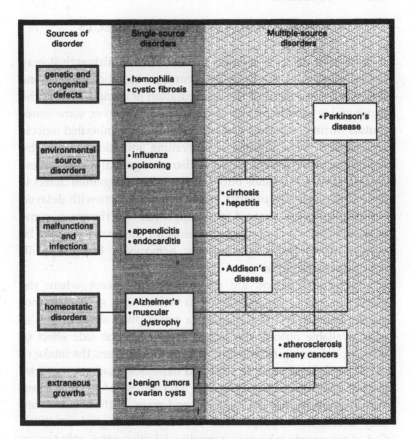

43. Sources of physiological disorders. Physiological disorders can originate from one or more of five categories of sources, as indicated by the examples in this chart.

region, or organ, for example, a gastroenterologist, thoracic surgeon, or cardiologist. Furthermore, regardless of the combination of sources that may have contributed to a disorder, the important thing in medicine is to eliminate or otherwise contain that disorder. Then, too, once most medical problems are correctly diagnosed—no small task in many cases—only a few specific remedies may be available.

GENETIC AND CONGENITAL DEFECTS

A congenital defect is a defect present at *birth*, although it may not be obvious at the time. By contrast, a genetic defect is one that is a consequence of the configuration or composition of the genes and is usually present at conception. However, some genes mutate during gestation, and conceivably some inherited defects are negated by a random contravening mutation. Hence, by definition, every genetic defect is also a congenital defect (unless by rare chance it washes out), but not every congenital defect is genetic in origin. (In practice, many genetic defects with delayed expression are not referred to as congenital.) Furthermore, some of these defects do not cause problems for the bearer yet may be passed on to future generations, hemophilia being a well-documented example.

Congenital defects are not widespread except where the mother ingests or exposes herself to various toxic substances (not always knowingly). Thalidomide (a prescription sleeping pill legally available outside the United States) has the side effect of severely deforming fetuses. But in terms of numbers, the intake of alcohol and drugs result in 1000-fold more deformations, e.g., fetal alcohol syndrome. Elsewhere, the freakishness of extreme cases attracts publicity, not the numbers, and many of these defects are only tendencies or propensities rather than biological directives. Out of ten people inheriting a specific defective gene, only two or three of them may develop symptoms attributable to it.[6]

Genetic research increasingly suggests that the incidence of many common disorders—including neoplastic growths such as lung and breast cancer—have a definite if not strong correlation with specific chromosomal abnormalities.[7] Yet because the rate of incidence per latent defect may be less than 10 percent and perhaps impossible to detect at birth, these defects are seldom considered to be genetic or congenital.

Whether a genetic/congenital defect affects most or only a small part of the body depends on where it occurs in the development of the fetus. If it appears in the blastula, it will likely replicate into every cell. If it crops up only toward the end of gestation, the defect will almost always be localized (though it still may be fatal). This is easily understood by comparing a 100-foot measuring tape with a 3-inch ruler, both of which are inaccurate by 1/16th of an inch. If you measure the length of a football field with the tape, your measurement will be off by 3/16ths of an inch. But if you use the small ruler, the measurement will be incorrect by 6 feet and 3 inches. That is, minuscule errors in any system that grows will be multiplied in direct and perhaps exponential proportion to the ratio of initial and final magnitudes.

But why are some defects more serious than others, and why are many if not most defects only propensities? The first question depends on the genes or cell mechanisms affected. A brain concussion is almost always far more serious than a stubbed toe. Almost. The brain concussion may be comparatively minor, whereas in rare cases a stubbed toe can lead to an infection that triggers a latent systemic defect which eventually causes death. Similarly, certain genes are responsible for development of the cerebrum and central nervous system, whereas others, if faulty, may result in barely noticeable webbing between two fingers. Granted, the webbing could extend to a complete fusion of the two digits (syndactyly), but that is still minor compared to Siamese twins.

The propensity issue is more complex in the sense that at least three explanations are possible, all of which probably occur at one time or another. First, the mutation of a gene can have many degrees for the same reason that different calipers can be manufactured with different degrees of precision.[8] Second, the manifestation of the defect may require a symbiotic reaction with another gene, enzyme, or other cellular element. Third, the symptoms may depend on the body attaining a threshold condition that does not occur in every bearer. Regardless, genetic and congenital defects can be arbitrarily subdivided into just four forms.

The first form consists of *visible deformities*. These deformities are the most obvious, though the dividing line between unusual growth and deformity is subjective. For example, when does short height become short stature (formerly *dwarfism*)? This form also includes the rare case of hermaphroditism—a person born with both male and female reproductive organs. (The name is a combination of the Greek mythical characters *Hermes* and *Aphrodite*.)

The second form covers *internal structural defects and deformities*. This ranges from a reversed configuration or placement of an organ (*situs inverses*), to a hole between the ventricles in the heart (*septal defect*), to the total absence of the cerebrum (*anencephaly*). Situs inverses is seldom harmful whereas a septal defect is usually fatal if not surgically repaired. As for anencephaly, which occurs roughly in one of every 200 pregnancies, the overwhelming majority miscarry, are stillborn, or die within a few hours of birth. An extremely small number, however, can survive for over a year.[9]

The third form includes *direct operational deficiencies*. These are not morphological deformities but functional or mechanical disorders (though in some cases the disorder may later result in irregular growth). Hemophilia appears in this classification. And the fourth form may be called *equilibrative deficiencies*. This category includes cystic fibrosis, Tay–Sachs disease, and all inherited autoimmune deficiencies.

ENVIRONMENTAL-SOURCE DISORDERS

An environmental-source disorder is one that can be traced *primarily* to an agent external to the body that is subsequently ingested, injected, or otherwise acquired by an organism, or conversely, an essential element the organism fails to obtain in sufficient quantity. This excludes internal conditions that may be exacerbated by increased exposure to such agents. However, some individuals may be more resistant to an agent than others or, the opposite, they may be especially susceptible. Also, these disorders may arise during gestation, in which case they are also considered congenital.

Fortunately, these disorders yield to preventive medicine more than all other categories of disorder combined. The reason is that once a harmful agent is identified, contact with it can be significantly curtailed if not blocked entirely. That failing, vaccines can be developed to enhance the body's immune system to destroy or nullify their effect. Unfortunately, intake of many of these debilitative agents is a matter of personal initiative and choice, and the choice is often made to continue the practice despite obvious dangers: smoking, alcoholism, drugs, profligacy and its venereal diseases, reckless driving, and so on.

The dynamics in every case consist of the organism contending with an invader or with damage inflicted by an external source, which ranges from inhaling dust that triggers a sneeze to a nuclear explosion that disintegrates organisms in a thousandth of a second. For the sake of discussion, environmental sources can be divided into five groups.

The first group covers all *organisms*, including subcellular viruses and rickettsias, cellular bacteria, other single-celled organisms, and multicellular parasites. Millions of different species exist, most of which are harmless. The others can trigger everything from colds to at least a few forms of cancer. Microorganisms are also responsible for the overwhelming majority of contagious diseases. And note that microorganisms contain genetic material, and that many of them invade cells and mutate the host organism's chromosomes.

The second group consists of *poisons and toxins*. Poison is the general term, while a toxin is limited to certain proteins produced by organisms. Arsenic is a poison; cyanide, a toxin. Virtually all poisons disrupt the structure or functioning of cells, though the time between ingestion and onset of symptoms can vary from less than a minute to years. Furthermore, most pharmaceuticals are poisonous if taken in excess dosage or under improper circumstances.[10] Other toxins include the venom of poisonous animals (and a few plants), some of which affect blood cells (hemotoxic), others of which affect the nervous system (neurotoxic), while still others destroy the effectiveness of epithelial cells. Some poisons may also operate as carcinogens, especially on lung cells.

Poisons offer yet another illustration of the power curve. Pharmaceutical dosages have little effect in any given set of circumstances until a threshold level (situational to each pharmaceutical) is reached. Slight increases beyond that level can sometimes magnify the benefits. Much beyond that level, however, the additional benefits diminish, eventually reaching the zone of negative returns by proving harmful if not fatal—a poison.

The third group comes under the heading of *radiation*. Radiation falls into two types: ionizing and nonionizing. Ionizing radiation is the stronger and can penetrate to any and all points in the body. *All* radioactive substances emit this type of radiation, which can easily destroy cells or mutate chromosomes. Controlled ionizing radiation is also used to eradicate malignant tissue. By contrast, nonionizing radiation is weaker and normally cannot penetrate below the subcutaneous layer of skin. Ultraviolet light and electromagnetic waves are good examples. Despite this weakness, nonionizing radiation exists in far greater quantities and therefore can cumulatively inflict considerable damage, especially skin cancer.[11]

The fourth group consists of *trauma and accidents*. In the United States, accidents claim as many lives between ages 1 and 24 as all other causes combined (though the incidence of death in this age range is small).[12] Trauma ranges from being hit by a vehicle, to wounds incurred in war or as a victim of crime, to swallowing a pin, to burns.

The fifth and last group covers *improper diet and fluid intake*. Dietary deficiencies can lead eventually to death, just as excessive intake of alcoholic beverages and drugs can. Conversely, proper diet has been shown to reduce the incidence of many disorders.

MALFUNCTIONS AND INFECTIONS

Most mechanical systems break down because a part breaks, wears out, or overheats. Or a system breaks down when some process no longer functions, say, the flow of material decreases below minimum required levels (e.g., a faulty fuel pump in an automobile). The same is true for biological systems. Without exception, every element of the human anatomy and every function of the human physiology has ceased to function properly simply by wearing out, breaking, or becoming infected. Nature is sparing with warranties.

However, it is often difficult to differentiate between a direct breakdown and a breakdown that results indirectly from another disorder. Or an individual with a weak heart can kill himself or herself by exertion that would be routine for, say, an Olympic athlete. Also note that the bulk of these breakdowns occur in the logistics functions if for no other reason than those functions are the most severely taxed by strain and continuous operation.

Regardless of the ultimate source, most of these disorders can be subdivided into four categories of malfunctioning: (1) rupture, (2) blockage, (3) pace, pressure, and related problems, and (4) biochemical destabilization. All of these categories have direct parallels in mechanical systems. A fuel line can rupture or become blocked, or the idle speed can be too slow or too fast for the current condition. The carburetor will certainly experience a disruption if diesel fuel is substituted for gasoline, or the gasoline breaks down chemically. In *Homo sapiens*, the parallels include aneurysms for rupture, embolisms for blockage, arrhythmias for pace dysfunctions, and infection for destabilization. Figure 44 is a matrix of these four categories versus specific organs and tracts, noting that myocardial infarctions are currently the single most prevalent cause of death in the United States.[13]

DEGENERATIVE AND HOMEOSTATIC DISORDERS

Without exception of any kind in more than 3 billion years of development, every species has been afflicted by at least one degenerative disorder. That disorder is aging. Every biological machine eventually reaches a point where no amount of intervention can prevent death. But when a more specific malady accelerates the process—more often than not by affecting a particular organ or subsystem—it is called a degenerative disorder. The disorder may progress so slowly that aging or some

	Breakdown	Blockage	Timing & Control	Destabilization
common generic names	accidents/trauma infarctions ruptures	aneurysms embolisms thromboses	arrhythmias hypertension hypotension	abscesses infections inflammations
Pulmonary system	• atelectasis • emphysema • pulmonary infarction	• pulmonary fibrosis • pulmonary stenosis	• asthma • cardiomyopathy • pulmonary hypertension	• alveolitis • bronchiectasis • bronchitis • pneumonia
Cardiovascular system	• cardiomyopathy • myocardial infarction • wounds	• arteriosclerosis • atherosclerosis • coronary embolism	• angina pectoris • cardiac arrhythmia • tachycardia	• anemia • endocarditis • myocarditis • septicemia
Gastrointestinal system and related organs	• esophageal diverticulum • hernia • peptic ulcer	• atresia • Crohn's disease • gallstones • stenosis	• colic • constipation • irritable bowel syndrome	• appendicitis • colitis • esophagitis • gastroenteritis
Urinary tract and kidneys	• hemolytic-uremic syndrome • nephrotic syndrome	• calculi (stones) • enlarged prostate • hydronephrosis	• enuresis • hyperuricemia • irritable bladder	• cystitis • pyelonephritis • schistosomiasis • urethritis
Nervous system and the brain	• Alzheimer's disease • multiple sclerosis	• cerebral thrombosis • hypoxia • neurapraxia	• cerebral palsy • epilepsy • myasthenia gravis	• encephalitis • meningitis • neuritis • poliomyelitis
Muscular and skeleton systems	• fractures • osteomalacia • Paget's disease • sprains	• cramps • muscle rigidity • prostratia couchpotatous	• gigantism • muscular dystrophy • spasticity	• ankylosing spondylitis • arthritis • gangrene

44. **The breakdown on breakdowns.** Many organic disorders arise from one of three mechanical problems from a common biochemical malady. However, some disorders arise from more than one type of problem, and some are the effect of another disorder.

other disorder wins the race to the graveyard, or it may progress with such speed that it is regarded as malignant.

Most of these degenerative disorders arise from flaws in the equilibrative or homeostatic dynamics of the body. However, not all equilibrative disorders are degenerative, and many can be controlled by a single pharmaceutical. In other cases, the disorder may generate symptoms only under severe stress or an unusual condition. Regardless, the causes of these disorders vary, though at least a plurality of them are thought to stem from a latent congenital weakness that is triggered by an invasive agent.[14]

Closely related to degenerative diseases are those conditions commonly grouped under *autoimmune disorders*. Although the eventual consequence of an autoimmune disorder is degeneration of tissue, the mechanics demonstrate a unique process. The system of antibodies mistakes some part of the host organism's own tissue for an invader (antigen) and begins to attack it. In the least troublesome cases, this process is a kind of internal allergy. At the other extreme, it is fatal.

Another common class of homeostatic disorders arises from the failure of a gland to produce (or overproduce) a specific hormone or other substance intended to regulate a physiological operation. Again, the end result may be degeneration of cells, but the mechanics are different and usually easier to treat. Sometimes the faulty organ or gland can be repaired; more often the necessary substance or effective substitute must be ingested or the overproduction chemically blocked.

Perhaps most of these disorders can be traced to genetic defects, but many require an infection or other malady as a trigger. Some of these disorders are uncommon yet heavily publicized because of famous individuals so afflicted. An example of this is Lou Gehrig's disease—amyotrophic lateral sclerosis (ALS)—which also afflicts the physicist Stephen Hawking. This motor neuron disease degenerates the nerves in the brain and spinal cord that control muscular movement. It is usually fatal within a few years because it tends to affect respiration but, as in Hawking's case, there are exceptions.

Far more common than ALS is the acquired immunodeficiency syndrome (AIDS). AIDS occurs when the human immunodeficiency virus (HIV) begins to infect the T-helper lymphocyte cells, which suppresses the body's ability to resist other diseases. Interestingly, there is some evidence that HIV existed before the outbreak of AIDS and perhaps always existed. This virus, like all organisms, probably underwent various mutations as it copied itself by way of a cell or bacterium until one particular variation became especially deadly.[15] We see this process in the attempt to control the spread of HIV in an infected patient. The chemotherapy destroys many of the parasitically reproducing viruses, which delays the onset of AIDS, but eventually a few especially virulent strains win out.

EXTRANEOUS GROWTHS

By definition, an extraneous growth is a growth the body does not need. That growth may be harmless, malignant, or harmless by itself yet indirectly destructive; but it does not include congenital deformities or other deformations such as *hydrocephalus* (an enlargement of the skull caused by the pressure of excess cerebrospinal fluid).

Extraneous growths can be divided into three categories: (1) inanimate material, (2) benign tumors or equivalents, and (3) malignant tumors. Examples of inanimate material include stones in kidneys, gallbladders, and bladders, and calcium buildups in the valves of the heart and in the aorta. Fundamentally, calcium buildups differ little from what so-called hard water does to pipes, and the equivalent of stones are found in oysters in the form of pearls. Mechanically, stones operate similarly to blood clots, but the latter have a tendency to block cardiovascular vessels, leading to death in many cases, whereas stones are seldom fatal.

The second type of extraneous growth—benign tumors and equivalents—can be further subdivided into three classes: (1) internal, encapsulated controlled growths; (2) external controlled growths such as cysts, warts, and hemorrhoids (but no boils, sties, and blisters); and (3) plaque, such as atherosclerosis in blood vessels. The encapsulated growths are commonly called benign tumors; the other two are not. Regardless of nomenclature, tumors or equivalents can form anywhere in or on the body.[16]

Benign tumors cause damage in one of three ways: (1) blocking essential flow, either directly as an obstruction or indirectly by exerting pressure on a vessel or organ, (2) transforming into a malignant tumor, or (3) bursting and releasing a toxin or other harmful substance into the blood or lymph vessels. Blockage is often fatal when the location of the tumor makes it inoperable or otherwise irradicable, especially brain tumors.

The third category—malignant tumors—are usually fatal in the absence of medical intervention and are often fatal even *with* such intervention. However, many embryonic cancers are destroyed by the body's immune system, and in exceptionally rare cases a patient may experience a spontaneous remission. Also, some cancers form late in life and grow so slowly that death results from unrelated causes. This is especially true for prostate cancer.[17]

Most cancers arise from a single cell and grow as a singular object by a chain reaction of cellular divisions.[18] Eventually, elements of this neoplastic growth break off and circulate within the body via the blood and lymph systems. It is this metastasis that is responsible for most cancer deaths.[19] The operational exception to this pattern occurs in leukemia. Here the bone marrow produces white blood cells in untoward but unattached numbers, thus crowding out the production of red blood cells, platelets, and normal white cells. However, the leukemia itself arises from a neoplastic transformation in one or more cells.

The fundamental cause of cancer is said to be unknown, although substantial evidence suggests that some cancers arise from viruses, from specific chemical carcinogens (especially cigarettes and asbestos), and from specific genes or mutations (often cumulative) of genes.[20] Despite this mélange, Sir Alexander Haddow once said that while cancer could be triggered by a wide spectrum of agents, it seemed probable that all of them operated by invoking a common "alteration." Furthermore, the potential for an eventual synthesis of understanding was supported by the fact that virtually all types of carcinogens were subject to mutation "of which carcinogenesis may be a specialized case."[21] Haddow further speculated that operation of every carcinogen deprived the host cell of its "growth-regulatory systems," which rely "on protein–nucleic acid association." The significance of this process was that it liberated the potential for growth "present all along but normally controlled."[22] Later references support this viewpoint, especially the September 1996 special edition of *Scientific American*, which was devoted to the subject of cancer.

Under the organic physiogenesis model, then, cancer can be described as a process whereby the configuration of the nucleus reaches a critical point and begins to reverse entropy in the form of a destructive catastrophic growth pattern. The scientific name for this process is *irreversible differentiation*.[23] And as mentioned previously, a cancerous growth most closely resembles a zygote and is therefore tantamount to parthenogenetic reproduction that subsequently fails to, and indeed cannot, differentiate as do the cells of a blastula as they chain-reaction replicate and grow into an organism. See Figure 45.

If this is correct, cancer is probably inevitable. Any machine that is as intricate and finely tuned as a mammalian cell is bound to get out of whack occasionally. The "miracle" is that this transformation occurs so infrequently, at least in terms of cancers that become established. True enough, nearly everyone who reaches the age of 90 (and possibly earlier) has a cancer growing somewhere within their bodies.[24] On the other hand, the body has or creates countless trillions of cells, and it only takes one oncogenic transformation to initiate a cancer.

The sharper questions on cancer focus on the specifics of the transformation, why certain chromosomes have greater propensities than others, and why different cancers have different incidence rates in different regions or cultures. For the obvious reason, skin cancers (basal cell, squamous cell, malignant melanoma) are highly prevalent (among Caucasians) in Queensland Australia, and the southwestern United States, but rare among blacks everywhere. Although most other distributions are not so easily explained, reasons must exist. In part, those reasons point sharply at environmental factors.[25] Still, certain cancers, including breast cancer, have a strong correlation with a specific defective gene or at least a differently configured one.

What, then, is the solution to cancer? A comprehensive analysis of the numbers, combined with the findings of the *Genome* project, will surely draw a clearer picture, but given the deep and perhaps inevitable roots of cancer, the hope of a

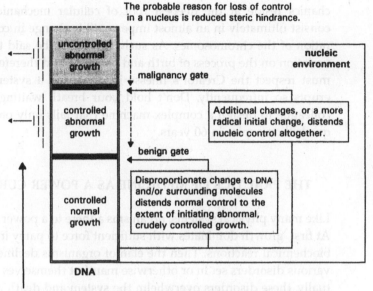

The probable reason for loss of control
in a nucleus is reduced steric hindrance.

uncontrolled abnormal growth

nucleic environment

malignancy gate

Additional changes, or a more radical initial change, distends nucleic control altogether.

controlled abnormal growth

benign gate

controlled normal growth

Disproportionate change to DNA and/or surrounding molecules distends normal control to the extent of initiating abnormal, crudely controlled growth.

DNA

In theory, it appears that any single DNA chromosomal chain can transform into a benign tumor or malignancy engine. Then once that transformation occurs the remaining chromosomes will likely be signalled to follow suit by way of symbiosis, similar to the probable sequence that occurs in mitosis.

45. Mechanical equivalent of oncogenesis. Despite 100 different cancers, the neoplastic transformation of a cell into malignancy arises from a common biomechanical process. This illustration depicts the equivalent of that process in engineering terms.

one-size-fits-all vaccine or other panacea seems naive.[26] Elsewhere, the efficacy of drugs that attack metastasizing cancer cells—which operate differently from healthy cells—may improve current mediocre performance, but that therapy will not likely solve the problem; it's too hit or miss. Regretfully, the only sound approach may be low-cost, extremely early detection of cancer on the grounds that any major change in the body's chemistry will *always* leave a signature, if only to activate the autoimmune system.[27] Sooner or later, technology will be able to identify that signature in every case; it is just a matter of increasing precision of medical technology.

If correct, this explanation is the worst possible—the me chanics reside at the deepest root of cellular mechanics an consist ultimately in an almost imperceptible change in configu ration of the chromosomes. As such, cancer can be said to be variation on the process of birth and life itself, and therefore on must respect the Creator's finesse in designing a system tha erupts so infrequently. Don't hold your breath waiting for colleague to design a complex machine that fails only once pe quadrillion copies in 60 years.

THE PRACTICE OF MEDICINE AS A POWER CURVE

Like many physical systems, organisms adhere to a power curve At first, growth dominates with sufficient force to parry inimica biochemical reactions. Then the élan of organisms declines, an various disorders set in or otherwise manifest themselves. Ever tually, these disorders overwhelm the system and death result:

The *theoretical* average life expectancy—in the absence c war, debilitating environmental factors, and malnutrition— estimated to be between 85 and 87 years, depending on th references one reads.[28] Actual life expectancy in develope countries now ranges from 75 to 79 years of age, and most of th difference between the theoretical and actual life expectancie can be traced to factors partially within the control of man. Tha is, in the absence of negative environmental factors and wit good nutrition and available medical care, life expectancy woul arguably rise to 83 or 84 years. A startling proof of this recentl occurred in Russia, albeit in reverse. When the former Sovi Union disintegrated in 1989, civil liberties suddenly expanded This new freedom included the "liberty" to drink even mo alcoholic beverages to excess and to face less prosecution fc violent crimes such as murder. Not surprisingly, this leniency le to a major increase in alcoholism and crime, resulting in a rapi decline of life expectancy between 4 and 6 years.[29]

The reason for this point of diminishing returns is tha medical science now deals successfully with most disorders unt the number of them in a person grows overwhelming. Nevertheless, medicine may have reached the point where its age-stretching ability will be limited to reducing negative environmental agents of physiologic disorders. To be sure, improvements in various treatments and therapies will continue, and those who suffer from chronic disorders will some day live more comfortable lives. It may even become possible to restore vision to the blind and to biogenetically engineer "original-equipment" replacements for faulty organs. Still, life expectancy in developed countries may already be as close to the theoretical maximum as practical.

Seemingly, the only way to raise the average significantly is selective breeding, a policy that raises the specter of Nazi Germany.[30] The use of amniocentesis to detect extreme genetic defects—which may encourage the mother to have an abortion—is controversial enough, but to expand this approach to form a superrace is repugnant at best. It is not that man is trying to play God. Rather, he has in fact attained some of that power by way of increasing knowledge. Wisdom is another matter.

CHAPTER 14

Blueprint of Evolution

"Where shall I begin, please Your Majesty?" he asked. "Begin at the beginning," the King said gravely, "and go until you come to the end; then stop."
—LEWIS CARROLL
Alice's Adventures in Wonderland

All we physicists strive to do is to trace His lines after Him.
—ALBERT EINSTEIN

The process of biological evolution can be compared to the evolution of musical instruments. Because the evolution of those instruments stretched over millennia, no single mind could oversee it, nor, as far as we know, did any person at any time predict how musical instruments would develop. Despite this lack of a unifying mind to *guide* development, a philharmonic symphony orchestra is today a wonder to hear, especially when playing Mozart's *Symphony Number 41 ("Jupiter")* flawlessly. Obviously, there must have been an underlying logic that substituted for an actively guiding hand.

Musical instruments are commonly divided into three categories: strings (including harpsichord and piano), winds, and percussion. The primordial antecedents were: rubbing objects together to generate a noise, blowing on a ram's horn, and striking hollow objects found in nature. From this came lyres, horns, and drums. Then over the next 2000 years, musicians gradually developed the superb instruments of the 17th and 18th centuries. They haven't changed much since then, and musicians today still wish for someone to craft violins on a par with Antonio Stradivari.

It seems that perfect harmony and polyphony resulted only by creating musical instruments exactly the way history did, and if so, the process would parallel the evolution of the species. But even if it did happen this way, the development required continual intervention on the part of mankind. That is to say, a perfect analogy with evolution would require the prototype instruments to reproduce themselves into ever more sophisticated models. That did not happen directly, though in a collective, sociological sense it did. What we need to do is reduce that sociological evolution of musical instrumentation into a pure, autonomic mechanical pattern for the evolution of the species.

EVOLUTION OF EVOLUTION THEORY

When the theory of evolution was first widely promulgated (1859), science lacked the tools and techniques to find *convincing* evidence. Eventually that evidence did accumulate, and more accurate techniques were developed to analyze it. Accordingly, no self-respecting scientist remotely doubts evolution, but many issues remain contentious. The least contentious of these issues "asks" if species evolved gradually or by punctuated equilibrium, i.e., by jumps or quantum leaps. On a graph, punctuated equilibrium resembles a staircase with each tread sloping upward slightly. More accurately, the resolution depends on the proportionality of each viewpoint because the evidence of some "jumping" is indisputable.[1] The implication is that nature is capable of near-catastrophic, constructive leaps in progression of the species.

The second issue looks at the comparative role of genetics versus interaction with the environment. The implications address the old question of which is more correct: (a) genetics is destiny or (b) the environment, hence society, is responsible for human conduct. But if the earth is an organism and no less physically deterministic than the physical aspects of any organism, then genetics would be destiny indirectly, in part, via its interaction with the earth, as might be negated by the exercise of human initiative.

The third issue is called the "Red Queen dilemma." It is named for a passage in *Alice's Adventures in Wonderland* wherein the Red Queen advises: "It takes all the running you can do to keep the same place," hence "if you want to get somewhere else, you must run at least twice as fast." If this is how nature works, it means that successful species must have been especially adaptive, else they would have been wiped out as the environment changed. The dilemma is the lack of evidence for *especially* adaptive species lower than man.[2]

PROBABILITY

Before we discuss evolution, the terms *determinism* and *indeterminacy* warrant reiteration. *Ontological indeterminacy* means that the outcome of a process is unpredictable with certitude; hence, it may or may not turn out the same way under two sets of identical circumstances. By contrast, *determinism* means that if all of the facts and logic are fully known beforehand, the outcome is predictable with absolute precision. If obtaining full knowledge beforehand is impossible or impractical, statistical techniques can be used to predict the likely outcome—*statistical determinism* (*actuarial determinism* with respect to time). As the experimental data or tests increase, actual outcomes will approach a theoretical distribution.

The consequence is that once you know how a system operates, you can configure it to produce the desired outcome the first time you set it in motion, assuming uncontrollable external factors do not intervene. As nature gives no evidence of consciously intervening in phenomena, her only alternative was to rely on actuarial determinism, even if it took four billion years. That is to say, nature did not "care" which *specific* elements combined into organic constructs, only that *some* of them did so.

The sociological parallel occurs in insurance companies and gambling enterprises, which habitually rely on actuarial determinism. For example, life-insurance actuaries do not know (or care) when Mr. Smith or Mrs. Jones will die, but for a million Mr. Smiths and Mrs. Joneses, they can predict the distribution of deaths as a function of age almost exactly.

In *Origin of Life on Earth*, A. I. Oparin stated that a sufficient number of trials of random letter type falling on the floor would eventually produce a page proof for Virgil's *Aeneid*.[3] Elsewhere, the actuarial–determinism thesis is illustrated by chimpanzees pounding on typewriters until one of them produces a play of Shakespearean quality. The chimp will on occasion type a few meaningful words, and less frequently will produce a profound thought (without comprehending its significance). Given enough chimps and sufficient time—say a trillion and an equal number of years—the play will eventually be produced.

Still, that procedure would be painfully slow. By comparison with those numbers, evolution proceeded at the speed of light. Nature relied on an enormous number of initial trials to generate a few basic molecular constructs, and from these she relied on subassembly development. Figure 46 compares symbolic models of element-by-element evolution with the subassembly alternative. Even in this symbolic representation, the advantages of the subassembly technique are clearly evident.

In terms of the chimps, this technique would put an untold number of them to work and gather up only the useful words and sentences. These words and sentences could then be given back to the chimps until by actuarial determinism they produced a requisite number of intelligent paragraphs or scenes. In turn, these larger units would be assembled until the play was completed. All things considered, four billion years should be adequate for the task.

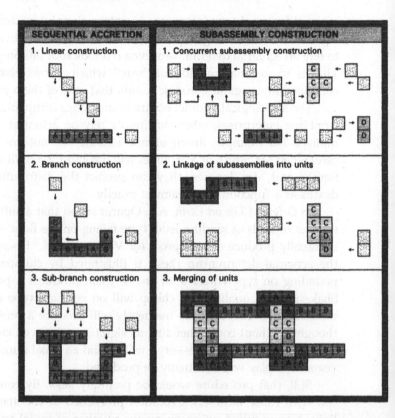

46. Sequential versus subassembly evolution. In any evolutionary or manufacturing process, a well-designed progression of subassemblies is inherently more efficient than sequential assembly of irreducible parts.

But who would judge the intermediate output and cull it for subsequent rounds? That question is irrelevant. Nature did not write plays; she built organic systems, and did so by phases: (1) peptides, (2) polypeptides, (3) encapsulated polypeptides, (4) subcellular organisms, (5) rudimentary cells, (6) advanced cells, and (7) multicellular organisms.[4] Each phase led straight to the next phase, and it is only the totality of the progression that obscures the underlying step-by-step, methodical progression.

This is not surprising when we try to comprehend in minutes what nature took four billion years to do.

NATURAL SELECTION

Natural selection is a euphemism for survival of the fittest. The "fittest" include roughly 13.8 million botanical and zoological species (more than half insects). That may be a large number, but they competed against an estimated 400 to 500 million other species. At that, the "laurels" signify only that the survivors had sufficient biological strength and resiliency to prevent extinction at the hands of a ruthless environment and fellow predators.

Furthermore, *fittest* does *not* mean that species adapted to the earth in any direct or conscious sense. The idea they could do so was known as Lamarckism and has been thoroughly discredited. What happened was that each generation of species produced a variety of specimens. In less than five percent of these species, a small number of specimens survived. Understandably, these few did most of the reproducing and, in turn, refined the species. Think of this as a kind of half-life, with the less-than-adequate species (and weaker specimens within a surviving species) tapering off to zero through a sequence of generations. By contrast, the fittest were able to reproduce at a ratio higher than 1:1, at least until other environmental factors (e.g., predators, reduced food supply per specimen) throttled the ratio back to unity. This implies a critical point (criticality), followed by a chain reaction that attenuated (reached a point of equilibrium with the environment) when the environment could not sustain any more of that species.

Also, keep in mind that *every* biological organism operates as if it were a cell in the metaorganism known as the earth. The ecology (read: physiology) of the earth is resilient enough to let many species try their luck by virtue of the enormous power demonstrated by the chain reaction of binary fissions unleashed during gestation. But nature also stacked the odds against success by destroying organisms exhibiting any but the most minor deficiencies. Only those few species—and specimens within a species—that survived long enough to reproduce in sufficient numbers avoided extinction.

Still, if the physiogenetic thesis is correct, accumulating those numbers was not a matter of chance. It was a matter of a design that ensured a few of the species would possess the appropriate configurational dynamics—patterns—to sustain continuous interaction with the environment, in a word, anthropy. By way of analogy, Stradivari's *summa cum laude* instruments resulted from painstaking care in *every* detail and aspect of his craft. Out of 1000 violin makers, one was statistically bound to achieve this level of perspective.[5] His name happened to be Antonio Stradivari. We use the same approach in sociological systems by inducing millions of students to seek higher education, knowing actuarially that half will drop out. And of the graduates, only a small percentage will return the investment in spades.

In short, nature permitted a high state of order for gravity because it is a very simple force. Moreover, cosmologists estimate that gravity, as well as atoms, formed within the first second after the big bang. Not so with life forms. The best she could do was to rely on actuarial determinism and billions of years, whereby a *few* productions achieved the necessary order and ability to contend with the environment. Many were called; few were chosen.

ORIGIN OF LIFE

The mechanics of evolution were identified by the Russian biologist A. I. Oparin in *Origin of Life on Earth* (1928)—peptides, polypeptides, encapsulated polypeptides, and so on. Yet until recent times, Oparin was among the least appreciated thinkers in science, though his theory has been enhanced and refined several times.[6] The most seminal of his followers was another Russian, Theodosius Dobzhansky, who wrote *Genetics and the Origin of the Species* (1941). With subsequent research and analysis, we now know that the evolutionary process proceeded, as mentioned, from peptides, to chains of peptides, to organized sets of peptides and other molecules, to semiencapsulated entities, to subcellular rudimentary organisms (most of which are extinct), and from there it branched to more sophisticated subcellular organisms (viruses) and rudimentary cells (precursor bacteria). The viruses fed (and still do) on bacteria, hence the bacteriophages.

However, nature was forced to use a shortcut. Recall the central dogma of genetics discussed in the chapter on genetics and physiology. DNA chromosomes can fabricate RNA chromosomes, various amino acids, proteins (from amino acids), and so forth, but the reverse does not occur except by random mutation.[7] To be sure, the prototype chromosomes were constructed from these components, but it took more than a billion years to do so. It is also true that the cumulative effect of random mutations over time—say a million years—can result in significant change. Not being *that* patient, nature combined slightly differing genes from two organisms to effect the necessary diversification. This reduced million-year-transition periods to mere generational spans, thus spawning the 400 to 500 million wanna-be species within three billion years.

We also know that cataclysmic events occurred at different times during the evolutionary process, which has led some people to conclude that the Creator intervened in the creation whenever evolution got out of hand—a kind of periodic megaflood. This is not likely because with one exception all genera survived these cataclysmic events. The exception was the dinosaurs, and even here one species is said to have survived, namely, the reactionary semanticist (*Brontothesaurus pedantus*).

BRANCHING AND DIVERSIFICATION

The most widespread misconception about evolution is that one species evolved from its immediate predecessor in a sequence of increasingly complex modifications. At the tail end of this misconception (no pun intended), man supposedly descended

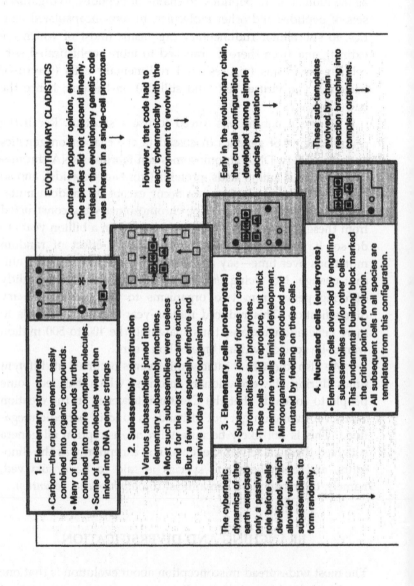

EVOLUTIONARY CLADISTICS

Contrary to popular opinion, evolution of the species did not descend linearly. Instead, the evolutionary genetic code was inherent in a single-cell protozoan.

However, that code had to react cybernetically with the environment to evolve.

Early in the evolutionary chain, the crucial configurations developed among simple species by mutation.

These sub-templates evolved by chain reaction branching into complex organisms.

1. Elementary structures
- Carbon—the crucial element—easily combined into organic compounds.
- Many of these compounds further combined into more complex molecules.
- Some of these molecules were then linked into DNA genetic strings.

2. Subassembly construction
- Various subassemblies joined into elementary subassembly machines.
- Most such subassemblies were useless and for the most part became extinct.
- But a few were especially effective and survive today as microorganisms.

3. Elementary cells (prokaryotes)
- Subassemblies joined forces to create stromatolites and prokaryotes.
- These cells could reproduce, but thick membrane walls limited development.
- Microorganisms also reproduced and mutated by feeding on these cells.

4. Nucleated cells (eukaryotes)
- Elementary cells advanced by engulfing subassemblies and/or other cells.
- This fundamental building block marked the critical point of evolution.
- All subsequent cells in all species are templated from this configuration.

The cybernetic dynamics of the earth exercised only a passive role before cells developed, which allowed various subassemblies to form randomly.

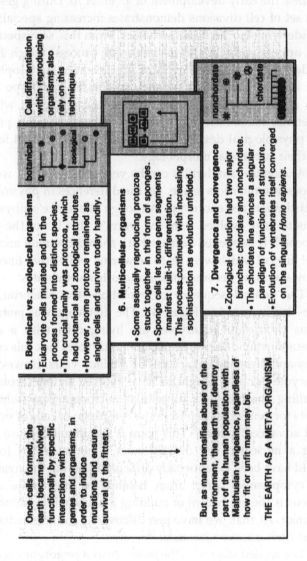

Once cells formed, the earth became involved functionally by specific interactions with genes and organisms, in order to induce mutations and ensure survival of the fittest.

→

But as man intensifies abuse of the environment, the earth will destroy part of the human population with Malthusian vengeance, regardless of how fit or unfit man may be.

THE EARTH AS A META-ORGANISM

Cell differentiation within reproducing organisms also rely on this technique.

5. Botanical vs. zoological organisms
- Eukaryote cells mutated and in the process formed into distinct species.
- The crucial family was protozoa, which had botanical and zoological attributes.
- However, some protozoa remained as single cells and survive today handily.

6. Multicellular organisms
- Some asexually reproducing protozoa stuck together in the form of sponges.
- Sponge cells let some gene segments manifest built-in differentiation.
- This process continued with increasing sophistication as evolution unfolded.

7. Divergence and convergence
- Zoological evolution had two major branches: chordate and nonchordate.
- The chordate line evinces a singular paradigm of function and structure.
- Evolution of vertebrates itself converged on the singular *Homo sapiens*.

47. The blueprint. The process of evolution was indeed complex, yet it appears to have relied primarily on permutations and combinations of repeating patterns.

from apes through some missing link. True enough, simpler organisms preceded complex ones, but the process of evolution parallels the early development of an embryo. During gestation, each set of cell divisions demonstrates increasing specialization of underlying biochemical mechanics. From this, semispecialized cells appear early in the development process, which further divide into ever higher levels of specialization and sophistication. In the evolutionary parallel, specialized prototype organisms developed early in the timeline, the science of which is sometimes called *cladistics*.[8] Cladistics stresses common, progressive physiological development instead of the external features and their superficial morphology.

Anatomical similarities can be very misleading. A worm is obviously similar to a snake, yet the differences in physiological sophistication are enormous. Although the snake's physiology seems to have evolved from the worm's, that is not the way it happened. It happened because both shared common ancestors, some lines of which went on to be worms while other lines went on, by stages, to become snakes and other vertebrates.

The general sequence of evolution—the blueprint, as it were—is illustrated in Figure 47. The first stage was *prokaryotes* versus *eukaryotes*. Eukaryotes are living cells with a central nucleus surrounded by cytoplasm. Prokaryotes lack this concentricity and preceded eukaryotes by a billion years. However, the eukaryote is now thought to have evolved by one prokaryote engulfing another (or a subcellular microorganism), which accounts for genes that persist in the cytoplasm of cells of even the most advanced species.[9] This point is significant because generating a nucleated cell directly from a nonnucleated forebear would have been exceptionally difficult. Directing the creation of the cytoplasmic "ring" from behind the membrane of the nucleus is the equivalent of building a house while sitting in an armchair. At that, we have just discovered an intermediate cell called *Methanococcus jannischii*.[10]

The second stage was the *protozoan* as a prototype organism. Most of the eukaryote line of cells branched off to become prototypes for the different phyla of botanical organisms, and they did this before the first true single-cell zoological creature was formed. The dynamics in botanical organisms are generally less active than in zoological counterparts, and therefore it probably took longer for a eukaryote cell to mutate successfully into the zoological scenario. But once that plateau was reached, nature attained a secondary critical point from which zoological specialization proceeded more rapidly, vaguely analogous to the role played by the DC-3 aircraft foreordaining all subsequent aircraft design.

The third stage was marked by *chordates* versus *nonchordates*. The animal kingdom is commonly divided into 10 phyla, ranging from one-celled animals (such as paramecia and amoebas) to the chordate phylum, which includes all vertebrates and predecessor subvertebrates such as sharks (which have a vertebrate-type skeleton made of cartilage). However, over the last 20 to 30 years evolutionists have recognized a twofold development pattern among all zoological phyla. One path led to chordates while the other path branched helter-skelter. As a result, the term *chordate* often connotes all species in the more-structured-development path while *nonchordate* refers to the balance.[11] The chordate path demonstrates comparatively greater homogeneity than nonchordates and culminates in a single species—*Homo sapiens*—that increasingly adapts the environment to its own needs insofar as the environment will tolerate it (as depicted in Figure 48). By contrast, the nonchordates are more heterogeneous, and shellfish are about as complex as they get.

EVOLUTIONARY DEVELOPMENT OF SUBSYSTEMS

The development of the principal physiological functions throughout most of evolution involved a series of increasingly complex variations on a few paradigms, and the same holds true for various organs associated with these subsystems. Two mutually reinforcing reasons account for this process. First, all four internal functions (logistics, reproduction, equilibrium/homeostasis, and control/

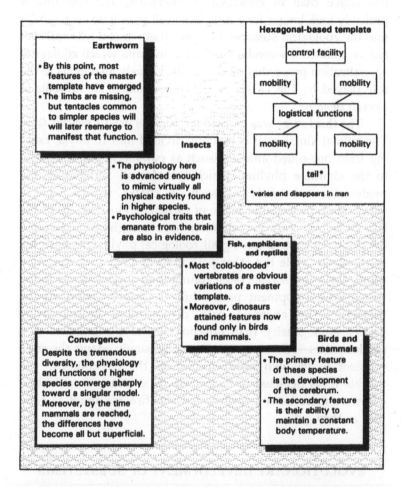

Hexagonal-based template

control facility

mobility — mobility

logistical functions

mobility — mobility

tail*

*varies and disappears in man

Earthworm
- By this point, most features of the master template have emerged
- The limbs are missing, but tentacles common to simpler species will will later reemerge to manifest that function.

Insects
- The physiology here is advanced enough to mimic virtually all physical activity found in higher species.
- Psychological traits that emanate from the brain are also in evidence.

Fish, amphibians and reptiles
- Most "cold-blooded" vertebrates are obvious variations of a master template.
- Moreover, dinosaurs attained features now found only in birds and mammals.

Convergence
Despite the tremendous diversity, the physiology and functions of higher species converge sharply toward a singular model. Moreover, by the time mammals are reached, the differences have become all but superficial.

Birds and mammals
- The primary feature of these species is the development of the cerebrum.
- The secondary feature is their ability to maintain a constant body temperature.

48. **Master template.** In the chordate line of development, the species adhere to progressively more complex variations of a common geometric heritage, with convergence toward a singular species—*Homo sapiens*—increasingly evident near the end of the scale.

locomotion) are essential for most lower-order zoological organisms and all higher-order forms. (Most forms of botanical life dispense with locomotion.) Second, if nature limited herself to only a handful of suitable chemical elements and the dynamics of their interactive configurations, she faced a strict limit on how many models could build themselves without intervention. Let us review the bidding:

Logistics To reverse entropy, all organisms without exception must be able to supply themselves with nutrients, or at least be able to absorb and process the nutrients that come their way. In practice, every living cell must have this ability, and multicellular organisms use the majority of their cells to obtain and route nutrients to every cell and rid themselves of waste products from processing. We see the same general configuration of organs and physiological processes in operation in all zoological species (especially the chordate line) from the earthworm forward: lungs, stomach, intestines, kidneys, and so on.

Reproduction The process of creating a new organism is not much different mechanically from binary fission, and the two operations coincide in one-celled organisms (asexual reproduction). Furthermore, reproduction is absolutely essential for evolution; therefore, the mechanics had to be developed before the advent of nucleated cells. The only fundamental difference between lower and higher species is in the complexity of the genes and consequently in the details of cellular differentiation.

Equilibrium and Homeostasis Homeostasis—the mechanics by which systems maintain equilibrium—supports the operation of all organisms; so it is not surprising that this function developed early in evolution. However, homeostasis occurs on three levels. The first, earliest, and simplest level is compartmentalization, i.e., the cell combined with separate development of various organs. The second level is the active controls that compensate for exertion and stress, which took a few phyla before they were necessary. The third level is the autoimmune system and this took much longer to develop.

Control and Locomotion   This function was perhaps the last to develop on the grounds that most simple organisms float or otherwise have little need to move to a source of nutrients. They simply absorb what they need from their immediate environment. Locomotion requires a refined nervous system interfaced with a muscular system that permeates the body, a complexity that was beyond the reach of simpler species.

ISSUES

The thesis of this book assumes that body and mind differ ontologically. That assumption resolves some difficult questions (e.g., roots of human consciousness), but it raises new issues. Some of these issues relate directly to evolution; others, to genetics and physiology in general, discussion of which before this point would have been disruptive to the continuing narrative on physical systems. We briefly review three of these issues here, which also serve as a transition into the chapters on sociological systems.

The first and most significant issue is the *definition of life*. Murder is all but universally considered to be a capital crime; swatting a pesky fly is not. Yet both *Homo sapiens* and the fly possess the same attributes of biological life and both were obviously products of the same evolutionary schema. Is there more than one valid definition of life? This issue is obviously related to abortion, especially in the case of anencephalia.

Very few anencephalias reach term, and almost all of the exceptions are stillborn or die within a few days. But in one recent case (1992), the baby survived for more than a year because the mother insisted it be kept alive on machines. The courts upheld her decision, but the one-short-of-unanimous opinion of medical ethicists—including ordained Roman Catholic priests—was to terminate the "life" of the baby.[12] The reasoning was that the baby was permanently comatose and could never exhibit any human attributes beyond those biological functions that exist to one degree or another in every organism starting with the amoeba. If this is correct, it follows *that psychical life cannot exist without a cerebrum*. The logical consequence, of course, is that biological and psychical life are two different things, despite any and all psychosomatic interactions. The cerebrum does *not* exist in an early blastula or early fetus, though a *healthy* blastula is obviously programmed to produce it.

The second issue asks: Could man have *accelerated the evolutionary process* by direct intervention? Probably yes, as evidenced by our ability to breed new or at least improved species of plants and animals. A related issue asks: Can man tinker with the evolutionary process to produce a "super" version of himself? Perhaps, but if steroid-enhanced athletic ability is considered repugnant, certainly the superman idea is to be condemned outright.

The third issue is that of *psychical–physical relationships*. If a person can exercise initiative, the cerebrum must have one *or more* focal points where a signal can *originate* in order to change an otherwise deterministic course of events. This minuscule signal must be amplified and relayed by the hierarchical nervous system, but it must originate somewhere and be other than the deterministic (or statistically indeterminate) result of cerebral mechanics at the instant it appears. The question: Where are those points? Descartes guessed the pineal gland, but that seems unlikely.

□ □ □

This concludes the discussion of the physiogenetic thesis as it applies to physical systems. The logical symmetry from the simplest to the most complex phenomena prevails, albeit in the form of complex permutations and combinations that increasingly obscure it. Obscure, not negate. So let us turn now to sociological systems.

Sociological Systems

PART V

Sociological Systems

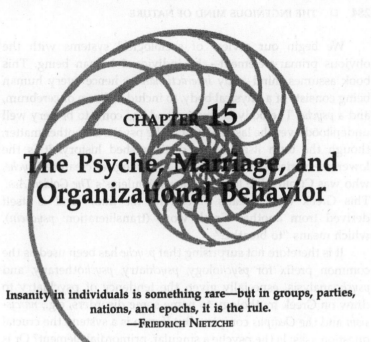

CHAPTER 15

The Psyche, Marriage, and Organizational Behavior

Insanity in individuals is something rare—but in groups, parties, nations, and epochs, it is the rule.
—FRIEDRICH NIETZCHE

Abraham Lincoln told the story about an asylum patient who claimed to be George Washington. The doctors were not sure how to proceed because they wanted to dispel the delusion without seeming to criticize the patient's choice of so admirable a role model. Fortunately, a week later he claimed that he was Napoleon—an easier syndrome to treat, or so they presumed. But to be certain, one of the doctors asked the patient if he was still Washington. He replied: "Yes, certainly, but that was by a different mother."

We begin our review of sociological systems with the obvious primary element—the individual human being. This book assumes mind–body *interactionalism*, hence every human being consists of a physical body, to include a brain or cerebrum, and a *psyche*. The body is obvious and has come to be very well understood over the last 400 years. The psyche is another matter, though the term itself has a distinguished history. It is the lowercase variation on the Greek mythological creature *Psyche*, who was Cupid's life love in Lucius Apuleius's *The Golden Ass*. This Greek word means "breathing *life* into," and is itself derived from another Greek word (transliteration: *psychein*), which means "to breathe."

It is therefore not surprising that *psyche* has been used as the common prefix for *psych*ology, *psych*iatry, *psych*otherapy, and *psych*oanalysis, especially given the tendency of psychiatry to draw on Greek mythology to name some disorders, e.g., *narcissism* and the *Oedipus* complex. But if man is a system, the crucial question asks: Is the psyche a singular, primordial element? Or is it comprised of discrete subelements, similar to an atom consisting of protons, neutrons, and electrons (as affected by various "fields" of force)? The answer to that question could elude us forever, but it is clear that much of human behavior arises from psychical aspects that often operate at cross-purposes with one another. Even the poets recognize this conflict via the mutually adversarial roles played by "head" and "heart."

If so, it follows that mental health and maturity, in part, are a function of a successful striving for equilibrium in some form among those conflicting aspects or subelements. To be sure, the effort often fails because one (or more) of those subelements can gain overpowering dominance, like the so-called workaholic. Or they may fail to operate or cooperate in sufficient tandem, common to various neurotic states. In a *figurative* sense, then, we may look on mental illness, or any substantive lack of maturity for that matter, as a misalignment of the interior lines among the psyche's aspects or subelements. And that misalignment is oft severe enough that professional intervention becomes the only recourse.

SUBELEMENTS OF THE PSYCHE

The aspects-versus-subelements issue is not to be dismissed out of hand, but for academic purposes the two terms may be considered to be equivalent. Apples and flour are different subelements of apple pie, but once on the table we don't think of them in those terms. Similarly, the "ingredients" of the psyche, so to speak, combine into character and personality whereby the synergism obfuscates them. In any event, one possible clue to the primordial elements of man resides in a book that was never intended for that purpose, namely, *Roget's Thesaurus*. In theory, all definitions are relative to other definitions, but Peter Roget, M.D., identified a hierarchical logic within etymology: a *logos* underwriting language. This point is crucial because all scientific theory depends on clear logical thought. And in *Roget's* we have an etymologically hierarchical, logical structure that has stood the test of time (145 years), translatable into any language without affecting its structure. However, bear in mind that Roget's categories are not limited words but rather umbrella concepts—logical reservoirs—from which all other terms flow in a hierarchy of tiered subconcepts.

Peter Roget devoted 50 years to this task before publishing his results (1852), and from that time until the Fifth International Edition (1992) his original structure for vocabulary had not been changed by subsequent editors.[1] More importantly, that structure suggests discrete psychical subelements, and even the revised fifth-edition structure only separates categories of human activities from the apparent subelements of the psyche identified in earlier editions. In more detail, and as depicted in Figure 49, Roget categorized all known words (in English) under a structure that indicates four psychical subelements: (1) the body (physiological host); (2) the mind, in terms of mental ability and processing of information; (3) the ego (or volition or the human executive function); and (4) the affections or affects, comprising emotions and feelings. (The numbers on the chart correspond to categories in *Roget's*). Keep in mind that the concept of ego in psychiatry is much more comprehensive than

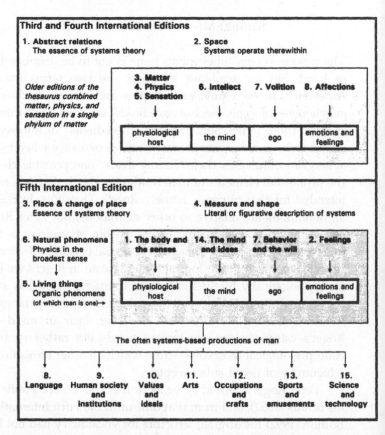

49. Roget's Thesaurus and psychical roots. The thesaurus was never intended as an analytical tool beyond the use of vocabulary, but its unbiased logical structure offers a model for understanding the underlying structure and roots of the human psyche.

the popular notion as the source of egotism. That it is, but ego is commonly regarded as the "executive function" of man—something that begs maturation, not disposal.[2]

First, *physiological body* included under the category of matter is common to all *Homo sapiens* and all other animals, which share a common set of appetites: oxygen (except for a few anaerobic organisms), water, nutrients, relative warmth and shelter, and (after puberty) the physical sex drive. Insufficient provision of any of the first four will cause death. The same is not true of the sex drive, but insufficiency here can threaten extinction, e.g., the giant panda (*Ailuropoda melanoleuca*).

Second, *intellect* (mind) superintends perception, intelligence, and logical processing. This subelement strongly correlates with the brain's physiology because deductive thought processes can be replicated on a computer, except for—and this is a very big exception—the power to create, both in terms of conception and implementation.

Third, *volition* incorporates ego. The existence of ego is not exactly controversial, but its role is another matter. As mentioned, many if not most psychiatrists and psychologists agree that the ego constitutes the executive function of the individual, but the relationship with underlying instincts and the social environment is not clear. Similarly, there is no agreement on how the superego or conscience forms: whether it is perception, a separate entity, an aspect of the ego, or a product of the individual mind interacting with parents and other social influences.[3] But beyond all doubt whatsoever, we know that the power of the ego can easily get out of hand, independently of or at least despite clear, logical thought.

Fourth, *affections* (affects) comprise a broad spectrum of *emotions* and *feelings*. These attributes are well known to all readers but are often ignored in scientific or professional analysis or, alternatively, they are seen as a mere transitory effect of physiological processes. So whether emotions and feelings emanate from a *separate* subelement—a reservoir of affects—or are an effect, surely their existence is an important aspect of the psyche. This book assumes emotions comprise at least one discrete subelement because of their overwhelming influence on the ego, which can nevertheless ignore them. At times. For example, mature people are said to control or throttle anger by morphing it into constructive behavior. By contrast, immature people often let their anger lead them into self-destructive behavior. Yet in both cases, the reality and intensity of the emotions associated with anger are the same.

If the above hypothesis is correct, or nearly so, human development and maturation must be, in part, a process of gaining control over the potential disorder generated by the action of inherently adversarial elements. Hence, gaining control over mental disorders may be considered as a process of psychical integration—a popular concept in psychiatry—or, that failing, stabilization or equilibrium among opposing psychical subelements. Similarly, sociological and/or organizational development can be perceived as the attempt to enlist contravening individuals into mutually beneficial if not more productive systems.

The need for psychical integration, to include control over physiological appetites, is not controversial. A controversy *does* arise over which force exercises the greater influence on behavior: genetic inheritance versus environment (and the processes by which these forces interact). Make no mistake; both sources can be overpowering. For example, the *proportion* of intelligence quotients (IQs) below 50 is identical among all socioeconomic levels, which strongly suggests that genes, not sociological environments, are responsible for profound retardation. But distribution of IQs between 50 and 70 are, in part, an inverse function of socioeconomic status, which suggests less-profound retardation is partially the consequence of social environment.[4] In support of environmental influence, Jane Healy concluded that excessive watching of television atrophies 20 percent of maturing brain cells and thus severely weakens the ability of the mind to concentrate.[5] Hence, the process of integration and/or psychological maturation must take the social environment fully but not solely into account.

CONJECTURED PSYCHOGENETIC MODEL

The various disciplines among the physical sciences may be ripe for a synthesis; a similar understanding of the psyche is not. Consider that doctors seldom debate the criteria for normality in the human body and physiology. From shortly after birth until age 45 or so, more than 90 percent of the population in developed countries have no symptoms of any physical disease or disorder at least 90 percent of the time.[6] When symptoms do appear, they tend to fall into a handful of categories: for example, trauma, complications of pregnancy, certain well-defined chronic disorders (diabetes, hemophilia), appendicitis, pneumonia.

We do not have this same neat situation when it comes to psychiatry. For starters, psychological maturity is something that takes years if not decades to achieve. Furthermore, psychological problems and dysfunctions are so common that people who demonstrate a high degree of maturity and integrity are, strictly speaking, abnormal. Lastly, we have little agreement on the dividing zone between psychiatric pathology and character defects. To be sure, the widely accepted manual of psychopathology, published by the American Psychiatric Association, classifies character defects as a form of psychopathology, and perhaps they are.[7] But sociological systems outside of psychiatry draw a distinct line between the two, especially when it comes to the issue of responsibility in court cases.

Nevertheless, if the physiogenetic model is correct, it will predictably have a figurative parallel in psychology. Ideally, we should develop a comprehensive model via a direct synthesis of the work of Sigmund Freud, Carl Jung, Alfred Adler, Karl Menninger, Jean Piaget, Erik Erikson, Karen Horney, Kurt Lewin, Arnold Gesell, Harry Stack Sullivan, M. Scott Peck, and others, not to mention an even larger number of noted psychologists and profoundly insightful philosophers. But it won't work. Each school of thought may be open to give-and-take, but only, it seems, if their particular school remains "first among equals."[8] The alternative is to posit a model and run it up the flagpole. In other words, the model may not be ready for prime time, but Figure 50 is a first cut at the task.

This model emphasizes psychopathology, because it is from disorder that we are led to investigate the fountainheads of human behavior and how they *should* operate. The central idea of psychiatric and psychological therapy, of course, is to bring subconscious elements, as applicable, to the fore, then to integrate

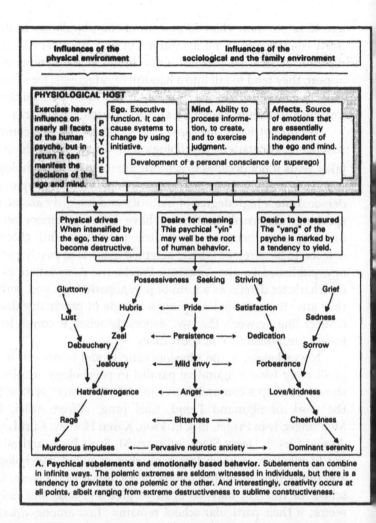

A. Psychical subelements and emotionally based behavior. Subelements can combine in infinite ways. The polemic extremes are seldom witnessed in individuals, but there is a tendency to gravitate to one polemic or the other. And interestingly, creativity occurs at all points, albeit ranging from extreme destructiveness to sublime constructiveness.

50. Conjectured psychogenetic model. The etymological structure of *Roget's Thesaurus* lends itself to this conjectured model of the psyche and its development.

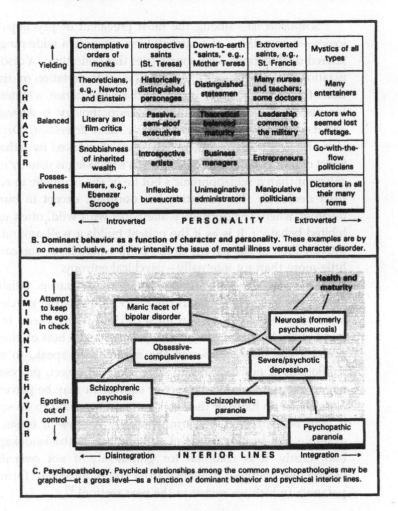

		Introverted ←		PERSONALITY		→ Extroverted
C H A R A C T E R	Yielding ↑	Contemplative orders of monks	Introspective saints (St. Teresa)	Down-to-earth "saints," e.g., Mother Teresa	Extroverted saints, e.g., St. Francis	Mystics of all types
		Theoreticians, e.g., Newton and Einstein	Historically distinguished personages	Distinguished statesmen	Many nurses and teachers; some doctors	Many entertainers
	Balanced	Literary and film critics	Passive, semi-aloof executives	Theoretical balanced maturity	Leadership common to the military	Actors who seemed lost offstage.
		Snobbishness of inherited wealth	Introspective artists	Business managers	Entrepreneurs	Go-with-the-flow politicians
	Possessiveness ↓	Misers, e.g., Ebenezer Scrooge	Inflexible bureaucrats	Unimaginative administrators	Manipulative politicians	Dictators in all their many forms

B. Dominant behavior as a function of character and personality. These examples are by no means inclusive, and they intensify the issue of mental illness versus character disorder.

C. Psychopathology. Psychical relationships among the common psychopathologies may be graphed—at a gross level—as a function of dominant behavior and psychical interior lines.

50. *(Continued)*

or otherwise "straighten out" the relationships among these subelements. This enables the bearer to deal more effectively with his or her desires and the environment in which those desires yearn to be manifested. Before looking at that process in the next section, let us briefly review the bidding on a number of well-known psychopathologies in terms of the conjectured psychogenetic model.

Schizophrenia—among the most prevalent of psychopatho-logies—is something of a catchall term covering a wide range of related dysfunctions. One source tries this definition: "a disorga-nization of the personality leading to misinterpretation or distor-tion of reality."[9] The key is the word *disorganization*, a synonym for disorder. Another common psychopathology is *depression*, and its cousin *bipolar disorder* (formerly *manic-depression*). Depres-sion also has many forms, but they are all marked by lethargy. Short of psychosis, a depressed person's thinking is usually more coherent than a schizophrenic's but he or she is unable to act or react decisively. The initiative lies dormant, except in bipolar disorder where the dormancy alternates with wild, often unin-hibited behavior. It is as if the patient builds a wall around the interior lines of the psyche to the point where he or she can do little more than rot internally—psychical entropy.

Compulsive-obsessive disorders, and their cousins fetishism and fixations, intensely focus the interior lines of the psyche on some object or condition that must be dealt with, at least in the mind of the beholder. By contrast, *paranoia* and a host of phobias position the interior lines of the mind, so to speak, to react irrationally if not radically to some external object, process, or one or more people. The object or equivalent may be perceived as intent to do harm or the agency of another mind to do so. Or it may be perceived as harmful simply because it exists, for example, the irrational fear of the number 13 (triskaidekapho-bia). Lastly, it is common for grandiosity if not *compulsive*, destructive madness to evolve from the more severe forms of paranoia (which we revisit in the next section).[10]

INTEGRATION OF PSYCHICAL SUBELEMENTS

The term *integration* when used in psychiatry can be misleading. As suggested earlier, even the most mature mind will experience conflicts between "head" and "heart," especially when it in-volves sacrificing one's own son. On this point, if the biblical Abraham got off easy, the Nobel laureate Max Planck did not.

During World War II, Planck elected to stay in his native Germany but refused to support the Nazi regime. Then in 1944, Planck's elder son was implicated in the plot to assassinate Hitler. Hitler offered to spare his son's life in return for Planck's political support. With heavy heart, Max declined, though he gave up the desire to live and died a few years later.[11]

Fortunately, few of us will be confronted with such choices, and therefore we can afford to spend some time looking at the process of integration, especially psychoanalysis. Karl Menninger described the crucial moment (turning point) of this therapy by noting that when the analysand had rooted out the source of "his misunderstanding and misinterpretations and mismanagements, having conceded his errors and forgiven those of others" and having admitted that many of his expectations were childish, he "begins to grow up." Menninger admits that the turning point is an enigma: "I presume that there is some balancing of forces in the way that any conflictual process becomes stabilized," adding that the existentialist concept of *kairos* might apply.[12] He defines kairos as a "crisis in the *sense of a dramatic moment* ... a particularly right time with various possibilities." He concluded that this moment was one "of heightened choice in that the individual is called upon to add his weight to the tenuous balance of forces."[13] It is difficult to find a better description of the psychical counterpart to crucial configuration and reversal of entropy, i.e., where a human being reverses his or her decline into disorder and begins the climb into psychological order.

The Dutch musicologist Walter Paap expressed somewhat the same thought when he described Beethoven's *Fifth Symphony* as "the record of a tremendous inner conflict which fills the listener with awe and trembling, and makes him share with joy and gratitude the final liberation." Interestingly, the final bars of this symphony force the conductor to draw imaginary [read: interior] lines in space with his baton.

The obvious benefit of restoring interior lines within the psyche is that it permits us to concentrate on life and goals. The renown physician Sir William Osler put the case this way:

Concentration, by which is grown gradually the power to wrestle successfully with any subject, is the secret of successful study. No mind however dull can escape the brightness that comes from steady application. There is an old saying "youth enjoyeth not, for haste"; but worse than this, the failure to cultivate the power of peaceful concentration is the greatest cause of mental breakdown.[14]

We find another example of integration in no less a person than Abraham Lincoln. One historian described Lincoln's almost sudden change from indecisiveness to decisiveness in February 1862 (10 months into the Civil War), saying that he "suffered a profound personal affliction in the death of his favourite son" (the *kairos*). Continuing, he admitted that "what caused his sudden emergence into self-confidence is a mystery" (similar to Menninger's admitted puzzlement). Then comes the key: "The difference between the earlier Lincoln and the later is not in the details but in the whole."[15]

Still, integration, or more accurately faulty integration, can lead to massive destruction. Herman Melville described this process in his novel *Moby Dick*—the tale of a whaling captain grown mad in his obsession for revenge against an albino whale:

> It is not probable that this monomania in him took its instant rise at the precise time of his bodily dismemberment. Then, in darting at the monster, knife in hand, he had but given loose to a sudden, passionate, corporal animosity; and when he received the stroke that tore him, he probably felt the agonizing bodily laceration, but nothing more. Yet, when by this collision forced to turn towards home, and for long months of days and weeks, Ahab and anguish lay stretched together in one hammock ... then it was, that his torn body and gashed soul bled into one another; and so interfusing, made him mad.... His special lunacy stormed his general sanity, and carried it, and turned all its concentrated cannon upon its own mad mark; so that far from having lost his strength, Ahab, to that one end, did now possess a thousand-fold more potency than ever he had sanely brought to bear upon any reasonable object.[16]

Captain Ahab lived only in the pages of literature; Hitler strode the earth. With respect to Hitler's recovering from wounds incurred in World War I, Sir Winston Churchill wrote: "as he lay sightless and helpless ... his own personal failure seemed merged in the disaster of the whole German people." Churchill went on to say that the intense trauma generated by the combined factors of defeat in war and the decline in law and order resulted in "an agony that consumed his being and generated those portentous and measureless forces of the spirit which may spell the rescue or the doom of mankind."[17]

Between these extremes of sober maturity and absolute evil, a rare individual comes along who, like the fictional Don Quixote, has lost his bearings but nevertheless charms us. For example, when this writer was serving in the Army's general-officer management office, we also had to deal with persons who claimed to be generals. One day, we received a letter from an American expatriate living on a kibbutz in Nazareth, Israel, who informed us—in a tone reminiscent of Clausewitz's writings—that he had been promoted to the rank of major general by General Patton in the thick of World War II, and now wanted to be hired as a strategist by the Army. Israel being outside our jurisdiction, we instead engaged in a long correspondence. His many letters were filled with discourses on strategic concepts and, except for a fixation on Dodge trucks, contained much food for thought. A scholar at the U.S. Army War College commented favorably on one letter, and senior officers would occasionally stop by to read his latest epistle. In time, we persuaded our man in Israel to abandon his imaginary stars and accept the status of "Honorary Lieutenant Colonel in the National Guard of the State of Alabama."

MARRIAGE

The simplest of all sociological systems is marriage: two human elements usually found under one roof. Simple? In the United States roughly half of all marriages end in divorce and arguably a significant portion of the balance are miserable. Why, then, do these systems break down? Conversely, why do at least a few marriages rise to enviable heights of enduring happiness? For that matter, how can objective scholarship approach the subjective and emotion-filled nature of marriage? Insanity is one thing; marriage, quite another. When Jack Benny was asked if he and Mary Livingston—his wife of nearly 50 years—ever contemplated divorce, he responded instantly: "No! . . . Murder, yes, but never divorce."

Actually, failure in marriage *can* be attributed to a number of distinct, sometimes quantifiable factors. As a corollary, success in marriage can be attributed to those same factors remaining below a critical threshold of danger. Plus—a very big plus—the husband and wife truly care for each other. Care is a subjective attribute, and it suffers by linking it with biological determinism— the consequence of physical monism. More than one marriage counselor has concluded that if a couple really care for each other, the chance of saving the marriage is good. In the absence of that attribute, failure beckons, no matter how well wired a person's physical brain may be. Let us, then, concentrate on the sources of failure.

First, despite the adage that in marriage two shall become as one, they don't. Any attempt to merge ego boundaries into a collective ego will destroy the psychical integrity of each partner. Sometimes the attempt is mutual; at other times, one-way, especially in what has come to be called *codependence*. Second, the marriage may lose any sense of purpose, and few systems endure in the absence of purpose (bureaucracies excepted). In the perspective of decades, procreation is transitory, and sexual pleasure is insufficient. No couple that happily reaches their golden anniversary attributes that milestone to still-yet-even-more-really-joyous sexual pleasure (despite an average of 4500 trysts of sorts).

Third, if one partner (for whatever reasons) proves dominant in too many aspects of the relationship, the other partner is likely to suffer unbearably. Successful marriages achieve a kind of stabilizing equilibrium in which the naturally dominant characteristics of each partner interact constructively, at least on the whole. Fourth, marriages almost always fail when destructive arguments reach a level of 10 percent of conversation but rarely fail when that level falls below five percent.[18] Fifth, adultery usually proves destructive because it introduces a third human element into a relationship that is meant to be one-to-one. In the popular novel *The Bridges of Madison County*, the character of Francesca engages in a four-day romp with a visiting photographer, only to return to life as usual, realizing that marriage to a somewhat dull husband has more to offer the innermost cravings of the human libido.

So we conclude this section with a vignette from the time when the *Titanic* went down to its doom. A woman, who had already been placed in a lifeboat, left it voluntarily to join her husband in death, quietly, hand in hand.

ORGANIZATIONAL FUNCTION VERSUS STRUCTURE

This book presumes that sociological systems are created, designed, or formed to accomplish one or more functions or purposes. To be sure, a few such systems arise serendipitously, but they don't stay around long unless they are seen to offer advantage, and that advantage becomes the presumed purpose or function. To manifest a function, a social system starts, or may be construed to start, with an initial configuration or structure subordinated to the function—what architects mean when they say "form follows function." In time, many of these systems emphasize procedures to the point where maintenance of the structure—the system itself—becomes the primary purpose of the system. At the extreme, our experience dealing with these systems seems like death warmed over—in a word, entropy.

Why? Three factors suffice: the effort required to sustain systems (logistics), the side effect of dominant roles played by a few individuals, and the overpowering influence of the *perceived* group ego. The factor of *logistics* exactly parallels the situation faced by organisms as described in the chapter on genetics and physiology. This needs no further comment except perhaps to say that the bureaucratic mind ingests, digests, and rids itself of waste in the sociological circulatory system known as interoffice memos.

The second factor of dominant individuals seems at first sight to enhance the life of sociological systems, and sometimes this happens. But to the extent any element in a system dominates, the other elements, be they ideas, atoms, or people, must accept an increasingly subordinate if not subservient role. In turn, these subdued human elements find their place in the sun, as it were, by integrating themselves into the organization. Go to any large social gathering. More than one guest will template his introduction on the well-known model: "Hello. I'm John Doe, Acme Widget Company." In point of fact, the annals of the U.S. Coast Guard yield the case where the helmsman of a yacht in immediate danger of sinking sent out the traditional SOS. On hearing it, a Coast Guard radio operator immediately acknowledged the call and asked, "What is your position? Repeat. What is your position?" The helmsman replied: "Chairman of the Board, Such and Such National Bank. Please hurry."

Most human abilities adhere to a normal distribution. This means that the majority of individuals possess abilities that evince a tight distribution. Only a few possess constructive ability in great measure. Conversely, another few lack any significant ability. Out of the 10 billion people born in recorded history, a mere 77 are represented in the first edition of the *Great Books of the Western World* set; 137 in the second edition (about 0.000001 percent). Fewer than 100 composers account for the overwhelming majority of Western classical music, and it is difficult to list more than 50,000 personages noted by history—literally one in a quarter-million. At that, a mere 10,000 dominate textbooks—one in a million or so.

In terms of organizations, these few individuals exert a disproportionate influence for several reasons. First, they are often in a position to take advantage of interior lines of communication, hence dominate the scene. Second, technology intensifies their influence by inflating publicity. Third, they tend to set standards, though not always for the good. Fourth, many of them will manipulate the organization to maintain power and influence.

The third factor of group ego is insidious and related to humans seeking meaning in a system in the presence of others dominant. The perceived group ego becomes synonymous with the system itself and is exacerbated by the well-documented tendency to render obedience to authority regardless of ethics and consequences, especially the 1961–1963 Milgram experiments.[19] Individuals from all walks of life were asked to administer increasingly higher-voltage shocks (ranging from 45 to 450 volts in 15-volt increments) to a person who gave incorrect answers or none at all to simple questions. The ostensible purpose of this shock treatment—which was faked—was to improve learning. The major finding was that *without exception* every participant obeyed instructions at least until the subject indicated extreme pain, and most participants continued to increase the voltage to the limit.

The need for a group ego is especially strong in the face of external threat. For this reason patriotic symbols were prevalent during the American Civil War, and on the part of democracies defending against aggrandizement during the two world wars in this century. The defeat of that global threat meant yielding individual preferences to the point of death for millions. The pain of sacrifice is not easily assuaged by theoretical principles and discourses. More visible focal points are essential, of which medals for heroism are conspicuous.

This symbolism of the group ego can sometimes merge with a nation's leader, especially if that leader possesses charismatic attributes. For the United States, the memory of Abraham Lincoln is closely associated with the success of the Union during the Civil War. Interestingly, he was secularly canonized at the moment of death by Edwin Stanton's remark "Now he belongs to the ages," and Lincoln's grave in Springfield, Illinois, is called a shrine on the road signs leading to it. Moreover, his memorial in Washington is openly labeled a temple by the words on the wall immediately behind the great white throne.

The group ego can be severely abused, for example, Hitler and his *fuhrer prinzip*. Even when not abused, this sociological umbrella can take on a life of its own and seemingly compel debilitating or destructive behavior. For example, patriotism is said to be the last refuge of a scoundrel, a tenet that became increasingly obvious during the McCarthy era and the sickening excesses of the House Un-American Activities Committee. Then, too, Colonel T. E. Lawrence (Lawrence of Arabia) admitted that his experience in rallying the Arab allies of the British against the Turks during World War I led to problems: "As time went by our need to fight for the ideal increased to an unquestioning possession, riding spur and rein over our doubts." Lawrence added that his mindset had become something of a religious belief and that "we had sold ourselves into its slavery, manacled ourselves together in its chain gang, bowed ourselves to serve its holiness with all our good and ill content" and that by their own act "we were drained of morality, of volition, of responsibility."[20]

ENTROPY AND ORGANIZATIONAL FORMS

The extent to which the above discussion operates to increase the entropy of sociological organizations depends in large measure on the form that organization assumes: relational, hierarchical, or integral. The *relational form* permits maximum individuality among members, who then cooperate to meet common needs. This form seems to have the best chance to avoid too much entropy or at least to delay its onset significantly. The *hierarchical form* mandates that members take direction from leaders, typically in some variation of the military chain-of-command environment. Obviously, and especially in the absence of crises, this form lends itself to autonomic strangulation. The *integral form* demands that members surrender or submerge individual identity for the good of the whole, which in practice usually leads to tyranny. Prognosis? Dead on arrival. However, bear in mind that any organization—any system—can (in theory) morph its structure from any of these forms into either of the other two forms. This is depicted in Figure 51, which constitutes a model for the discussion on nations and nationalism in a subsequent chapter.

Many organizations try reorganizations to preclude or delay the onset of entropy, and this approach has led to numerous jokes, dating back to the Roman Empire. Elsewhere, we sometimes look for new organization models based on organic systems. Much has been written on this theme, especially by extending the concept of homeostasis to political science. This obviously has some truth to it, for example, checks and balances to keep an organization from collapsing in on itself or, alternatively, erupting into discordant behavior and activities. We can find this approach in three variations or models.

The first model emphasizes *balance of power*. This model configures dominant but antagonist or adversarial elements so that they mutually negate excess force—dynamic equilibrium—yet combine to excise especially unruly elements (external or internal). This approach comes closest to an organic system and, in fact, appeared in sociological systems before it was recognized in physiology. The U.S. Constitution is the archetypal application.

The second is the *semi-independent satellite* model. This model grants maximum independence to members, holding them only to obligations on which they agree beforehand. A good example is the doctrine of states rights in the Tenth Amendment to the U.S. Constitution (though the import of that amendment has been gradually usurped by the vast increase of federal power). Elsewhere, this model is sometimes employed by large corporations to free various divisions of stifling bureaucratic control, for example, the initial relationship between the General Motors Corporation and the Saturn Company.

The third is the *selected-controls* model. This is a variation on the semi-independent satellite model. The superintending tier relinquishes detailed control (micromanagement) of lower tiers but retains explicit control over one or more focal points. For example, when the federal government's attempt to regulate an activity by direct regulation is thwarted in the courts by the states-rights doctrine, legislators often use economic incentives

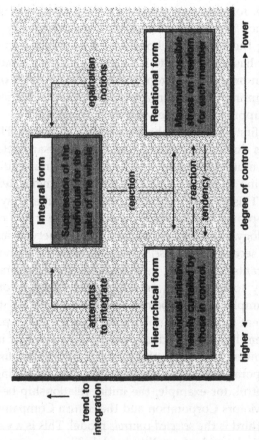

51. Relationships among sociological forms. Most organizations and nations adhere to a structure somewhere on the relational–hierarchical range of systems, but the notion of egalitarianism leads to the integral variation that can evolve from either polemic.

or unfunded mandates to effect the same purpose, e.g., environmental restrictions.

These three models all stress structure and dynamics at the expense of leadership. In terms of systems, this situation translates to an emphasis on equilibrium, homeostasis, and autonomic control (reducing leadership to an administrative or managerial role). One reason for this emphasis is to dampen the human tendency toward corruption. Indeed, some historians have argued persuasively that the only reason the proposed Constitution, with its strong executive powers, gained legitimacy was the presumption that George Washington would become the first president and set the standard of conduct for future incumbents.

But what is leadership? Leadership can be defined as the ability to motivate people to do what they cannot or will not do on their own volition. On the other hand, a study by the Franklin Research Institute (1967) determined that the only characteristic common to successful *military* leaders (where leadership is at a premium) is the ability to assimilate massive amounts of information from diverse sources and, after digesting it, disseminate the results in useful form.[21] If this is correct, the decisive interior lines of communication in a successful organization must radiate primarily from the leader. This makes it difficult to compensate for poor leadership, and thus the key to reducing organizational entropy is to ensure that leadership skill percolates to high levels without being felled by corruption. That goal being unrealistic, you can only roll the dice and then, if necessary, roll heads.

One last point here. Riots and revolutions are destructive catastrophes. Can we generate a controlled, constructive sociological catastrophe (as that term is understood by science) complete with a chain reaction of events that obviously reverses entropy? Yes, and perhaps the best example in modern times was the restoration of Colonial Williamsburg in Virginia. One person—the Rev. Dr. William Goodwin, rector of Bruton Parish Episcopal Church—envisioned the project despite Williamsburg having fallen into a sleepy, backwater town, whose economic existence hinged on a snake-pit-type mental institution. Not having funds, he tried numerous times—the last with success—to enlist the support of the Rockefeller Foundation, including the personal interest of John D. Rockefeller, Jr.[22] From that point forward, the chain reaction—extraordinarily well documented—led to results far exceeding the most wild imaginings. At one point, though, an elderly parishioner confronted our down-to-earth prophet: "Father Goodwin; come Sunday, will you be preaching the Gospel? Or will you be preaching wallpaper? If it's wallpaper, I'm not coming."[23]

URBANIZATION AND SLUMS

Cities exemplify the focal points and interior lines of sociological systems. Most cities arose, like ancient Troy, because they were situated at commercial crossroads. In time, cities take on a life of their own to the point where most operations support their own residents rather than external trade. Further in time—a point of diminishing returns—growth tapers off and a kind of rigor mortis sets in. Thomas Jefferson noted this atrophied tissue when he commented that cities do as much for the health of a country as sores do for the body.[24] Two centuries later, in the film *Defending Your Life*, the Saint-Peter-at-the-gate defense counsel explains to his client that while hell doesn't really exist, Los Angeles was coming close. Actually, the idea of lawyers running heaven is an even more hellish thought.

The existence of urban slums is not in dispute and, even outside the slums, urban life is marked by higher rates of violent crime, illegitimacy, and depersonalization of the individual. Furthermore, at least in the United States, a disproportionate percentage of people living in slums belong to minority groups. But why does this happen? Does systems theory in any of its many forms offer insight? Can the obvious urban rot— sociological entropy—be reversed? In theory, the answer is yes, but the problem may stem from socioeconomic status and the associated levels of literacy and conduct, not race or other discriminatory dividers.[25] Worse, the commendable experiment personally funded by Oprah Winfrey, to the tune of nearly a million dollars focused on just five families, suggests that the "entitlements mentality" had become too entrenched, even among the most motivated, free-of-drugs poor.[26]

In any event, the poor get poorer, and the territorial need for living space is further curtailed. The density of crime, illiteracy, illegitimacy, drugs, alcoholism, and other failings of human conduct rises to intolerable levels. Accordingly, despite support now given such as Aid to Families with Dependent Children

52. Slums as a euphemism. The plight of urban slums is worse than the word *slums* indicates, and the rate of decay is accelerating. Note the Capitol in the background.

(AFDC), food stamps, and the earned-income tax credits, the situation continues to deteriorate. What, then, is the answer? One proposal advocates centers to create economic opportunity within inner cities, so as to give inhabitants the necessary impetus to improve their lot. That idea is noble, but aside from pumping in even more money, how many bona fide jobs can be created in the absence of a demand for more goods and services? Another solution proposes integration by permanently subsidizing rents and mortgage payments in wealthier suburbs. Unfortunately, the cost is prohibitive and different socioeconomic classes seldom mix well. Would that it be otherwise, yet this attempt at forced integration partly explains why after three decades of school integration, many if not most public schools remain highly segregated.

Still another solution proposes reintroduction of the successful Civilian Conservation Corps (CCC) from the Depression era. That idea has both merit and obstacles: the costs are high, the parks are in better shape, alternative work on physical infrastructure would take work away from the current labor force, young men (there were no women in the CCC) from inner-city environments fared poorly and often had to be dismissed from the program.

So is there any solution? Doesn't at least some model of "social engineering" work? Probably not. Why? True enough, sociological systems exhibit the same operational characteristics as physical systems, to include equilibrium, homeostasis, culminating points, crucial configurations, reversal of entropy, and the like. But this picture ignores the tremendous, incomparable significance of human emotions and the exercise of initiative. In a word, then, people are *not* subassemblies to be shoehorned into some organic-like, social structure. What might work is an approach that gives primacy to individuality. Unfortunately, neither capitalism nor socialism seems prone or capable to give that emphasis, and the power of indifferent economic forces—discussed next—may prove ultimately unmanageable. But if there is a solution, the individual must be elevated to a higher level than a drone.

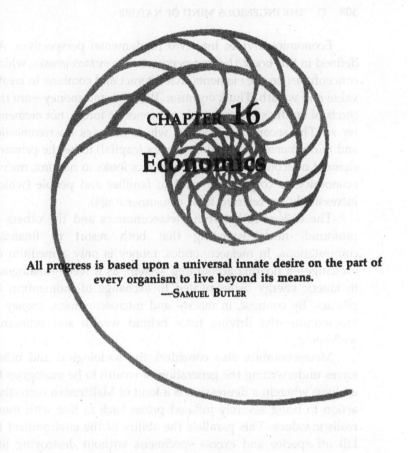

CHAPTER 16

Economics

All progress is based upon a universal innate desire on the part of
every organism to live beyond its means.
—SAMUEL BUTLER

Economics is a pervasive sociological system, touching every
individual, organization, and nation. However, the perception of
economics varies with the era, often because of the proportional
weight given to production and free enterprise versus ethical
considerations and the welfare of people. And if the love of
money is the root of all evil, then the study of it may be the root
of all confusion. Alan Greenspan is reported to have said that the
Federal Reserve Board specializes in precision guessing. If so,
perhaps the chief accomplishment of economics has been to
make astrology a respectable science.[1]

Economics divides into two fundamental perspectives. As defined in this book, the first perspective is *metaeconomics*, which concentrates on the elements that interact and combine to create value and wealth. Think *organism*. Wealth—and money—are the *effects* of an underlying system of economic forces, not elements *per se*. The second perspective, which includes *macroeconomics* and *microeconomics*, considers money (capital) to be the primary element of economics. Macroeconomics looks to nations; microeconomics, to corporations and to families and people (which increasingly are referred to as consumer units).

The difference between metaeconomics and the others is profound, notwithstanding that both resort to financial computations. In metaeconomics, money is only a medium of exchange—a floating accounting system, as it were—analogous to kinetic energy being related to exchange of momentum in physics. By contrast, in macro- and microeconomics, money *is* momentum—the driving force behind wealth and economic well-being.

Metaeconomics also considers the sociological and other forces underwriting the generation of wealth to be analogous to ecology, wherein a depression is a kind of Malthusian corrective action to bring severely inflated prices back in line with more realistic values. This parallels the ability of the environment to kill off species and excess specimens without destroying life across the board. By contrast, macroeconomics (and its microkin) views depression as an unmitigated disaster to be staved off at almost any "cost." Same depression; two legitimate viewpoints.

Elsewhere, other terms from the physical sciences have seeped into metaeconomics. One analyst questioned the wisdom of bank mergers. He said that "big deals" lead to more "big deals," which in turn generate even more such "deals." Then he asked: "Has the *chain reaction* gotten out of control, and the financial *meltdown*" [started]?[2] It's a safe bet that when major corporations experience "meltdown," so too will the general economy.

METAECONOMICS

Wealth is sometimes regarded as having the means of production or service which, in turn, creates even more wealth. Alternatively, wealth may be considered as having sufficient financial resources and opportunities to produce wealth. As such, *wealthy* (the adverb) usually connotes *excess* wealth compared to needs. This wealth is typically expressed in terms of capital or value, and those terms are usually measured in units of currency. Consequently, the meaning of capital and value all too often reduces to price. Since price is easy to visualize, money quickly becomes the focal point of economics.

But price and value are not the same thing. Excess supply compared to demand will drive prices down, just as excess demand will drive prices up. Neither dynamic changes an item's contribution to work and the formation of more wealth. The economic history of civilization can be indexed to a steady accretion of wealth as man learned to reconfigure resources into more productive systems. True enough, nations and empires go "belly-up" on occasion, but to metaeconomics—which, like ecology, ignores national boundaries—this merely signals a change in ownership of goods and the means of producing them.

However, the graph of this accrual is linear only on average. Many factors operate to distort the "curve" of that graph with respect to different nations and different eras. The most prevalent factor is the control nations have over resources or the capital to acquire them, combined with the knowledge to link or otherwise transduce the elements into more favorable configurations (technology). Other factors include the ratio of the population with respect to production capability, government policies and purposes, which are usually enforced by laws and taxation, and the degree to which entrepreneurs are given freedom of action. Think also of the prevalence or lack of fraud, greed, and overdependence on welfare, the competition (if not economic warfare) posed by other countries, and the degree to which a nation attempts to aggrandize its neighbors or, the opposite, the degree to which it may be compelled to expend major resources defending its own prerogatives.

The apparent consequence of these factors is significant disorder in the best of times and catastrophe in the worst, i.e., uncontrollable depression, worldwide at times. Yet in *The Wealth of Nations*, Adam Smith identified an underlying system, which he called the "invisible hand." By way of analogy, physicians sense an "invisible hand" in the body that seeks to defeat various disorders, often culminating in a moment of crisis that marks the turning point to death or a return to health.

Similarly, the "invisible hand" of economics seems to be guided by the collective recognition of value versus price, competition or the power of governments to break up monopolies when they appear inimical (or to regulate them when they are unavoidable), rebellion or revolution in its many forms when too many needs go unmet, and accumulation of knowledge and the progression of technology that lead to more technology. But understanding the "invisible hand" and controlling it are two different tasks. We understand earthquakes and volcanos very well; we have not been able to control them.

Still, we *can* summarize the basic operation of the economic organism, starting with production. Production means to arrange, refine, transform, or transduce physical resources into more useful configurations, or at least into configurations that people are willing to buy (pet rocks, for example). The root elements are: raw material or lesser subassemblies, human labor, machinery or other means to assist in production, and the infrastructure essential to manage the first three elements. Infrastructure includes (but is not limited to) leadership, adequate financing, technology, favorable markets and transport thereto. The only dispensable element (in part) is labor because it can be, and often is, replaced by machinery in the form of automation to the extent production reduces to time-motion activity (see Figure 53).

Supposedly, service differs from production because existing goods are delivered, sold, maintained, repaired, or perhaps improved instead of new goods being produced. Aside from

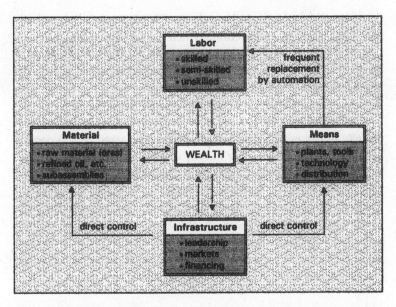

53. **Contributing elements of wealth. Wealth results from favorable combinations of material, labor, means and technology, and infrastructure. The labor component, in part, is being replaced by technology, especially by automation.**

delivery, service is: less amenable to assembly-line mechanics, generally requires a greater knowledge and ability on the part of the labor force, and for the most part requires less investment and capital costs. Still, the elements of service are nearly the same as production. The key differences are that *configured* material is substituted for raw material, and labor is less easily dispensed with.

Economies in developed countries are changing emphasis from production to service, and no scarcity of reasons confronts the analyst. First, the so-called information age means that more and more service workers are required to keep tabs on all of the documentation associated with work. Second, increasing productivity per worker means that fewer workers are needed for production, and more workers are needed to maintain and repair increasingly sophisticated products. Third, automation substitutes for labor in production, which drives down the price of goods via competition, which means the population can acquire more of them, which intensifies the demand for service. Fourth, the leisure time created by increased productivity has hosted the development of a vast recreational service market.

The problem with a service-based economy is that *proportionally* fewer new goods are produced compared to the cost of services, yet the clamor for money continues. Hence, an increasing percentage of the population survives by charging higher fees for services—a parasite economy as it were. Gouging by lawyers, stock market brokers, and physicians attests to this, not to mention systemic fraud in the tens of billions of dollars annually.[3] Inflation beckons, especially on the price of investments. During the period 1990–1996, the Dow Jones average nearly tripled, far in excess of the increase of goods and wages.

MACROECONOMICS

Macroeconomics addresses the ebb and flow of money within a nation and, for that matter, the world. The impetus behind this focus is specialization—the ever-increasing division of labor. Specialization demands an ever more fungible (fluid) medium of exchange. In former times, precious metals filled that role, but today the common media are negotiable instruments and currency (most of which are now reduced to magnetized bits in computer databases). The accumulation of money—or effective control over it—places individuals, corporations, and governments at the various focal points within the interior lines of commerce. Because almost everything has its price, either directly or in lost opportunities from *not* investing elsewhere, nations measure their economic health in terms of money rather than the means of wealth. The two common rulers for this measurement are *gross national product* (GNP) and *gross domestic product* (GDP), both of which measure the sum of a nation's transactions. The *national* product takes international trade balances into account; the *domestic* product does not.[4] Still, for many reasons both measures are misleading.

The first reason is the *production/population ratio*. If country A has twice the product of country B, and four times the population, A will probably be poorer. The doubled wealth must be distributed to fourfold the population, i.e., A's per-capita share is only half of B's. The second reason concerns *turnovers per cycle*. Money flows. When you earn a dollar and spend it, that same dollar will undergo a number of transactions before it winds back in the pot that pays your salary. This does not mean the same physical dollar changes hands a particular number of times in each cycle, but that the nation's gross product equals the value of all wages times a factor (ranging from 2 to 10). Capitalist countries typically have a higher turnover ratio than do socialist countries because capitalism relies more on free enterprise than direct, one-shop government redistribution.[5]

The third reason—*government share of the gross product*—is another way of looking at the second reason. In the 1930s, federal, state, and local governments in the United States consumed or redistributed about 11 percent of the GDP. Today that figure is 45 percent and rising. The bulk of this increase goes to various welfare and income security benefits, commonly labeled entitlements. Furthermore, many nongovernmental transactions are mandated, regulated, or influenced by government decree.

The fourth reason looks at *distribution*. Production of wealth requires a higher density of capital than does day-to-day living; therefore, wealth and the sum of transactions for corporations and investors must necessarily be higher than average. However, while increasing concentration yields more wealth, it does so only by intensifying poverty among perhaps the majority of the population. This explains why governments seek an optimum trade-off. Too much egalitarian redistribution of wealth dries up the economic fountainhead. Conversely, too little redistribution means that too few can drink from that fountain (which sometimes leads to rebellion and revolution).

The fifth reason is the degree of *bartering, self-support, and underground transactions.* The value of produce grown and consumed by a farmer does not always appear in the GDP nor does the value of barter and other underground transactions. Hence, developing countries with a low gross per capita product may be better off than the numbers indicate. For example, the *per capita* GDP of Tanzania is about 1 percent of the U.S. GDP, yet surely a Tanzanian has more than 1 percent of the U.S. standard of living.[6]

Macroeconomics from an international perspective is even more complex than this. The "have-not" nations request and sometimes demand aid, support, and concessions from the "have" nations, but here sovereignty comes into play. There is no *legal* obligation to provide these funds or equivalent nor is there any agency effective enough to enforce such obligations if they did exist. Also, multinational companies reduce tax liabilities by farming out production to other countries. Lastly, international crime has become big business because of the ease by which proceeds can be laundered through international banks. Granted, life imposes moral obligations, but from the perspective of economics, moral obligations plus 75 cents or so will buy a cup of coffee at a fast-food restaurant.

ECONOMIC HOMEOSTASIS

If you draw a graph and plot the daily sum of all your assets as a function of time, you will likely see a splatter of points that collectively go uphill or downhill as time passes. Draw a line through these points so that roughly half the points are above, and half below, that line. That line is called the *line of regression.* Most long-term analyses of macroeconomics discern lines of regression about which the value of all goods minus liabilities (net worth) fluctuates. In turn, this "regression" suggests that metaeconomics is a zero-sum game. This means that every gain is offset by a loss somewhere else, despite the fact that wealth gradually increases. The reason for ignoring this growth is that it is very small compared to wealth; the annual increase in global net worth rarely exceeds 2 percent of the value of all transactions occurring in a year. Although manipulations of this zero-sum game hasten or retard the rate at which total assets accumulate, these manipulations have severe limits whereas, in theory, the entire world capital above local subsistence farming and barter can and does gravitate toward comparatively few nations. A mere seven countries of more than 200 control 67 percent of the world's gross domestic product.[7]

As shown in the upper frame of Figure 54, these mechanics result in the most common economic cycle, namely, periods of sustained growth alternating with recessions. Recessions are often regarded as "invisible-hand" (homeostatic) corrections to excess growth. When excess growth is too severe or some economic factor is allowed to progress too far, the correction takes on the malignant attributes of a depression. But how can this excess growth reach dangerous proportions undetected, and why isn't there any middle ground between recessions and depressions?

The answer to the first question is that excess growth does not normally generate unfavorable circumstances that are immediately apparent. On the contrary, the economy gives the appearance of being robust. This is analogous to *some* cancers that grow and metastasize without generating obvious symptoms until shortly before death is inevitable.

The answer to the second question is that most recessions are comparatively short-lived, rarely exceeding two years and seldom causing unemployment levels to exceed 10 or 11 percent. By contrast, depressions can drag on for 10 years and evoke unemployment rates of 25 percent or higher. The difference seems to be that in potential depressions, some aspect of the economy is allowed to hyperventilate out of control until the whole economic organism collapses from exhaustion.

For example, in the 1920s the value of the stock market increased at a giddy pace, far in excess of bona fide increase to wealth—its value in terms of being able to create even more wealth on a sustained basis. Eventually, this artificial growth reached a point where buyers were no longer willing to pay

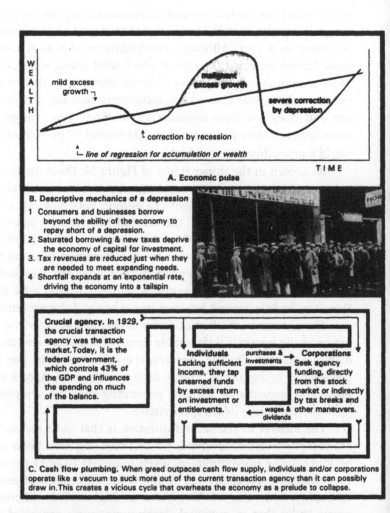

A. Economic pulse

W E A L T H

mild excess growth

malignant excess growth

severe correction by depression

correction by recession

line of regression for accumulation of wealth

TIME

B. Descriptive mechanics of a depression

1. Consumers and businesses borrow beyond the ability of the economy to repay short of a depression.
2. Saturated borrowing & new taxes deprive the economy of capital for investment.
3. Tax revenues are reduced just when they are needed to meet expanding needs.
4. Shortfall expands at an exponential rate, driving the economy into a tailspin

Crucial agency. In 1929, the crucial transaction agency was the stock market. Today, it is the federal government, which controls 43% of the GDP and influences the spending on much of the balance.

Individuals Lacking sufficient income, they tap unearned funds by excess return on investment or entitlements.

purchases & investments

wages & dividends

Corporations Seek agency funding, directly from the stock market or indirectly by tax breaks and other maneuvers.

C. Cash flow plumbing. When greed outpaces cash flow supply, individuals and/or corporations operate like a vacuum to suck more out of the current transaction agency than it can possibly draw in. This creates a vicious cycle that overheats the economy as a prelude to collapse.

54. Economic cycles. Cycles of economic growth and recession are common in all developed countries, but when *perceived* growth vastly outpaces *substantive* wealth, the culminating point of this overstretch can lead to catastrophic depression. The chain-reaction mechanics of this process constitute an economic cat-chasing-its-tail game.

inflated prices, which forced sellers to lower their sell price. This haggling set off a chain reaction, partially motivated by fear, as large holders of stock tried to cut their losses. As prices fell, buyers were increasingly unwilling to buy in; hence, the supply exponentially outstripped demand. The market lost 80 percent of its former value, which dried up capital for investment, which closed down industry, which put people out of work, who could no longer buy goods, which forced more businesses to quit, and so forth. Yet until the bubble burst, the accelerating, *robust* Dow Jones average prior to late 1929 had little direct effect on the vast majority of the population.[8] See the lower frame of Figure 54 for a semigraphical picture of this process.

Could this catastrophic reaction happen again? Are there not stock-market controls that curtail selling when prices drop too rapidly? Are there not safety nets such as unemployment compensation, Social Security, welfare, the earned-income tax credit, as well as pension plans and individual retirement accounts? Yes, but these safeguards feed on contributions, taxes, or dividends from investments. In addition, the enormous federal debt continues to increase by deficits exceeding $100 billion per year despite the 1993 tax increase. Worse, the government has spent all of the money (about $700 billion) in the Social Security and 21 other trust funds in addition to "official" annual deficits (further increasing the national debt), as reported in every annual edition of the *Statistical Abstract of the United States* (published by the U.S. Government).[9]

We downplay the malignancy of the national debt. It is an abstraction, or so goes the theory. Yet the accelerating interest payments—if nothing else—will eventually increase to the point where the government cannot borrow sufficient funds to cover the deficits and therefore must curtail entitlements, which will curtail spending, which will force companies out of business, which will reduce total tax intake, which will further reduce benefits, which will erode consumer spending, which will reduce taxable business income, and so on, until perhaps half the country is unemployed. Think *chain reaction.*

CORPORATE MICROECONOMICS

Analogously speaking, microeconomics describes the organs and cells of the metaeconomic organism. However, the proportional influence of each such unit varies considerably, and corporate economics are more dependent on the national economy than nations are on the world economy. Furthermore, the range and distribution of corporate operations (and their volatility) are much greater; hence, it is harder to generalize on microeconomics than the macro variety. Then, too, corporations vary in size and clout from the largest controlling hundreds of billions in assets, to millions of so-called mom-and-pop shops or single-proprietor operations.

Proprietary corporations are by definition driven by the profit motive while the nonprofits exhibit a much wider range of motivations: from profit disguised as excess income, to the genuinely altruistic environment of sheltered workshops. The nonprofits also include municipal corporations and many quasi-government agencies, which in theory are bent on providing services. Elsewhere, some corporations and agencies are natural monopolies, at least regionally, and are more highly regulated. Still, corporations and agencies of whatever kind are subject to the marketplace in some form, even if it is only the voters' concern for keeping if not increasing entitlements.

Because of this variety, business schools emphasize case studies. Each case is unique in one or more respects, though collectively they provide underlying lessons—analogous to the principles of war—that apply to microeconomics in general. These principles have never been formalized, but they abound as rules of thumb. One well-known and easily documented generalization is that most new business ventures fail for one or more of four common reasons: overestimation of marketplace demand, failure to understand the competition, insufficient capital, and lack of perseverance. Occasionally an entrepreneur takes the market by storm with a new product (say, concrete that never hardens and therefore never cracks), but the majority of new businesses focus on existing types of products and services.

Hence, they must take business away from the competition or generate a new demand for the same product or service.

Corporate microeconomics also offers a study in optimum configurations. Some companies reach equilibrium, for example, a business that does well with a one-plant operation but fails when it expands into a multiple-plant operation to accommodate increased sales. The reason is that management of multiple plants entails higher administrative and overhead costs, at least compared to the marginal cost of adding shifts to a single plant. Elsewhere, corporations have learned that micromanagement is counterproductive. The interior lines become too rigid, and too much effort is expended trying to satisfy every node in the system. When the Otis Elevator Company, which had cultivated decentralized management, consolidated their computer operations, senior executives discovered that middle managers at the headquarters used the consolidated data to reexert micromanagement.[10]

Perhaps the most difficult problem of microeconomics is the conflict between short-range and long-range profitability; or for nonprofits, between short-range and long-range efficiency. This is especially true when major investment and extensive borrowing is called for, because most corporations think and act in terms of short-range profitability. A telling amount of vision, courage, and decisiveness is essential to modify the future. When these traits are scarce, most companies will stick to the *status quo*. In the process, markets change and the internal bureaucracies increase. The problems faced by IBM in the 1991–1993 period are an excellent case in point. Despite accelerated use of computers and automation, IBM survived only by cutting its labor force nearly in half from its high point in the 1980s, and by decentralizing internal controls in order to increase its sensitivity to market conditions.

Lastly, let us consider the role of leadership and exceptional business judgment on the part of a chief executive officer. Until the advent of computers, probably no industry was more subject to volatile trends than automobile manufacture. There is only so much you can do with oil and steel before another corporation buys it, but consumer taste in automobiles is another matter. It is not surprising that the two most renown case studies in business involve automobile manufacture: Alfred P. Sloan (1923–1963) at General Motors and, more recently, Lee Iacocca at the Chrysler Corporation (1978–1994).

But why do comparatively few firms grow large to the point whereby five percent of them account for 90 percent of all sales?[11] Is it not the right concentration—the right configuration—of resources and leadership ability at the right time in the right place—the right timing? Among other attributes, these exceptional leaders know how to coax more production out of the same resources than do their competitors, and sometimes a small difference in configuration can result in a major difference in outcome. Thus operates the power curve.

CONSUMER-UNIT MICROECONOMICS

Most singles and families (consumer units) strive to develop some form of reliable economic system for their own financial needs. The common name given to these systems is *budget*, in which little attention is paid to future needs beyond an item marked savings (typically the first to suffer when expenses increase). Still, most consumer units that have disposal income try to accumulate wealth in excess of current need in order to contend with unforeseen-but-predictable contingencies and with predictable-but-distant expenses, including college education of children and retirement. In *David Copperfield*, Charles Dickens wrote:

> Annual income twenty pounds, annual expenditure nineteen nineteen six, result happiness. Annual income twenty pounds, annual expenditure twenty pounds ought and six, result misery.

When Dickens wrote this, annual expenses for most families held steady, mortgages were less common, and the vast majority of people died before they grew too old to work. Thus, Dickens's point was easy to fathom. Today, variable debt and long-term, high-value mortgages are common, as is fluid household income.

Then, too, the ever-changing tax and insurance picture, and the volatility of Social Security benefits, exacerbates the analysis. Still, the idea remains for people to create "wealth" by accumulating capital in excess of needs.

This "profitability" can be fed by increased income, reduced expenses, "sweat equity," and leveraged investments. The first two are self-explanatory. Opportunities for "sweat equity" (which keeps money in the bank by substituting personal labor) are less frequent, the most common being do-it-yourself projects around the home and possibly for some of the minor construction and finishing work on a new or rebuilt home.

The fourth—leveraged investments—operates by borrowing money in order to make more money. Borrowing money at a lower interest rate than can be earned in a safe investment is all but impossible, but capital appreciation is another matter. Many people have done well by buying and selling their own homes, or by investing in the stock market, more often than not by way of mutual funds. Others have lost small fortunes.

Most people overlook the role played by dominant elements. As developed in previous chapters, dominant elements operate by way of crucial linkage to other elements or by sheer mass (or both), hence they tend to dictate how a system will operate. Similarly, a handful of decisions in a household budget will generally determine the overall financial picture. These dominant "elements" typically include automobiles and the cost of college education, but the single most dominant item is the choice, size, location, and cost of a residence. Too much house means high mortgage payments, higher taxes, empty space to fill up with furnishings, and the subtle pressure to spend like the neighbors—a low-level kind of chain reaction. In the long run the capital return on the investment *may* compensate for the strain but only after many years of financial misery.

Another interesting aspect of microeconomics is the effect of diminishing returns, especially as it applies to second incomes. When a spouse works for the sake of increasing income, and young children are at home, the amount of additional *disposable* income is typically a small fraction of the gross additional wage.

Federal, state, local, and FICA taxes can claim 50 percent (or more) of the gross, while the cost of earning that income—from day care to transportation—can absorb another 15 to 20 percent, perhaps more. Then, too, the strain of the work reduces the time for money-saving by careful shopping and the like.

CYCLE OF POLITICAL ECONOMICS

Until the end of Medieval times, economics was of interest primarily to popes, kings, and other rulers, who relied on taxes and "voluntary" benefices to support the activities of state. Elsewhere, the charging of *any* interest on loans was deemed usury and evidence of moral degradation. Lastly, merchants and bankers were considered to be one step removed from social outcasts.[12] But as manufacturing technology and international trade increased, this semimoralistic perspective evolved into *mercantilism*—the accumulation of capital and precious metals by the state government to fund state functions. In turn, mercantilism evolved into the free enterprise common to modern times.

This summary view overlooks the influence of communism, which has failed almost totally, and the partially successful inroads of socialism. As Will and Ariel Durant point out in *The Lessons of History*, socialism dogged even the Roman Empire.[13] Thus, it is more accurate to say that metaeconomics almost alternates between capitalism in its many forms, which is strongly associated with democracy, and socialism (typically under the euphemism of welfare capitalism), which requires a pervasive hierarchical form of government.

This cycle is difficult to perceive because the pace and magnitude differ radically among different countries. For example, the United States seems ready to delve—some would say sink—ever more deeply into socialism by controlling the provision of national healthcare, which now consumes 15 percent of the GDP. By contrast, Sweden, Great Britain, and perhaps Canada are giving serious consideration to reducing the welfare functions of the state. Elsewhere, socialist Russia and many of the former Eastern Bloc states are at least attempting to change over to free-enterprise systems.

But why do these two incompatible schools of thought take turns on center stage? The answer seems to be that socialism dampens the entrepreneurial spirit and the work ethic, which reduces wealth, which strains a nation's finances, which leads to a renewed emphasis on free enterprise, which generates larger gaps between the proverbial haves and have-nots, which encourages a restoration of government programs, which leads to more socialism, on which the cycle begins anew. That is to say, we see a kind of homeostasis operating in the economic machinery, with the middle class serving as the balancing agent or ballast that keeps economic swings within manageable bounds. Regardless of the current emphasis on free enterprise or socialism, if the majority of a nation's citizens lead a comfortable life economically, they have the votes to keep it that way. But when excess socialistic redistribution of income hits too hard, they balk and take it back by way of their voting power. Similarly, when poverty grows malignant, they use that same voting power directly or indirectly to support higher taxes and greater redistribution.

At one time in the United States, the highest federal tax bracket was 77 percent (increased to over 90 percent during World War II). This rate fell to 50 percent and then to 28 percent. Now the maximum rate has turned back upward by two stages to 41 percent (39.6 percent plus the 1.45 percent no-limit Medicare tax), with the looming prospect for an even higher rate as the national debt continues to mount. But in the absence of a viable and substantial middle class, this metacycle could take on a more violent swing.

TECHNOLOGY AND THE SHRINKING MIDDLE CLASS

Some economic policies seek to maintain free enterprise and a large middle class, most notably the antitrust legislation of the Roosevelt–Taft era. However, a relatively new factor operates to shrink the middle class, namely, technology. Technology exerts itself in two forms. One form—medical advances—has created a large and still growing retired class. In the 1940s, only a handful of the population lived past 65 years, and few people retired before that age. Today, 15 percent of the population exceeds that age while millions of others retire early (not always willingly). The second form—automation and robotics—reduces the need for human labor and divides the balance into a comparatively small group of well-paid technocrats and a much larger pool of lower-paid workers, who are increasingly subject to layoffs and longer periods of unemployment.[14] Consequently, society in most if not all developed countries can now be divided into five classes: (1) the wealthy, (2) professionals, executives, and technocrats, (3) resource workers, (4) most retirees, and (5) absorbers.

The wealthy may be few in number—no more than 10 percent of the population and probably less—yet they provide the majority share of investment capital, taxes paid, and charitable contributions.[15] This group exerts more economic influence than any other group and perhaps nearly as much as all of the other groups combined.

The professionals, executives, and technocrats amount to perhaps another 10 percent of the population. Members of this group exert high influence by virtue of their jobs, their ability to project their thoughts, and their role in nurturing others. Not all of them gain this influence by way of technology but, in general, technology enhances their leverage.

Resource workers constitute the largest of the five groups and include so-called white-collar and blue-collar workers in an economy where service and information increasingly dominate. They exert little influence beyond their immediate work environment, but they pay their own way, contributing as much in taxes as they draw directly or indirectly in benefits. However, they are increasingly dependent on jobs and backup government programs such as unemployment insurance. And as the demand for labor shrinks with respect to supply, they cannot afford to be choosy where they work or the salary they agree to accept.

The fourth group—*retirees*—number about 40 million in the United States (roughly 15 percent of the total population and 20 percent of the adult population). Most draw Social Security, and many add pensions and interest from savings and investments to that. Few retirees contribute to the *generation* of wealth, but they sustain a great deal of the economy by way of expenditures and some taxes. The point to note—at least in the United States—is that Social Security (since 1939) and Medicare (since 1965) are funded entirely from present taxes. Hence, from the perspective of the U.S. Treasury and annual budgets, retirees constitute a large group that absorbs more benefits than they pay in.

The fifth group are the *absorbers*. Absorbers are unable to contribute anything to the economy, but they absorb a disproportionate share of its resources. To be sure, many people drift in and out of this status, and the numbers fluctuate with the health of the economy. Still, the government must subsidize them to an increasing degree. Put another way, developed nations will be called on to fund more and more entitlements from a shrinking capitalistic base without resorting to a pure socialistic economy that dries up available wealth even faster.

□ □ □

So this chapter ends on the same pessimist note as the previous chapter, but the intent is not pessimism. The intent is to describe how economic systems operate. All systems can and do degenerate into lethargy and disorder—entropy. Economics is no exception. Chaotic configurations can be reconfigured to restore some semblance of order, but only if the "players" understand how the underlying forces and elements interact or transact.

Systems theory and its various kin can help, but it takes more than wishful thinking, or political "sound bites," or socialistic diatribes, or mere tinkering with the Federal Reserve prime rate. Given human greed and other shortcomings, the cycle of excess growth followed by recession is probably inevitable. And avoiding depression may require governmental mandates that are inimical to democracy. For example, how can a "free" nation regulate stock-market prices? Still, the choice remains to deal with the insidious factors of metaeconomics or pay the price of its homeostatic ability to trash excess valuation with Malthusian-like depressions. Again, think: organism.

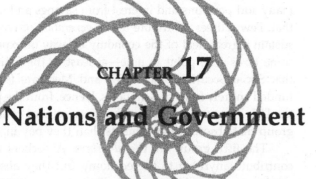

Nations and Government

A portion of mankind may be said to constitute a nationality if they are united among themselves by common sympathies which do not exist between them and any others—which makes them cooperate with each other more willingly than with other people, desire to be under the same government, and desire that it should be government by themselves or a portion of themselves exclusively.
—JOHN STUART MILL

At times, politics and disorder seem to be synonymous. Aristotle concluded a long time ago that man is by nature a political animal. In systems terms, this means that each individual will exercise initiative—usually to his or her own advantage—in a sociological group that strives to mold if not subdue that initiative for the benefit of the group and its other members. Strives. The political animal usually finds a way to avoid being caged.

Still, tribes were established, which combined into cultures and regional entities, which led to the development of nations, which, however, seem to be the endpoint. Despite a growing world population, land is essentially a fixed quantity, and there is little prospect of any supranational authority to which nations will yield sovereignty. Moreover, all empires to date have disintegrated except for a few small-island possessions of various countries, the continued occupation by Russia of the Kuril Islands off Japan, and the suzerainty that China exercises over Tibet. Lastly, most of the current world turmoil occurs in those countries where different cultures were forcefully brought under a single jurisdiction, most notably the states in the former Soviet Union, the ethnic regions in the former Yugoslavia, and in many African nations. Elsewhere, national boundaries have achieved a remarkable stability compared to the musical-chairs statecraft of centuries past, especially in Europe.

The upshot of this paradigm is that the world has become something of a geopolitical chessboard—a metascale variant of game theory. In systems terms, the elements (nations), and their interrelationships, have become more stabilized as have relations among them. The sporadic violence that we continue to see lacks the force to upset this relative global equilibrium. However, civil wars, rebellions, and revolutions continue to provide ample evidence of critical mass and catastrophe in regional bodies politic. In that vein, this chapter looks at nations as systems while the next chapter focuses on how those nations vie for influence and power within the larger system of the world.

THE POLITICAL SEED BED

To survive and endure, nations must overcome many barriers that work to their detriment. One barrier is the sheer size of population. Population may not be a problem for Canada or Australia; it is a severe problem for China, India, and Indonesia; and may eventually prove equally severe for many other countries. Another problem is the seeming political indifference of perhaps a majority of a nation's citizenry. It has been said that less than 5 percent of an electorate makes the sustained, semischolarly effort to understand issues and candidates. At that, in the United States only 60 percent or so of the electorate bothers to exercise the right to vote. Furthermore, the general tendency to obey authority, despite obvious wrongdoing, virtually ensures indifference to wrongheaded government.

We must also consider the extent of human greed, criminal intent, and other character flaws among citizens. The Internal Revenue Service estimates that $130 billion in tax liability goes unpaid annually.[1] That is, on average each of the 130 million tax-filers (individual and joint returns) cheats the government of $1000 annually. If 75 percent of tax-filers are honest, then the remaining 25 percent fail to report an average of $18,000 of taxable income. Another consideration is the extent of collective mores, ethical standards, and the level of education among citizens. Still another is the ancient maxim that power corrupts and dements. Edward Gibbon, in *The Decline and Fall of the Roman Empire*, defined corruption as "the most infallible symptom of constitutional liberty."[2]

It is not surprising, therefore, that a few enterprising individuals—dominant elements in the organism of a state or nation—take advantage of this situation to the detriment of the majority. Some are content to use the system anonymously while others seek the political limelight for the sake of power, often to sate some inner psychological need. The first movement toward democracy in England arose from feudal lords demanding preservation of their own prerogatives vice King John. That was the significance of the *Magna Carta*, and it took another 600 years for the benefit to diffuse to the common man.

From time to time, a few individuals seek to improve government or at least keep it from becoming worse. Of these few, some are so naive that the system readily chews them up. Others are subsequently corrupted by the experience of power. Still others push their views to the point of intransigence and lose all influence, of whom Woodrow Wilson is probably the best example in U.S. history.[3] The minuscule balance go on to earn the thanks of history if not of their peers. In this century, General George C. Marshall, Sir Winston Churchill, and Mahatma Gandhi were archetypal examples of competence and integrity. Still,

Marshall was repeatedly castigated by Senator Joseph McCarthy, Churchill was voted out of power before World War II ended, and Gandhi was assassinated.

Arguably, the ideal is to create a system of government that is not dependent on exceptional leadership (inherent equilibrium) yet can call on it in times of great crisis (homeostasis on tap). This ideal must also place checks and balances on those who would usurp power while providing at least a modicum of redress for the bona fide complaints of the normally silent and acquiescent majority (autonomic homeostasis).

NATIONS AND NATIONALISM

A *nation* is an organization that has significant sway over the decision-making of its members. This includes the authority to send those members to their deaths in defense of the common good, as if they were little more than antibodies. Restated, and within the limits set by the polity, a nation exercises the power of life and death. The sustainment of that power requires a strong collective conscience or group ego which psychologically binds members to their nation. That collective conscience is called *nationalism*. However, nationalism means different things to different people. When a state aggrandizes its neighbors or otherwise coerces them to conform to its own perspective, this nationalistic spirit is typically labeled *jingoism*, and for this reason nationalism sometimes carries a pejorative connotation.

Regardless of any ethical issues, nations seem here to stay. Attempts to create a world authority elicit about as much enthusiasm as watching paint dry. The few times the United Nations has been effective in ameliorating international conflict are attributable to voluntary support by affected nations, who more or less used the United Nations as a cover to support national interests. Examples include the U.S. counterinvasion of South Korea in 1950 and the counterinvasion of Kuwait subsequent to the occupation by Iraq. Elsewhere, the indecisiveness of the United Nations has sometimes been used as a scapegoat when one or more nations wanted to steer clear of some regional conflict, e.g., Bosnia and Somalia.

In the absence of supranational authority, a nation requires a group ego in some form to survive if not flourish. That ego requires sufficient commonality as perceived by a nation's human elements to override personal differences. Think of this as group-psychical interior lines linking the otherwise divisive effect of individual initiative. If the differences are controllable, dynamic equilibrium will tend to keep the group or nation in comparative order by way of mutual back scratching. When those differences grow out of bounds, and malicious biting begins, the culminating point of political equilibrium is passed, resulting in revolution or rupture in some form.

This situation leads to the question: What are the factors or common bonds of nationalism? Do those bonds preexist, or can they develop (or disintegrate) over a period of time? From a polyglot of human sources, some nations develop a strong political fabric that defies conventional logic, for example, the United States. Elsewhere, force was used to unite various regions that remain nations to this day, e.g., Germany. Still, the tendency among empires and nations alike is to break up into more culturally homogeneous entities more so than to voluntarily merge into superentities. The recent collapse of the former Soviet Union attests to this. Elsewhere, Africa is now divided into 55 countries, with civil war in one form or another infecting at least 10 of them at any given moment. About the only exception to the disintegration of empires is the attempt of Europe to create an economic if not a political superpower, and the jury is still out on that endeavor.

SOVEREIGNTY AND THE POLEMICS OF GOVERNMENT

The group ego of nations has a formal name, and that name is *sovereignty*. Roughly speaking, sovereignty equates with the party or parties at law that have sufficient authority to resolve internal conflicts and decide how to counter external threats. In a pure monarchy, sovereignty and the king coincide. Louis XIV supposedly intoned *L'etat, c'est moi*. Conversely, in a pure democracy sovereignty resides in the citizenry, as Lincoln put the case at the dedication of the cemetery at Gettysburg: "that government of the people, by the people, and for the people, shall not perish from the earth." Between these polemics reside intermediate forms of government known as oligarchies, aristocracies, and republics, all of which vest power—de facto or de jure—in a relative handful of individuals.

When sovereignty condenses into one individual at a focal point of power—a point from which interior lines of authority neatly radiate—maximum efficiency ensues. If power did not corrupt or give reign to corrupt mind-sets, the monarchical form of government would continue to dominate history. That not being the case, polities seek to diffuse power in order to restore political equilibrium. Unfortunately, the loss of neat interior lines ensures inefficiency. Will and Ariel Durant concluded that "freedom unlimited is chaos complete."[4]

In systems terms, the totalitarian model exemplifies the hierarchical form of system. At the other polemic, the democratic model distributes political power equally among citizens, thus exemplifying the relational form of system. Western history evinces a general progression from monarchy and totalitarianism to a kind of democracy in the form of constitutional republics. In practice, however, as populations expand to the hundreds of millions, political power is reconcentrated de facto into oligarchies or aristocracies. Thus operates the primary give-and-take of political power.

SOCIAL CONCERNS AND THE THIRD POLEMIC

At either polemic of the governmental spectrum, a substantial minority of citizens—sometimes the majority—suffer. This fact needs no amplification at the totalitarian polemic, while the looseness of democracy provides maximum opportunity for the most able citizens to achieve dominant status at the expense of peers. It is not unusual for chief executive officers to earn 200 times the minimum wage, and for some celebrities (with or without much talent) to rake in 1000 times that wage.[5] Enter socialism, with its tenet that everyone is somehow equal and should therefore share income and benefits more or less equally regardless of contributions. In the extreme form, enforced socialism becomes communism or even Nazism (a name derived from the German words for *national socialism*).

This socialist polemic is offset from the baseline spectrum of governmental forms, because socialism can be instituted in any model of government system (as depicted in Figure 55). However, to the extent that a system of government presumes all

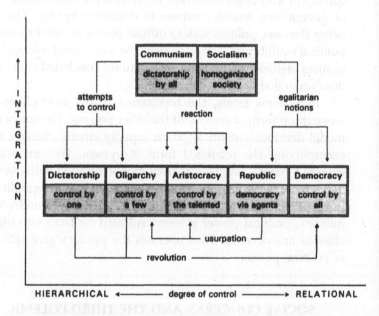

55. Forms of government as systems. Most governments hew to one of three forms: relational (democracy), hierarchical (monarchy, oligarchy, or aristocracy), or integral (socialism or communism). Government structures at the polemics almost always fail, but the attempt to find a self-sustaining equilibrium among the polemics faces equal difficulties.

people are equal and must be treated as equals, the government must to one degree or another superimpose that "equality" over the obvious differences and ranges of human capability. At the extreme, this means that each person must submerge her or his natural differences for the sake of the common good in order to receive an "equal" wage on which to survive. This approaches the integral form of system, in which the identity of the elements (citizens) are alloyed into the overall picture: what Marx and Engels called "the dictatorship of the proletariat." Most socialistic theoreticians abhor this extreme, but all of them stress massive redistribution of income.

In any event, a socialist-leaning state should be studied from the polemic on the baseline spectrum from which it evolved. When imposed by a dictatorship, socialism typically degenerates into preserving the status of the bureaucracy in power, perhaps sustained by ensuring the masses are guaranteed at least a subsistence-level standard of living. By contrast, when socialism evolves from a democracy, the motivation is usually more altruistic, though in the end, it too generates massive bureaucracies that tend to suffocate human initiative.

REBELLION AND REVOLUTION

Revolutions occur when government practices become intolerable for a major segment of a nation's population. In systems terms, the operation of that nation reaches a critical point of conflict among its elements—its citizens and its institutions. The discontent is shared among many in a classic case of sociological interior lines. This phenomenon can be literally heard in small riots and mutinies, especially in dramatized yet historically accurate accounts such as *Mutiny on the Bounty*. In larger-scale rebellions, the stabilizing influence of political equilibrium breaks down, and chaos enters the picture. When the uprising is local and short-lived, the ensuing disorder is usually called a rebellion, for example, Shay's rebellion in 1792. When an uprising infects an entire nation, the disorder transforms into revolution.

The ultimate objective of a revolution is to create a more effective configuration of government in the eyes of the revolutionists or at least to reduce abuses of power. That reconfiguration may seek internal reform or attempt to expel an occupying power. In either case, revolutionists often introduce their own violence. It is not surprising that revolutions often fail, or the objective may take several generations to resolve itself, or that the new configuration might be worse than the one it replaces. The ancient saw is that the oppressed become the oppressors. Figure 56 summarizes the outcome of 10 well-known revolutions.

Interestingly, the process of upheaval operates in science similar to the way it does in political environments. For example, when an existing theory no longer explains a growing array of contravening facts and relationships, the replacement theory is often called revolutionary, and it may take from 20 years to a century to gain acceptance. Furthermore, revolutions in science—especially in the form of technology—can eventually lead to political upheaval. No greater example exists than the Industrial Revolution. The economic chaos arising from that technological upheaval spawned political revolutions by the score: from the back-to-nature Luddites (1820–1850) to the disastrous, long-term consequences ensuing from Marx and Engels's *Communist Manifesto* (1848), namely, the death of some 90 million Soviets and Chinese (not counting war deaths).

A more intriguing case is the history of Great Britain and members of her former empire. The tribes residing on this island were conquered by Rome, and a millennium later (1066) by Normandy. Then England combined with its northern neighbor Scotland to form Great Britain (1707) and later absorbed Northern Ireland to create the United Kingdom (1801). It also developed the largest and most far-flung empire in history.

In the process of building a nation and empire, England experienced dozens of revolutions and civil wars, of which the War of the Roses and the Cromwellian epoch are the most noted. Still, Great Britain is no longer a major power and will probably lose Northern Ireland, if only because the Roman Catholic population has a higher birth rate than the Protestants. Its royal family has been muddied of late, and its experiments in social-

	Description	Outcome
Reformation 1520–1600	This was the great rebellion against the practices and the theology of the Catholic Church.	Protestantism gained permanence, followed by two centuries of even more virulent religious genocide.
English Civil War 1643–1660	Radicals in Parliament rebelled against Charles I, who had often usurped their prerogatives.	Cromwell as Lord Protector beheaded the king, yet monarchy was restored in a decade and regicide condemned.
American Revolution 1775–1783	The American colonies (aided by France) rebelled against Great Britain to attain independence.	The U.S. prevailed and established a government based on the principles for which the war was fought.
French Revolution 1787–1799	Primary cause was the growing influence of the *bourgeoisie* vis-à-vis monarchical traditions.	The outcome was a Reign of Terror, restoration of monarchy and, later, Napoleon's slaughter of Europe.
European revolutions of 1848	Concurrent revolutions in Italy, Germany, Austria, and France flared over economic concerns.	They all failed, although the reaction subsequently opened the door for German and Italian unification.
American Civil War 1861–1865	The legal issue was the right of states to secede from the Union, but the real issue was slavery.	The Confederacy was defeated and slavery abolished, but the aftermath evoked a century-long bitterness.
South American revolutions 1810–1815	Argentina, Chile, Colombia, and Peru waged prolonged war against the Spanish occupation.	Spain was defeated in each case, yet control passed to ruling elites that remain in power to this day.
Chinese revolutions 1909 and 1949	Nationalists ousted the emperor, grew corrupt, and in turn were defeated by the communists.	Mao Zedong replaced Chiang Kai-shek with an ideological dictatorship that slaughtered eighty million people.
Bolshevik Revolution 1917–1918	This revolution was caused by social conditions intolerable even by Russian standards.	The monarchy was overthrown but Lenin's successor Stalin devolved into a genocidal dictator.
Indian emancipation 1920–1947	The Indian people, led by Gandhi, resisted British rule by massive civil disobedience.	Britain granted full independence in 1947, but civil war between Hindus and Muslims flared immediately.

56. Well-known revolutions and their outcomes. The revolutions highlighted in this chart are well known to history, and most of them were marked by violence and war. However, the revolutionary goals were not always achieved or only with a bitter aftertaste.

ism have been accompanied by high unemployment and a national healthcare system that almost no one likes and is often cited as government bureaucracy at its worse.

To be sure, some historians point out that the majority of revolutions are relatively peaceful, and that only a few turn radical. That statement depends on the definition of "radical." In political science, "radical" implies actual or impending violence and bloodshed. Certainly a few well-known revolutions did transpire without destruction or maiming of human life, for example, the Glorious Revolution of 1688. Still, if some revolutions are not explicitly violent, they are all inherently radical. They impose, or attempt to impose, a different way of doing business within a jurisdiction. The interior lines of authority—of sovereignty—are remapped. During the French Revolution of 1789–1796, the new map was simple—the route from various kangaroo tribunals to the guillotine, along which some 20,000 Frenchmen took a one-way trip. A larger number were spared the formalities of a hearing before being massacred. And in the region of Vendée (along the Atlantic seaboard), a half-million Frenchmen were slaughtered for rebelling against the new central authority in Paris.[6]

The mechanics of revolution may lie in the obedience-to-authority syndrome. The prerequisite for a revolution is massive discontent, combined with a leader who demands a new configuration of government—a new set of interior lines of authority. As the majority of citizens are muddle-through followers, the moment when allegiance changes hands from the existing authority to the would-be revolutionist marks a critical point, at which time the normally dormant energy of the citizenry is unleashed. But an entire population rarely switches allegiance at the same time; hence, the two sets of contravening interior lines (establishment and revolutionary) ensure the catastrophe of revolution will be accompanied by chaotic conditions. This was evident in Vietnam (a revolution and a civil war), if we take note that North and South Vietnam were artificially created as separate states after World War II.[7] The government in Saigon attempted to radiate its authority to the countryside, while the Viet Minh (and later the Viet Cong) had other ideas.

CIVIL WAR

Civil war is a variation on revolution in which the contending forces have sufficient clout and resolution to engage in a more or less conventional war until one side prevails, or both sides grow weary from the effort and mutually sue for peace. However, unlike wars of aggrandizement, civil war is usually impelled by an ideological force or moral issue that bears no satisfactory compromise, for example, the American Civil War. That conflict was fought over a distinctly moral rather than a political issue, namely, slavery (constitutionally legal at the time). Outwardly, the Union and the Confederacy contended over the right of a state to secede, but that conflict would never have arisen to the height it did without the slavery issue. Moreover, the ideological nature of conflict in civil war tends to transcend the bloodshed to the point where nations often glory over the battlefields as if they were shrines.

On the other hand, civil war is not quite the same thing as isolated guerrilla warfare or terrorist ambushes that may endure for decades, though practitioners of civil conflict may rely heavily on such tactics. For that matter, such tactics have been common in many international wars. This leads to the question: What is the *realpolitik* difference between civil and international war? The difference is that in civil war one side lacks sovereignty or international recognition of claimed sovereignty, whereas in international war the contending parties almost always have equal standing in the eyes of the world community.

This distinction makes little difference on the immediate battlefield; it can have a major influence on the outcome of the war. For example, France recognized the United States during the Revolutionary War and provided military assistance on several crucial occasions (though the motivation was more to hurt Great Britain than to aid the new Americans). By contrast, the Confederacy was not able to obtain recognition from a single country, and that lack severely hampered her ability to trade for arms abroad. As it was, Lincoln went to great lengths to assuage Britain over the *Trent* affair, lest the mother country extend recognition to the Confederacy.[8] In systems terms, the side lacking international recognition is "behind the power curve," whereas the side enjoying that recognition "rides the power curve." That is, an internationally isolated contender must strive much harder to exert the same power as does its opponent drawing on de facto (if not de jure) alliances.

Another difference in civil war is that the war-initiating or rebelling side typically lacks the infrastructure and experience of established government, though this lack does not necessarily ordain failure. On the contrary, consider the total overrun of China by the insurrectionists under Mao Zedong in 1949. The Nationalist government was rotten, did not have effective leadership (despite the public stature of Chiang Kai-shek), lacked an effective armed force, and were up against a masterful leader who developed a formidable army over the preceding 16 years. Similarly, the Falange party under General Francisco Franco defeated the ruling Loyalist government in Spain in less than 3 years, albeit with considerable if clandestine international aid. Still, while successful revolutions seem to be landmarks of history, at least five and probably more attempted revolutions have failed for every one that succeeded.

In short, civil war is a case of a homeostatic political reaction carried to an extreme degree, but in order to succeed the rebelling opponent must have a sufficient mass of public support (clout), or the interior lines of existing national authority must be so weak or distended that the mere leverage exerted by the insurrectionists is sufficient to bring the house down.

RADICALS, REVOLUTIONISTS, AND REFORMERS

Although the descriptions in this section are arbitrary, they recognize a progression in types of leadership with respect to implementing major change in the way a government operates. The key to this ladder is the form of authority the individual possesses as a means of implementing change—the degree to which he can harness, impose himself, or create new internal lines of authority. The first rung is the venue of the *radical*, who advocates massive change but lacks the authority to implement those changes except by moral suasion and the force of his words. Henry David Thoreau is the archetypal example here.

The next rung up is occupied by the *revolutionist*. A revolutionist is a radical who assumes or otherwise gains de facto power by way of personal leadership. This situation requires unusual circumstances if not a critical point in the chaos of the polity. And because the overriding human tendency is obedience to authority, a radical will *not* gain obedience unless his potential followers see no other recourse. The revolutionist must embody the aspirations of those whom he would lead or at least give that appearance. He may mold those aspirations or otherwise cause them to shift direction, but first he must lay hold of his erstwhile followers. Mostafa Rejai put the case this way:

> Revolutionary ideology has three components: (1) a thorough-going critique of the existing order as inhuman and immoral; (2) a depiction of an alternative, superior order embodying positive values of liberty, equality, and fraternity; and (3) a statement of plans and programs intended to realize the alternative order. Deploying such an ideology, the professional revolutionary undermines the confidence and morale of the ruling regime, rationalizes and legitimizes the need for revolution, politicizes and mobilizes the masses, enhances the followers' sense of cohesion and solidarity, and elicits commitment and devotion.[9]

Of such revolutionaries, Mahatma Gandhi is probably the best known and probably the most benevolent example. Though Gandhi never possessed a whit of formal authority, he clearly exemplified the description cited above. In so doing, he gained the freedom of some 400 million people from British colonialism.

On the highest rung we find the *reformer*, who may or may not be a revolutionist at heart. In either event, the reformer always gains formal political office legitimately or comes to be recognized as having formal authority by those in a position to sustain it. This is not quite the same thing as the revolutionary who overthrows a government and assumes power himself. Rather, the reformer operates within the existing system. Understandably, the system must have degenerated significantly to tolerate that much change, but it does happen. The history of Great Britain is replete with bona fide reformers and, for the

United States, Presidents Andrew Jackson, Abraham Lincoln, Theodore Roosevelt, and Franklin D. Roosevelt were all reformers of the ideological bent. Lincoln in particular imposed his will on the entire nation at a cost of 660,000 lives.

□ □ □

In summary, most nations strive to stabilize their internal polity (equilibrium), and rely on various techniques (homeostasis) to keep the most unruly and recalcitrant elements under reasonable control. But sometimes the dynamics of this political homeostasis cannot contend with excessive malcontent, leading to the political-science application of catastrophe, namely, a revolution. And if a nation learns from its experience, as demonstrated by a more lasting and resilient form of government, then it could be said to be acting cybernetically. As George Will concluded, nations are organisms, not machines.[10]

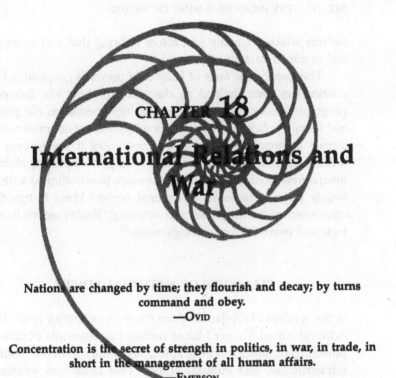

CHAPTER 18

International Relations and War

Nations are changed by time; they flourish and decay; by turns
command and obey.
—OVID

Concentration is the secret of strength in politics, in war, in trade, in
short in the management of all human affairs.
—EMERSON

This chapter concentrates on the relationships among nations.
In the absence of an effective supranational power, and given the
frequent appetite for aggrandizement of neighbors, arguably the
most prominent theme in modern history has been the attempt
to maintain an international balance of power. Technically, then,
every nation comprises a system in itself existing in the larger
system of the geopolitical world. Though that larger system
exerts little direct control over its elemental nations, some kind
of *equilibrium* is necessary, typically through balance-of-power
treaties. To be sure, these treaties often fail to prevent war and its
ensuing chaos, but it is equally true that once war breaks out, the

341

nations affected usually join forces to bring that war to an end and restore a balance of power.

The inextricable facet of balance of power is geopolitics. Geopolitics has been defined as the study of how the factors of geography, economics, and demographics operate on the politics and foreign policy of a nation. In practice, many if not most nations pursue a foreign policy that underestimates if not ignores the salient facts of national power. They attempt to impose an abstract, international configuration of relationships poorly aligned with the largely physical elements of national power.[1] Hans Morganthau once summarized this misfit by concluding: "Reality asserts its own logic and overrides legal arrangements."[2]

ELEMENTS OF NATIONAL POWER

In the seminal *The Influence of Sea Power Upon History 1660–1783*, Admiral Alfred Thayer Mahan outlined six elements of national power: geographic, demographic, economic, military, political infrastructure, and cultural ethos.[3] That book was written in 1890, and since that time his leading contemporary advocate added only the category of technology.[4] However, and like all nonlinear systems, the magnitude of a nation's power is not the sum of its elements but rather the effect of their configuration. This also means that national power will vary with time as the elements change, or even when the significance of an existing element changes, for example, oil. Furthermore, a comparatively minor change in one element can result in a radical increase or decrease of international influence, e.g., the opening of the Panama Canal. Let us review these elements in more detail.

The element of *geography* includes size or mass, terrain, weather, access to seaports, position with respect to other nations, degree of insularity (if any), and natural resources. It is the least changeable element and arguably the single most influential one in world affairs.

The *demographic* element refers to the size and distribution of the population but also implies the ratio of population to land mass and resources. For example, China's population is roughly four times that of the United States yet it occupies approximately the same land mass. The consequence is that China must use the bulk of her resources for survival. An aging population is also a strain because seniors absorb enormous funds and resources without contributing to the production of wealth.

The *economic* element—financial clout—is measured by gross national or domestic product, but the percentage of those products exceeding survival-level requirements is more important, as is a country's financial position in the world market and its balance of trade.

The *technology* element is scientific prowess—the ability to elicit more wealth from available resources. It may include sophisticated innovations like a supercomputer or something as simple as an interstate highway that ensures efficient distribution of resources.

The *military* element is the easiest to visualize, but in this instance it refers to available force above the constabulary level and the ability to project that force if necessary. Keep in mind that a nation is limited in the amount of military force it can field without weakening the underlying economy. During World War II, Washington made the decision to form no more than 90 army divisions even if it meant prolonging the war. The rationale was that a higher number would have impeded civilian industry (which was furnishing roughly half of all war materials used by the Allied powers).[5]

The *political infrastructure* element refers to the interior lines of politics and the laws within which the polity operates. Dictatorships tend to be more efficient but proscribe human initiative. Democracies, by contrast, are inefficient but put a premium on the exercise of initiative. In the short run, a dictatorship easily exerts more influence, whereas in the long run democratic infrastructure typically proves more resilient.

The *ethos* element is the collective or cultural mind-set—the group ego—of a country beyond the authority of the government as an artificial person at law. It is the most abstract and elusive element of national power, though sometimes the most influential. The most common example of ethos occurs in countries dominated by a single religion, e.g., Islamic nations and Israel. Elsewhere, the Japanese (until 1945) were so fanatically loyal to their country and their emperor that soldiers readily committed suicide rather than submit to capture.[6] By contrast, the United States lacks a strong, singular ethos, except by encouraging pursuits of self-interest. This situation creates an enormous reservoir of potential capability to draw on in crises, but otherwise aids and abets social chaos.

DYNAMICS OF INTERNATIONAL RELATIONS

Every nation has its own interests, and it will do whatever it can do—or what it thinks it can get away with—to advance and protect those interests. Whether a nation succeeds depends on how well it plays the game. In the 19th century, Lord Palmerston said: "Great Britain has no permanent friends, she has no permanent enemies, she has only permanent interests."[7] That is to say, nations align themselves with or against other nations to the extent such alignments serve national interests. When those interests change, so too will her alignments. The *manner* in which a nation advances its interests depends on the ratio of complementary to conflicting interests. The ancient formula is:

$$\text{propensity to use force} = \frac{\text{conflicting interests}}{\text{complementary interests} + \text{conflicting interests}}$$

To be sure, this formula does not lend itself to precise numerical calculations, but the idea remains valid: The propensity to use force is a function of the degree to which conflicting interests begin to overshadow complementary interests.

When that ratio indicates a preponderance of conflicting interests, the decision of a nation to actually *use* force (military or economic) to advance its own interests depends on the comparative national strength of the contending parties (including any alliances) and the likely cost of exercising that power. Montesquieu said that the true strength of a prince is not his ability to conquer his neighbor but the difficulty his neighbor finds in attacking him.[8] So if a nation is fortunate enough to realize when would-be aggrandizement would position itself out on an international limb (read: culminating point), it may think twice before proceeding. Otherwise, she might invite counterinvasion or devastating, retaliatory economic sanctions.

Paul Kennedy's *The Rise and Fall of the Great Powers* suggests that this kind of insight is rare. The reason is the underestimation of what it takes to subdue or otherwise browbeat other nations into subservience.[9] Military force is effective in defending against aggression, the general rule being that the attacker must have *at least* three times the defender's military strength in order to succeed, or what is the same, a defender needs only one-third of an aggressor's strength to hold him off.[10] Of course, the actual numbers also depend on terrain, fortifications, and so forth, but the assumption is that the 3:1 ratio is a minimum. In systems terms, the team with the home-field advantage, as it were, possesses interior lines while the invader must pay the exponentially increasing price of maintaining forces as the distance from his logistical base increases. Moreover, if the invaded country possesses great land mass, it can trade space for time until the ratio of comparative effective forces favors them. This is exactly how Russia got rid of both Napoleon and Hitler.

The attempt by the three comparatively small Axis powers to conquer the world (1931–1945) was ludicrous. They succeeded initially because they invested their resources primarily in the production of military force, their opponents were largely unprepared, and they had efficient dictatorial political infrastructures headed by especially effective leaders (assuming that leadership ability has little correlation with ethics). And there is no doubt that Japan's sneak attack on Pearl Harbor in late 1941 was effective, but not the way Japan intended. Instead, the attack merely aroused the United States to war, in which she applied her vastly superior resources. By late 1944, the disproportionate national strength of the combined Allied forces became strikingly evident. The Allies started to sell off surplus material before the war ended, whereas the Axis nations were reduced to wreckage and poverty.[11] Twenty-five hundred years earlier, Macedonia relinquished the spectacular conquests of Alexander the Great shortly after his death.

Furthermore, the geopolitical chessboard is now more stabilized. Despite the skewed distribution of land mass and other elements of national power among 200 countries, none have the power to prevail much beyond their borders. In 1843, Alexis de Tocqueville observed that the United States and Russia were headed for superpower status.[12] And in 1904 Halford J.

57. Leverage versus clout in geopolitics. The Japanese empire badly miscalculated when it declared war on the United States at Pearl Harbor. The United States controlled 10 times the resources and had the political infrastructure to manage those resources with a vengeance.

Mackinder predicted that Russia's land mass would permit her to dominate the Eurasian continent.[13] The United States did become a world power, but she lost a few shirts in the small countries of Vietnam and Somalia. Similarly, Russia became a superpower and exerted a kind of global hegemony but no more.

This global stability (read: equilibrium) is enhanced by the collective, nonaggressive-but-fiercely-defensive ethos of nations formerly within the British Empire. Great Britain misjudged the culminating point of leveraging its navy and industrial strength to colonize much of the world, and in time that empire disintegrated. Still, Great Britain, the United States, Canada, Australia, and India (officially) all speak the same language, facilitating the interior lines of mutual communications. Moreover, English has become the international language of science and foreign affairs, vice French. Lastly, English jurisprudence and England's emphasis on higher education still exerts a subtle (or not so subtle) international influence. Mahatma Gandhi learned about civil disobedience while studying at Oxford.

THREATS AND ALLIANCES

The obverse side of aggressive or other untoward conduct is that for every such act by one nation, at least one other nation will be placed on the defensive. Still, defense requirements depend on the nature of the threat, which themselves can be categorized into four levels.[14] The first level is *defense of homeland and territories*. This is the clearest danger because land will be occupied or at least rendered unusable. When a state cannot defend against this level of threat, it will not long remain a state.

The second level consists of *economic threats*. Economic warfare can take many forms, ranging among: cartels, maintaining artificially high prices (on specific commodities), trade wars, subsidized industry leading to trade imbalances, and gaining effective control over another nation's economy (at least a distinct slice of it).

The third level concerns *imbalance of geopolitical power*. Th situation almost always begins as a potential threat long before erupts into an actual one. The genesis is typically the act of tw or more nations combining in alliance to the point of exercisin hegemony over their neighbors. In reaction, the affected sove eignties attempt to form counteralliances.

And the fourth level is the subjective one of *perceive ideological threats*. Ideology has led to more human destructio than all other causes combined. In former times, the emphas was on religious dogma. In modern times, one only nee consider the wars—hot and cold—engendered by communis and fear of communism.

All threats can be countered, if not defeated, by force some kind, and the efficacy of that force is almost alway strengthened by alliances, at least with respect to the objectiv for which the alliance was formed. For example, NATO succes fully deterred any potential Soviet threat during the so-calle cold war, but that alliance has become something of an albatros In the absence of an overbearing Soviet threat, NATO can on try to persuade member nations to support military objectiv elsewhere that they may not want to support, and often refuse do so (e.g., Bosnia).

Alliances trace back to ancient times, and they gaine impetus during the time of Napoleon. Nations that considere themselves adversaries were threatened by the prowess of th bona fide military genius and were therefore obliged to jo forces to stop him. When Napoleon's reign was finally broug to an end at Waterloo, various statesmen—especially Austria Metternich—sought to create a more permanent balance power in what became known as the Concert of Europe. No withstanding the Bismarckian wars of German consolidatic and several other peripheral wars, the concept held for 99 year

Kaiser Wilhelm II had different ideas. He initiated the first the global world wars of this century, wherein two allianc battled one another: the Triple Alliance (Germany, Austri Hungary, and Turkey) and the Triple Entente (France, Gre Britain, and Russia). The Triple Alliance eventually broke u and the Triple Entente changed when the defeated Russia dropped out and, several years later, the United States joined with France and Great Britain, which gave the former Entente the necessary edge to force Germany to capitulate.

Twenty years later, the business of wartime alliances expanded to global dimensions and, on one occasion, impelled Sir Winston Churchill to remark that he intended to get in league with the Devil (Stalin) in order to drive out Satan (Hitler). After the war, the alliances re-formed as defensive treaties, e.g., the North Atlantic Treaty Organization (NATO) and the Australia–New Zealand–United States treaty (ANZUS). In 1990–1991, some 28 nations allied themselves briefly to expel Saddam Hussein from Kuwait, especially as he was in a position to invade Saudi Arabia and tie up much of the world oil supply.

CRUCIAL CONFIGURATIONS FOR THE ONSET OF WAR

Major wars give the appearance of sudden onset, but the onset is only a critical point reached after a long period of unfavorable change in part of the international geopolitical configuration. This formative period may take a generation or more, for example, World War II (at least in Europe) was said to have arisen from the reparations forced on Germany at the end of World War I. On hearing of the terms of the Peace Treaty at Versailles, Marshal Foch remarked: "This is not peace; it is an armistice for twenty years."[15] In the Pacific theater, Japan's sneak attack on the United States at Pearl Harbor was foreseen by General "Billy" Mitchell and others as early as 1925. Moreover, Fleet Admiral Ernest J. King—the Chief of Naval Operations during World War II—staged two mock attacks on Pearl Harbor in the 1930s to prove the point.[16]

This same formative process also applies to many if not most civil wars. For the United States, the break with the mother country brewed for 20 years, while in the following century the antislavery mood of the northern states was well entrenched by the 1820s. Consequently, it is all but impossible to prevent major war once the brew has passed a certain point; the configuration takes on a life of its own and is bound to be resolved even at the cost of tremendous bloodshed.

In this regard, Sir Winston Churchill concluded that World War II had been the most preventable of wars if wise decisions had been taken a generation earlier at the end of a conflict in which the combatants had lost millions of men and did not want to think about future war.[17] That is a very big *if*. Philip of Macedonia once threatened Sparta that if he ever entered Laconia, he would grind Lacedaemon into dust. Sparta replied with a single word: "If."

Still, we must ask how the *catastrophe* of war operates. Why does the harmful configuration of elements take so long to form, only to explode in a comparatively short time? Why can't we see it coming if the signals are so clear? The second question is easier to answer. The catastrophic process is ill understood, even when it occurs in natural disasters. It does not seem logical that a seemingly minor incident can trigger a momentous epoch. Instead, we think in terms of sequential cause and effect. From that arithmetic perspective, war can only be the endpoint of a long, steady progression of events that can, in theory, be stopped at any intermediate point. Also, the human mind prefers not to dwell on the potential of great tragedy. In a scene in the television miniseries *The Holocaust*, the wife of the protagonist refuses to believe that Hitler intends to persecute the Jews. She exclaims such a movement could not occur in the land of Goethe and Schiller. He responds that, unfortunately, those people were not then the ones in power.

Answering the other question—why it takes so long for potential war to brew silently before erupting suddenly—is no different than the explanation for all catastrophic phenomena. It took time for nature to align—configure—disparate physical elements to the point where they could initiate evolution, and, at that, the process seems possible only on a small percentage of planets in the universe. In the sociological parallel, major wars involve a massive realignment of the elements of national power. But to effect that realignment requires a major sticking point—

typically some irreconcilable issue on which the opposing parties refuse to negotiate. All the what-if doubts must be overcome.

Irreconcilable issues tend to arise from greed when a nation wants to possess or otherwise control more than encompassed by its internationally recognized sovereignty. It follows that most international wars arise from attempts by one or more nations to aggrandize others. Hence, the leaders of a nation and its citizens align themselves with respect to the issue at hand. The interior lines of communication bespeak a more singular message. Physical and economic resources are sharply focused on the tools of war. In time, the momentum of this activity overrides caution until some seemingly minor event becomes "the straw that breaks the camel's back."

On this matter, it is interesting to note just how helpless the participants of World War I were to control events after Archduke Ferdinand was assassinated at Sarajevo on June 14, 1914. The story of the famous international chain reaction has been told many times, though some interpretations posit that this war could have been prevented anywhere along the chain. On the other hand, Germany had planned to conquer France since 1905, and strong nationalistic antagonisms against neighboring states abided in the minds of most of the ruling monarchs.[18] Only Great Britain initially stood outside the stew-pot, and she was obliged to come to the aid of France once war commenced.

NATIONAL PURPOSE VERSUS THE CONDUCT OF WAR

Although Clausewitz's most famous dictum was that war is a means of carrying out national policy by other means, he also observed that the conduct of war tends to run amok.[19] This means that once a nation is engaged in war, that nation will tend to pursue hostilities even when the conflict is no longer consistent with the national purpose for which it was initiated or joined, if in fact a clear purpose had ever been stated.

What, then, accounts for this collective malignancy? It is not enough to state the obvious, namely, that human emotions often

erupt out of control. Why do they lose control? The explanatio
seems to reside in a number of mutually reinforcing factors. Th
first of these factors is the *relationship of purpose and militar
objectives*. This means the degree to which military capabilitie
and national purpose fail to mesh. For example, when a countr
is invaded and the purpose of war is to expel the invader b
force, that purpose can only be effected by military means. A
the other extreme, military force has never been particularl
effective in curbing the spread of what is perceived as a
inimical ideology.

The second factor arises from the *virtue of heroism*: admir
tion for courage on the battlefield. This attribute is a true virtu
but in war it is mass that prevails over time, not the occasion
leverage obtained by the action of genuine heros. But quitting
battlefield or otherwise withdrawing from a war that cannot b
won gives the appearance of cowardice. Yet if the continuanc
leads only to more casualties with nothing to show for it, of wh
value is the heroism? Masada was a rare exception; few othe
qualify.

The third factor is *perseverance*, which is strongly tied t
courage. Perseverance is the momentum of the human psych
When coupled with heroism, it may easily cause a nation t
continue a conflict that is no longer winnable.

The fourth factor is *ethnocentrism*. Belief in one's country
probably essential for survival, but this loyalty is easily carried
extremes in the face of ideological tenets. Combining this and th
previous points, B. H. Liddell Hart said that peaceful natior
tended to "court unnecessary dangers." Once into the fray, Ha
continued, such nations were more likely "to proceed to extrem
than predatory nations." Predatory nations made "war as
means of gain" and have been inclined to back out "when the
find an opponent too strong to easily overcome." He then co
cluded that is was "the reluctant fighter, impelled by emotion an
not by calculation, who tends to press a fight to the bitter end."

The fifth factor comes under the rubric of *external chang
The strategic configuration of a war can change suddenly wit
out a nation recognizing it; hence, that nation continues its o

strategy despite factors that mandate a new strategy. The classic case of this occurred during the Korean conflict, 1950–1953. This conflict was actually two distinct wars. The first war entailed the expulsion of North Korean forces from South Korea, an objective that was achieved by the end of September 1950, roughly 3 months after the war began. This rousing success encouraged both President Truman and the Joint Chiefs of Staff to extend the war into North Korea in order to subdue that country permanently. Once the war extended to her borders, China felt threatened and accordingly threw millions of her own men into the fray. At that point, U.N. forces—primarily the United States— were engaged in a new war with China. The only thing that was the same was the terrain, but Washington failed to take this radical change into account. The result was that the United States endured the war for another two and a half years with little to show for it.

Elsewhere, independent changes in the international environment or the internal polity can negate the purpose of war or, alternatively, make pursuit of an existing war all the more imperative. The most significant case of this was the rapprochement of the United States and China—*détente*—that deflated the justification for continuing the war in Vietnam, i.e., stopping the spread of communism. True enough, the increasing lack of internal support had a bigger influence in bringing U.S. participation to an end, but that is besides the point.

TERRORISM AS A FORM OF WAR

The practice of terrorism is a form of warfare that depends almost exclusively on leverage. Furthermore, it is about the only option available when the initiating party has no significant strength except stealth with respect to its adversary. Accordingly, the ultimate target is *not* the object struck. Far from it. Short of using clandestine weapons of mass destruction, the actual physical damage or casualties inflicted are minuscule compared to conventional warfare. By way of another comparison, the human toll from terrorism seldom exceeds 10 percent of the innocent victims killed by drunk drivers during the same period of time, and the percentage is usually much lower.[21] Instead, the real targets of terrorists are the interior lines of the political infrastructure and cultural ethos superintending or otherwise having a vested interest in the physical targets. The terrorist's goal is to persuade a polity to change its attitude in favor of whatever he wants.

Aside from the use of terrorism as an instrument of crime (typically extortion) or as a variation on tactics within a conventional conflict, acts of terrorism can be divided into three categories. The first category consist of acts committed by so-called *radical groups*. The intent of these groups is to gain international recognition of various causes and/or obtain funds. Tactics include assassination, kidnaping, hijacking, and sabotage. The common denominator is that its practitioners believe such conduct is the only means to advance their cause. Still, they rarely succeed. International order poses an inherent resilience against radicals and, moreover, acts of terrorism inspire loathing because they are usually directed at innocent victims.

The second category consists of acts under *state-sponsored terrorism*. This category operates similar to a radical group, except that it is controlled and usually funded by a sovereign power and acts under the aegis of that power or is permitted to do so without prosecution. The aims of the government supporting it may or may not coincide with the aims of the terrorists, but that subtle point is secondary to the fact those members serve as a de facto military arm of a state. Their effectiveness is greater than occurs with a radical group because the ultimate target is usually a specific country, the focus is stronger, and they can more easily obtain resources and shelter.

The third category of terrorism is the product of *insurrectionists* or *rebels*. Organized rebellion and revolution against a sovereign power is as old as recorded history. In terms of terrorist activities, the tendency of the insurrectionist is to use guerrilla warfare as a prelude to more conventional tactics. Even when he does not, the success rate tends to be the highest among the three categories because many neighbors will prove sympathetic to the cause or can be cowed into submission. For proof of this, consider the endurance of the Irish Republican Army (IRA) in Ireland and the Shining Path insurrectionist movement in Peru.

POTENTIAL CATASTROPHE OF OVERPOPULATION

Nature has no qualms about triggering catastrophes, for example, natural disasters, famines, and epidemics. In the 14th century, the Black Plague wiped out a third of Europe's entire population. Supposedly, man now has the technological wherewithal to reduce the effect of such catastrophes. For example, we have virtually eliminated smallpox, a disease for which no *known* case has been documented in 15 years. On the other hand, the exponentially increasing world population presents a more difficult challenge, namely, that the resources of the earth are barely able to sustain the present five and one half billion people. What will happen if this growth continues and reaches a level of, say, 15 billion, by which time we will have severely depleted existing resources? Clearly, this would constitute a crisis in international relations.

And should this happen, billions of people would die prematurely until the earth could again sustain what was left of the human population. The carnage would make genocide seem like a misdemeanor. Imagine *five* large football stadiums, each filled to the brim with coffins (the entire space, not just the seats), a new set every hour of every day for an entire year. That is the magnitude of reducing world population from 15 to 5 billion.

How this potential catastrophe might pick up steam, as it were, is in part speculation. Certainly, malnutrition would play a major role, as would the evolution of viruses and other microorganisms resistant to most if not all known antibiotics and vaccines. Even if viable antidotes could be developed, the mass of deaths would overwhelm medical providers and available supplies. But can't we see this coming and do something about it before it is too late? In theory, yes, but catastrophes seldom give advance notice in terms that will be heeded. For the most part, people seem to have inured themselves to human suffering, which means that moderate increases in mortality would not sound an alarm clear enough to motivate governments to take necessary corrective actions (despite the notable success of Rachel Carson's *Silent Spring*). Hence, as in all catastrophes, the critical point will eventually be reached, followed by a predictable chain reaction of deadly epidemics and malnutrition.

Ironically, the commendable global efforts to reduce childhood diseases and malnutrition will only serve to accelerate the population growth rate. By contrast, birth control *could* reduce the growth rate to zero, but that approach does not seem to be a priority, except in China. China, which is faced with a population of 1.2 billion, gives birth control very high priority (often via Draconian measures).[22]

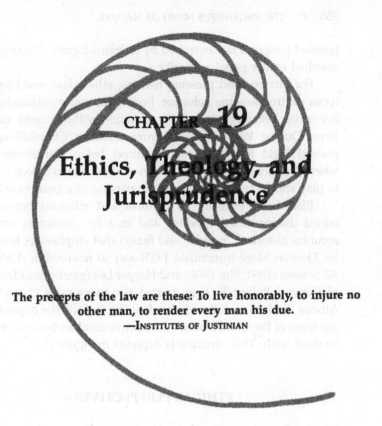

CHAPTER 19

Ethics, Theology, and Jurisprudence

> The precepts of the law are these: To live honorably, to injure no
> other man, to render every man his due.
> —INSTITUTES OF JUSTINIAN

Jurisprudence strives to bring order to the chaos of human
behavior that arises whenever two or more persons live or work
within proximity—an abstract system imposed over groups, nations, and other sociological systems to the point where it is also
considered sociological. Actually, jurisprudence encompasses two
distinct abstract systems: ethics and law. Ethical systems address
human conduct but rely only on moral suasion or peer pressure.
Reliance on moral suasion usually leads to anarchy; hence, legal
systems codify portions of ethical perspectives, which can be
enforced by executives. Over this, we find theological systems
insofar as theology prescribes ethical behavior, and, in times past,

justified temporal law exercised by religious figures. No king ever matched earlier papal authority.

For obvious and pressing reasons, ethics, law, and theology focus on troublesome behavior. People recognize extraordinarily evocative behavior—especially courage—without need of systems. During World War II, Prime Minister Churchill openly praised Field Marshal Erwin Rommel before Parliament, and when President Franklin D. Roosevelt died, Radio Tokyo paused to play special music "to honor the passing of a great man."[1]

Elsewhere, virtually all cultures and religions honor sustained distinguished conduct, and in a few instances produce accurate historical "novels" and fiction that emphasizes integrity. Sir Thomas More (canonized 1935) was so honored in *A Man for All Seasons* (1960, film 1966), and Harper Lee (great-granddaughter of General Robert E. Lee) created the memorable character of Atticus Finch in *To Kill a Mockingbird* (1960). Still, the overwhelming mass of literature stresses destructive conduct because it must be dealt with. This situation is depicted in Figure 58.

ETHICAL PERSPECTIVES

People do not, and perhaps cannot, agree on ethics. At the extreme, some observers postulate that ethics do not exist or, alternatively, that ethics are solely a matter of personal choice. Accordingly, philosophers and ethicists have quarreled for millennia over how fixed or fluid ethical systems can be. Many of their arguments arise from differing sociological perspectives. For example, *individual ethics* focus on personal responsibility for one's own conduct and behavior, holding that a sense of ethics is necessary in order to avoid drifting into hedonism. The law alone does not suffice and, indeed, there are moments when personal ethics compel, or should compel, civil disobedience.[2] Elsewhere, *family ethics* come close to individual ethics, though in practice conflicts between and among parents and children occur frequently. Moreover, children do not have the same rights as adults, and adults are obliged to provide a level of care for their

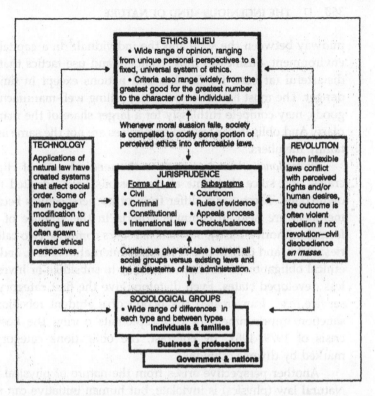

58. System aspects of ethics. Ethics is an abstract system superimposed over the mass of sociological systems, from which the conduct of men and nations can be evaluated and judged. When conflicts among people and sociological systems cannot be resolved by moral suasion, ethics are codified into enforceable systems of law.

children that they are not similarly obliged to offer to strangers and, in many regions and cultural groups, to more distant relatives.

Under *ethics of state*, every state is obliged to care for its members to one degree or another. Therefore, the state must seek a compromise between ideal notions versus the implied obligation to provide the greatest good for the greatest number. Next, presume that *corporate, organizational, and marketplace ethics* lie midway between the state and the individual.[3] In a capitalistic environment, businesses must compete and use tactics that are distasteful (at best) for families and nations except in time of danger. The most honest corporation, selling well-manufactured goods, may compete ruthlessly for a larger share of the marketplace. And obligations toward employees are not the same as for family members.

Professional ethics date back to the medical oath of Hippocrates, and since that time all professions have adopted their own codes. These codes differ from other perspectives because members are held to common criteria in the absence of any formal authority. Lastly, *international ethics* comprise two categories: rights and obligations created by international law, and the ethical obligation of developed nations to subsidize or invest in less developed states. Even dictators give the first category lip service, e.g., Iran created the fiction of a student rebellion to sanction imprisonment of U.S. diplomats during the hostage crisis of 1979–1981. By contrast, the obligations category is marked by diverse opinions.

Another perspective arises from the nature of physical law. Natural law (physics) is inviolate, but human initiative can rearrange physical systems (technology) and thus change outcomes. Human initiative can also modify any sociological system, including a system of law intended to regulate and throttle individual behavior. Still, all changes in systems incur consequences. A notable passage from the film *Inherit the Wind* tells the story:

> Gentlemen, progress has never been a bargain. You've got to pay for it. Sometimes I think there's a man behind a counter who says, "All right, you can have a telephone; but you'll have to give up privacy, the charm of distance. . . . Mister, you may conquer the air, but the birds will lose their wonder and the clouds will smell of gasoline.[4]

That is to say, any person with the necessary physical freedom can do anything he wants short of repealing a natural law, but he must suffer the consequences of bad judgment inflicted by the homeostatic reaction of the physical or sociological environment as the case may be. Edward Tenner's *Why*

Things Bite Back: Technology and the Revenge of Unintended Consequences (1996) describes this process in some detail, with special emphasis on medical technology, the modern office, sports, and environmental initiative.[5]

HUMAN INTENT AND ETHICAL CRITERIA

We evaluate—and judge—human behavior by mandate in legal cases and oft by necessity outside the courtroom. When Emmanuel Sieyès was asked what he did during the French Reign of Terror (1792–1795), he replied "I survived." (*"J'ai vécu."*) By what criteria should we evaluate individual behavior? One popular criterion is the personal *intent* (or its lack) to do wrong. In fact, a defendant's intent is a crucial element in most criminal trials. Elsewhere, we can arbitrarily postulate four levels of behavior as a function of intent.[6]

The first level is the *unavoidable side effect*, which evolves from doing what seems right or profitable in an honest way. At times, participants may be aware of those effects but can do little about them, and in general it is all but impossible to hold individuals responsible for any consequences. When you drive a car, you pollute the air, but you won't go to jail.

The second level is *opportunistic misuse*. This arises when an opportunity has a legitimate purpose but is used under questionable circumstances, for example, checking credit reports (legitimate) for the purpose of hiring (illegal or questionable). Opportunistic misuse also occurs when acts are committed simply because the opportunity presents itself with little chance of being caught, for example, taking money from a wallet found in the street when that wallet contains the owner's name and address.

The third level is *systematic exploitation*—misuse expanded to systemic proportions, for example, an employer who lowers wages to a subsistence level when jobs are scarce, solely to increase profit. Although this is not a crime, many observers see it as unethical and blameworthy. The intent to do wrong is clear, but the conduct remains entangled in gray areas between law and ethics until the consequences rouse public sentiment to the point of drafting new legislation, e.g., the first minimum wage law (1940).

The fourth level is *malicious intent*, whereby wrongdoing crosses the line into conduct that is unmistakably criminal and/or subject to punitive damages in civil court. Virtually all felonies involve, or are construed to involve, malicious intent. Willful and culpable negligence also come under this heading.

Intent, by definition, is attributable only to individuals. By contrast, sociological systems are concerned with the common good and must evaluate ethical rules and guidelines from that perspective. To do so, we must have criteria. One of these criteria focuses on the *consequences of an act*. When a person sets off a false alarm and, just as the firemen arrive, a flash fire breaks out in a nearby building so fierce that only the presence of the firemen saves the occupants, the consequences of the wrongful act are favorable. Conversely, when a person sacrifices his life trying to save another person but in the process leaves his family destitute, the consequences of an inherently good act are mostly unfavorable, especially if the attempted rescue fails. Still, sociological systems must consider numbers. For every false alarm with a happy outcome, hundreds of others put innocent people at risk. Therefore, society deems false alarms as unethical—and illegal.

The second criterion is *character*. When a great person does a little wrong, we typically overlook it in extenuation of his or her reputation or it is presumed to be the inevitable consequence of living. When wrongs accumulate or an especially bad judgment leads to ruination, the saving grace of character loses its effect.

The third criterion is sometimes called *fixed versus situational morals*. A fixed moral defines behavior as right and wrong. By contrast, a situational moral implies the partial absence of standards. Obviously, fixed morals lack flexibility to accommodate human nature, while situational morals can be used to rationalize any act of human conduct. Yet, we cannot logically alloy these two dichotomous polemics.

The fourth criterion is the well-known *greatest-good-for-the-greatest-number* idea, which earns credence because by definition it has mass on its side. In an imperfect world, decisions that render the most good while incurring the least "bad" earn wide acceptance as a compromise. This is especially true if the interests of the minority are protected.

The fifth criterion, *means to an end*, determines if wrongful means are justified if the outcome is favorable. When the means are not especially wrongful and the benefits are significant and lasting, the issue is moot. Social etiquette even praises it as the so-called white lie. When the opposite situation prevails, we may severely criticize it. Between these polemics, most cases are a matter of degree and are often decided on the consequences.

The sixth criterion is called *dilemma ethics*, and applies when *all* options in a situation have serious side effects, yet that situation mandates a decision. For example, if a country has a fixed budget for healthcare, should expensive treatment for the aging be curtailed to provide a much greater amount of preventive care for the young?

Despite the broad sweep of these factors, we can see a common thread running through them all: the consequences of acts and decisions. *Any* change in *any* system results in a different configuration—a consequence. In purely physical systems, consequences are deterministic, hence are *not* subject to evaluation from an ethical viewpoint. In sociological systems, human initiative is a factor, hence consequences *are* subject to ethical review.

THEOLOGICAL SYSTEMS

Several theologians have written multivolume treatises with the same title: *Systematic Theology*, of which the most recent is Paul Tillich's (1951–1963).[7] Understandably, while these theologians employed logic extensively, their efforts could not be objective in the scientific sense of that word. Theology—and religion—is an intensely personal experience. People who differ from this do not write theological treatises, except perhaps to criticize religion, for example, Sigmund Freud's *The Future of an Illusion* (1927). If we can find someone who is willing to take on the task, he or she might begin with the obvious subject, namely, God.

If God (as a Creator) exists, that God would likely have attributes and some relationship with the creation. The relevant scale would declare *pantheism* as one polemic, whereby God and "reasoning" are equated, consisting primarily of the logos underwriting nature, as exemplified by Spinoza's *Ethics* (1648). The other polemic would presume a monotheistic, monolithic God who predestines human behavior and foreordains each person to heaven or hell from the moment of birth, for example, John Calvin's *Institutes of the Christian Religion* (1541–1555). Arguably, every religion and denomination in the world can anchor itself to some point on this scale before veering off on details, except for purely ethical religions (e.g., Confucianism) that do not acknowledge deity in any form.

More importantly, many religions constitute major sociological systems in their own right, especially Roman Catholicism. Then, too, some nations are theocracies, for example, Israel and many Islamic countries. Law in theocratic countries is heavily based on documents declared or presumed to be sacred. Other countries may lack a theocratic basis for law but policy remains heavily influenced by a dominant religion, e.g., Catholicism in Ireland and Italy, and Lutheranism in Sweden. Still other countries are subtly influenced by a pervasive if not dominant religion. The Church of England (Anglicanism) *is* the state church of Great Britain, and the King or Queen is the head of that church. The subtle influence arises from the fact that the plurality if not the majority of Members of Parliament (MPs) are Anglicans.

Elsewhere, some religions have extensive bodies of moral codes. Judaism is without peer because its system enabled the most persecuted minority in history to survive pogroms over several millennia, then reemerge as the state of Israel in 1947. Among Christian denominations, Roman Catholicism has an extensive ethical code accompanied by its own legislative system (canon law). Similar situations are found in other world religions.

Then, too, the role of theology in ethical systems is a popular theme in great literature, for example, "The Grand Inquisitor" in Dostoyevsky's *The Brothers Karamazov.* Elsewhere, consider Will and Ariel Durant's imaginary dialogue between Pope Benedict XIV and Voltaire, which illustrates the problems of freewheeling ethical codes in light of the Church's strictures on human behavior.[8] And George Bernard Shaw gave the virtues of religion short shrift in his play *Major Barbara,* suggesting that entrepreneurship has done more for the betterment of mankind than all religious piety.

Also consider that the so-called monkey trial of John T. Scopes in 1925 did *not* settle the evolution issue. As late as 1985, a federal appeals court (Judge William K. Overton) had to overturn an Arkansas statute prohibiting evolution texts in the classroom.[9] And in March 1996, the state legislature of Tennessee seriously debated a bill that would prohibit the teaching of evolution in public classrooms as anything more than a theory.[10]

We cannot understand sociological systems unless religious beliefs (or their lack) are considered. Even Sir Isaac Newton devoted more of his time to the study of theology than physics, and at one point the Archbishop of Canterbury remarked that he (Newton) knew more about theology than all his clerics and prelates put together.[11]

ROOTS OF LAW

Conflict in all cultures and nations is inevitable: conflicts between constituents (civil law), conflicts between a constituent and the state (chiefly criminal law), conflicts with a system of law itself (constitutional law), and conflicts between nations or sovereignties (international law). Each form of conflict led to a different body of law, but the common element of all conflict resolution is *justice.* The ultimate meaning of justice hinges on an understanding of rights and obligations essential to the structure and functioning of a state. Thus, the only way to achieve a collective good is to restrain some of the freedoms of the individual and impose obligations on him to boot. Similarly, the operation of the state finds that *its* rights are also circumscribed by obligations. More than one king has literally lost his head by overlooking those obligations. Thus, the ultimate goal of jurisprudence is arguably to keep rights and obligations in dynamic equilibrium.

This equilibrium must take the contending forces of three points into account: the right of an individual to pursue his or her interests, the right of others to be protected from the excesses of the first right, and the obligation of the state to adjudicate conflicts. The first two points ensure continual conflict, and that conflict must be adjudicated if a culture or nation is to endure. The process of maintaining equilibrium was once illustrated by Supreme Court Justice Oliver Wendell Holmes, Jr., when he ruled that the right of free speech does not extend to yelling "fire" in a crowded theater.[12] Absolute rights do not exist; there is a give-and-take to all rights, which means that the pushing and pulling of various parties pursuing their own interests is the stuff of legal equilibrium.

LEGAL SYSTEMS AND THE PROCESS OF LAW

The development of law uses a cybernetic process, or what scholars call an ongoing synthesis. Figure 59 depicts this process, which shows that when cases conflict, resolution interacts logically with legal precedence.[13] Conflict arises from: changing values within the polity, the consequences of new technology, and new findings in science that affect human conduct. Resolution of any and all legal conflicts hinges on several dominant factors.

The first factor is the degree of stability. Law, at least in free-world republics and equivalents, demonstrates high stability and evolves only gradually. Because law applies to the entire citizenry, any radical or sudden change is typically disruptive as

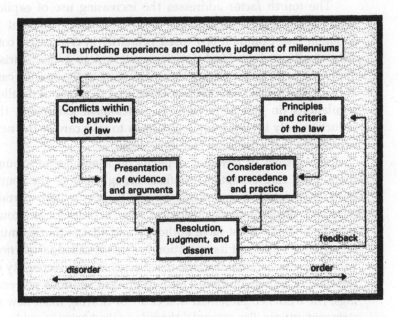

59. Jurisprudence as a dynamic system. Law is perhaps the most profound accomplishment of civilization, but its development can be understood from a simple chart depicting the dynamics of conflict and resolution.

occurs in any stabilized system. The second factor is the role of dissent. When jurists disagree, the minority almost always write one or more dissenting views. Some of these views lodge themselves in jurisprudential thinking and slowly evolve into prevailing opinion.

The third factor concerns the usurpation of legislative and executive roles. Jurists dissatisfied with social problems often issue rulings that go beyond the resolution of conflict and into the prerogatives of the legislative and executive branches of government, which upsets the balance of power among the branches of government (as intended by constitutional law). A well-known case in U.S. jurisprudence comprises the ongoing dictates from the bench on how to run and manage school districts, primarily to enforce racial integration.

The fourth factor addresses the increasing use of explicit statutes and legal codes in lieu of the flexibility inherent in common law. Common law (used primarily in civil cases) consists of all previous legal decisions relevant to a case. Jurists apply the logic underwriting these decisions to the specific case at hand. But when common law is reduced to tomes of detailed regulations (in lieu of relying on professional judgment), the reduction only invites more regulations to take care of the new exceptions that arise from statutory detail.

The system of law also depends on a number of ground rules. With rare exception, conflicts brought before the courts must have resulted in damage or loss to one or both parties: monetary, physical, or psychological. Also, in order for any court to have effect, it must have authority and that authority must prescribe the limits of its operation in what is called *jurisdiction*. Except in international law, the usual aegis is the sovereignty of the state or its constitution. Third, to prevent spurious proceedings, the evidence presented in a court or hearing must hew to stringent criteria. For example, the rule against hearsay evidence outlaws most secondhand testimony.

The courtroom is the focal point of all jurisprudence. In most countries somewhere between 100 and 200 cases of high significance dot the jurisprudential landscape and become part of general history.[14] For the most part, however, the courtroom subsystem is severely restricted and the structure is rigidly codified. For example, the law never fully trusts the judgment of a single individual. The vast majority of lower courts are under the control of a single judge or magistrate, but any decision or action by each judge is subject to appellate review, which *does* provide a panel of jurists to adjudicate the point or points of contention. Then, too, lay juries are charged primarily with deciding matters of fact. That task is usually limited to weighing the evidence presented to determine if it supports the contention of one of the litigants in a civil trial or the indictment in a criminal trial. If a legal point arises during deliberations, the jury is obliged to seek the judge's ruling on it. Lastly, personal freedom is restricted in a courtroom. A person who acts beyond prescribed limits usually finds himself or herself held in contempt.

Courts also recognize the legitimacy of circumstantial evidence. In the everyday world, the term *circumstantial* connotes peripheral if not specious arguments, but in a legal case circumstantial evidence can lead to a conviction when that evidence admits of no other *plausible* explanation, and that includes capital crimes.[15] In physiogenetic terms, think interior lines. The facts, as brought out by evidence, lead to only one, usually simple, logical pattern. Furthermore, courts are quite tolerant of advantage wrought from the timing and sequencing of the presentation of evidence, e.g., tripping up witnesses during cross-examination. In jurisprudence, the path from chaos to truth is not a straight line.

Of special significance are the cases in which momentous social issues are brought before the bar, albeit technically on a matter of lesser weight. In these rare situations, judges tend to overlook the rules of evidence if what is said sheds light on the social conundrum. The most well known of these cases in the United States took place in 1925: the so-called Scopes "monkey" trial. Scopes was brought to trial for violating a Tennessee law that prohibited the teaching of evolution in a public classroom, and his "guilt" was never in doubt. The famous William Jennings Bryan was brought in to conduct the prosecution, and the equally famous criminal lawyer Clarence Darrow was hired as the defense counsel.[16] The play (and subsequent film) *Inherit the Wind* accurately dramatized the court proceedings.

CONSTITUTIONAL AND INTERNATIONAL LAW

From a systems viewpoint, constitutional law is the most interesting facet of jurisprudence because it comprises the dynamics by which a legal system stabilizes itself (a form of homeostasis). That is to say, constitutional law is the process of refining and adapting the legal relationships among the rights, prerogatives, and obligations of its citizens versus its elected and appointed representatives before any conflicts erupt into rebellion and revolution. In the United States, the Supreme Court contributes to the body of constitutional law in almost every ruling. In a few cases, the decision may explicitly address a constitutional issue, but in others the decisions typically add weight to existing interpretations (of the Constitution) or nudge those interpretations slightly.

International law is nearly as interesting as constitutional law because it has evolved over more than 3500 years. No effective supranational authority exists, which means that nations will abide by certain agreements only if it is in their interests to do so. The surprising thing is the large number of points that nations agree on. Even the most destructive dictators usually give the appearance of adhering to international law to avoid international condemnation—a high-level peer pressure. Hitler, for example, set up a "showcase" camp for Jews at Theresienstadt, which the International Red Cross was permitted to inspect.[17]

Early international law was subjective at best, and existed primarily in the form of edicts directing how a state was to conduct business with other states. By medieval times, jurists sensed an unwritten natural law that governed or should govern relations among nations. The most notable book in this era was *On the Law of War and Peace* (1625), written by Hugo Grotius and still in use today. This book set down legitimate reasons for war and argued that the means of war should never exceed the cause to be remedied. In time, these doctrines led to the Geneva Conventions, which in turn provided the authority for the war crimes trials following World War II.

In time, economic interdependence among nations led most sovereignties to realize that they could no longer rely solely on their own authority. Not only environmentalism but a kind of geopolitical ecology operates in international relations that demands more cooperation in the form of adherence to various bodies of international law—a stab at a global equilibrium. Without loss of national sovereignty, of course, which leads to a logical impossibility. Any international law or treaty enforceable by a supranational authority impinges on national sovereignty by definition. In systems terms, international law is an attempt to impose a hierarchical form of control over the distinctly relational form of system inherent among sovereignties competing on a globe from which there is no escape save by extinction. One form or the other must predominate, and that is likely to remain relational.

To be sure, the excesses of various nations will continue to erupt and create lasting problems, not to mention the Malthusian threat of the exponentially increasing world population. Similarly, civil wars will continue in less developed and in ethnically divided countries while the major powers jockey for comparative influence and favorable trade ratios. It is equally certain that advocates of world government will continue to plead for a true supranational authority, unaware that success for a complex system depends as much if not more on homeostasis than hierarchical control. For example, the destructive potential of thermonuclear power motivates many nations to agree to treaties reducing or regulating nuclear weapons, e.g., the Strategic Arms Limitation Treaties (SALT). These treaties cannot guarantee that such weapons will be laid to rest. On the other hand, neither could a world authority, especially if and when such weapons fall into the hands of terrorists.

DETERMINISM VERSUS INITIATIVE

If there is any point that has been emphasized over the last five chapters, it is that sociological systems differ radically from their abstract and physical cousins because of human initiative. But how much initiative an individual can exercise remains an unanswered question. How often does our exercise of initiative lead to frustration? Can man reorder society to reduce the frustration? In *The Fountainhead*, Ayn Rand said that the purpose of civilization is to set man apart from his fellow man. In that vein, let us review the bidding on two factors: historical trends and the entropy of organizations.

60. The issue of initiative. This scene—which evokes memories of Thomas Gray's *Elegy Written in a Country Church-Yard*—seems to ask how much initiative an individual can exercise in the face of the great movements of history. The answer is not always clear, but if initiative is a delusion, we are all Frankensteins disguised by nature's neoplastic surgery.

The unfolding of history seems too massive an enterprise for an individual to exercise much influence beyond personal decisions, except perhaps for a few leaders in times of great crisis. Leo Tolstoy addressed this factor at some length in his second epilogue to *War and Peace*. In that final essay, he wrote that power is a force that moves nations, a power that "is the collective will of the people transferred, by expressed or tacit consent, to their chosen rulers."[18] In turn, this power—so he argued—compelled obedience or was itself compelled to react to external problems. Hence, individual freedom held little sway on the course of history and made the issue of initiative nearly moot. Each man may be a degree of freedom—an independent human being—in the sociological system of the state, but the power of the state, combined with the necessity for group survival, often negates that freedom.

The second factor—organizational entropy—surfaced in the film *Schindler's List*. During World War II, Oskar Schindler's conscience led him to save 1100 Jews from the gas chambers by employing them in a factory, and he made many attempts to save their children too. It is an inspiring story in the midst of the Holocaust, and there were several others like it.[19] Yet these heroes who risked their own lives to save others were rare. In the massively greater scope of the horror they were able to protect only a very few. In short, it is only the unusual individual who will "buck the system." As Justice Holmes put the case:

> I know of no true measure of men except the total of human
> energy which they embody. . . . The final test of this energy
> is battle in some form—actual war—the crush of Arctic
> ice—the fight for mastery in the market or the court. Many
> of those who are remembered have spared themselves this
> supreme trial, and have fostered a faculty at the expense of
> their total life. It is one thing to utter a happy phrase from a
> protected cloister, another to think under fire—to think for
> action upon which great interests depend. The most power-
> ful men are apt to go into the mêlée and fall or come out
> generals.[20]

Carl Jung echoed this thought in writing: "Resistance to the organized mass can be effected only by the man who is as well organized in his individuality as the mass itself."[21]

The bottom line seems to be that in order to beat the "system," one must build his or her own system. Building systems always risks disorder by underestimating the difficulties. In this vein, bear in mind what Sir Winston Churchill said about disorder versus order, namely, that the human story did not always proceed "like a mathematical calculation on the principle that two and two make four." At times, Churchill continued, the human sum of two and two added to "five or minus three, and sometimes the blackboard topples down in the middle of the sum and leaves the class in disorder and the pedagogue with a black eye."[22]

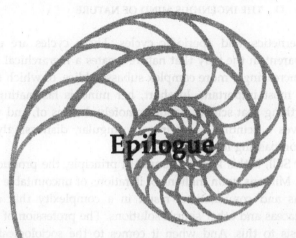

Epilogue

One universe made up of all that is; and one God in it all, and one
principle of being, and one law, the reason, shared by all thinking
creatures, and one truth.
—Marcus Aurelius Antoninus

The mind of nature is not only ingenious but parsimonious. She
starts with a configuration of elements in space, sets them in
motion, then sits back and waits patiently for billions of years
until the program inherent in the original configuration develops
into the organism known as the earth, accompanied by nearly 14
million species that survived the evolutionary trial by fire.

The development of organisms, and all natural physical systems, arises from patterns efficaciously programmed to reach out
into their environment and take what they need to grow further
and then fission in order to replicate themselves—computer science
practiced on biological looms. These patterns both stabilize and
grow by various mechanics, including equilibrium, homeostasis,

cybernetics, and aperiodic cycles. These cycles are especially apparent in the way that nature creates a hierarchical sequence of increasingly more complex subassemblies, of which the cell is the most important. In short, her mind is fascinating beyond anything that science fiction remotely dreams of, and yet never moves a scintilla away from a singular, distressingly simple, encompassing logic—her *logos*.

Still, if this logos is simple in principle, the productions are not. Millions upon millions of iterations of uncountable permutations and combinations result in a complexity that defies all panaceas and other simple solutions. The profession of medicine attests to this. And when it comes to the sociological systems spawned by nature's most sophisticated creation—*Homo sapiens*—the complexity goes ballistic. Each person can *re*configure or *re*align any system he or she sees, at least in theory. And as our knowledge increases, so too does the reach of that capability.

To be sure, the consequences of tinkering with a host of physical and sociological systems may not be immediately apparent. Malthusian reactions to overtaxing the carrying capacity can burst catastrophically on mankind. Elsewhere, superficial manipulation of economic parameters, ignoring the insidious patterns of sustained human greed and excess valuation, leads to equally catastrophic depressions. Gloom and doom face us at every turn.

But there is hope. As we gain insight into the mind of nature, we concurrently gain the ability to apply that insight, and can do so at a much faster rate than nature. We can head her off at the pass as it were, for example, someday defeating cancer every time by extremely early detection. For no matter how complex a system may be, we can eventually come to understand how it operates. From that, and providing we recognize *all* of the factors that influence that system, we can delineate the consequences of various options available for remedy. However, sometimes all options come packaged with undesirable side effects.

Still, let us end this journey on a cheerful note. The team of Laurel and Hardy were forever solving problems with methodologies that provoked lovable disaster. We call this passing a *culminating point*, whereby a person, or system, or nation extends itself so far beyond its resources that the slightest mishap begs defeat. This turning point was never more evident than in the 1932 film, *The Music Box*. In that episode, our heroes struggle to move a piano up a long, long outdoor flight of steps. This does not require a rocket scientist to foresee the consequences.

Sometimes this hilarity occurs in real life, even to St. Joan. Joan met her end through the most insidious treachery, but earlier in her brief, and brilliant, career she lost a battle of sorts—fair and square—with her favorite and most respected lieutenant, General La Hire (pronounced *La Heer*). Joan repeatedly badgered her generals not only to ensure all liturgical rites were observed, but to pray openly before the troops. The latter, La Hire was not willing to do. Joan persisted to the point where he realized she was not going to give up, yet he remained equally adamant about avoiding the role of deacon.

With cheerful malice aforethought, he bided his time whilst she grew bolder and began to harangue him—ever politely, of course—in earshot of his colleagues. At that moment, he uttered the French equivalent of "Oh, very well," clasped his hands, turned his head to the heavens, and intoned in full treble: "Pray, Sir God, that you will do by La Hire as he would do by You, if You were La Hire and he were God." Joan erupted into uncontrollable laughter to the point of tears and had to retreat into a tent until she regained her composure. For some strange reason she never thereafter pursued the matter.

And so it is with the patterns-are-everything thesis of this book. Theories should jettison their authors, so I will pursue the general idea no further. It has been a heady experience, taking the giants of science to task on a few points here and there, only to realize, as was said centuries ago, that even a dwarf can see farther by crawling over such shoulders. Instead, I will pick up a few threads and run with just those. Said Justice Holmes: "This, too, is the faith of the soldier; having known great things, to be content with silence."

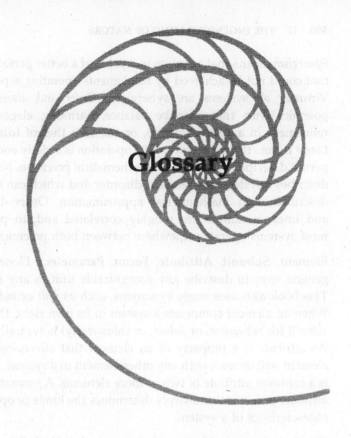

Glossary

This glossary groups related terms into paragraphs. As far as possible, the definitions adhere to those used in the relevant scientific disciplines, but in cases where the definitions and/or connotations differ between disciplines, yet remain relevant to the physiogenetic thesis, the simplest definitions are used. Sometimes they are the vernacular.

System, Machine, Order, Disorder, Synergism, Linear, Nonlinear. A *system* is a set of elements (or their equivalent) organized to accomplish the purpose for which the system was designed or which can be inferred. *Machine* is a common name given to many compact, usually man-made physical systems.

Synergism means that a system is capable of a better performance that could not be achieved by its elements operating separately. Virtually all systems are synergistic. *Order* and *disorder* are polemic terms that describe balance, harmony, elegance, or refinement in a system (order), or the lack thereof (disorder). *Linear* means that a system and its operation is orderly enough to permit description with classic mathematical precision. *Nonlinear* describes a system that leans to disorder and which can only be described with mathematical approximation. Order–disorder and linear–nonlinear are roughly correlated and, in practice, most systems operate somewhere between both polemics.

Element, Subunit, Attribute, Factor, Parameter. *Element* is a generic term to describe any recognizable unit in any system. This book also uses many synonyms, such as *unit* or *individual*. When an element comprises a system in its own right, the term *subunit* (or *subsystem*, or *subset*, or *subassembly*) is typically used. An *attribute* is a property of an element that affects how that element will interact with any other element in a system. A *factor* is a common attribute of two or more elements. A *parameter* is an attribute or factor that sharply determines the limits or operating characteristics of a system.

Physical System, Abstract System, Sociological System. A *physical system* is one composed solely of physical elements with no evidence of psychological attributes. An *abstract system* consists solely of ideas or concepts and their logical relationships (such as mathematics). A *sociological system* consists primarily of individuals or higher-order animals that evince strong psychological attributes.

Arrangement, Configuration, Pattern, Fractal, Art, Strategy, Focal Point, Symmetry. *Arrangement* is the geometric position of all elements in a physical system, or the logical schematic of relationships in an abstract or sociological system. *Configuration* is any arrangement combined with the attributes of each element or individual. When a configuration has a purpose or goal, that configuration is concurrently a system. *Pattern* describes any sequence of changes to any dynamic configuration. A *fractal* is a configuration or pattern that facilitates change in the direction from disorder to order or, alternatively, maintains a near-constant state of order despite interactions that would otherwise degrade the system at issue. *Art* is the action taken to rearrange, add to, reduce the number of, or otherwise modify elements in a system to improve its efficiency or appeal. The process of doing so is defined as *strategy* (though this term is avoided when referring to works of art). A *focal point* (alternatively, a *center of gravity*) is an element, or a point within an element or a system, that when acted on generates a significantly greater change or offers more advantage than occurs when the same action is applied to other elements or points in the same system. *Symmetry* describes common attributes between or among systems, theories, bodies of law, and so forth. Symmetry may or may not be visibly apparent.

Morphology, Linkage, Network, Interior Lines, Relational Systems, Hierarchical Systems, Integral Systems. *Morphology* is the subscience of categorizing configurations. *Linkage* and *network* are two of many synonyms that describe the means or logic by which elements of a system relate to each other. (In communications, a network and its system frequently coincide.) *Interior lines* means an efficient internal network or equivalent among elements in a system, e.g., the interstate highways. A *relational system* emphasizes the individuality or semi-independence of its elements. A *hierarchical system* imposes overriding lines of authority over constituent elements. An *integral system* alloys its elements to the point where those elements have little utility outside the system.

Model, Template, Paradigm. A *model* is a generic arrangement or configuration from which, or on which, systems can be tailored. When so tailored, the model is also called a *template*, especially if that model can be copied and the copy tailored. A model that offers exceptional versatility or widespread (sometimes global or universal) application, or is the only arrangement that seems to work for any given type of system, is a *paradigm*. The simplest example of a paradigm is the wheel.

Space, Time, Mass. *Space* is a dimensioned void that extends to infinity. In practice, systems have a finite number of elements; therefore, the space occupied by those elements is also finite. (In most abstract and sociological systems, space is figurative; hence, synonyms are often used, such as *sphere of influence*.) *Time* is the framework within which changes occur. Because time continues to transpire with or without systems, it is independent of them. *Mass* is the fundamental basis of physical matter.

Purpose, Design, Inferred Purpose, Serendipitous System. *Purpose* is the objective or aim of a system, and *design* means the initial arrangement and attributes of the elements of a system. *Inferred purpose* is the purpose that a system seems to have, and is used in cases where lack of evidence precludes certainty. This can occur with a conglomeration of loosely related systems that were never intended to be a system in its own right, for example, the presumed healthcare "system" in the United States. *Serendipitous system* describes a chance arrangement of elements that may be used as a system despite a lack of forethought or intent.

Operation, Process, Effect, Interaction (or Transaction), Efficiency. *Operation* describes how the elements in a system interact or cooperate in order to manifest the system's purpose. *Process* is nearly synonymous with *operation* but emphasizes the pattern of operational changes with respect to time. The distinction is not rigid. *Effect* is the measure of work performed or influence exerted by one system on another. An *interaction* (or *transaction*) is an action or act within a system that alone may or may not result in usable output but which contributes to an ultimate outcome. *Efficiency* measures the economy—mechanical, fiscal, or figurative—by which a system achieves its purpose. This measure is often expressed as the ratio of usable output to input (or equivalent ratio).

Causality, Determinism, Indeterminacy, Initiative, Degrees of Freedom, Controllability, Probability, Statistical Determinism, Actuarial Determinism. *Causality* means that one or more events occur as a consequence of preceding events or actions. *Determinism* means that the unfolding pattern of configurations in a physical system is inevitable, given the attributes of each element and universal law on the outcome of elemental interactions. *Indeterminacy* means that outcomes under apparently identical situations will fall into a distributed pattern, and that no one outcome can be predicted with certitude. However, the distribution may arise from minute varying differences in underlying deterministic mechanics too small or too impractical for observation or measurement, or—according to some theoreticians—because of a fundamental lack of causality. *Initiative* is a psychological attribute that enables an individual to initiate a change to a system by way of his or her cerebrum and central nervous system (analogous to the terms *free will* and *volition*). *Degrees of freedom* refers to the degree of physical or logical independence among elements in a system (or set of observations) with or without the exercise of initiative (and is roughly correlated with nonlinearity). *Controllability* refers to the extent that, or the ease with which, external influences or internal initiative can modify what a system otherwise seems programmed or destined to do. *Probability* is the statistical likelihood a system will achieve the purpose for which it was designed or how it will react when acted on. Deterministic systems are incompatible with probability, but lack of knowledge of operational details mandate probabilistic calculations. By contrast, outcomes in sociological systems, which include the attribute of initiative among members, are inherently probabilistic. *Statistical determinism* describes that practical necessity, and *actuarial determinism* is statistical determinism measured or gauged over a period of time, typically in years or longer.

Momentum, Inertia, Force, Potential Energy, Kinetic Energy, Entropy, Power, Dominant Element (or Dominant Force). *Momentum* measures an element's mass multiplied by its velocity.

Unless acted on by an external force, that element will continue to move at its current speed and direction (velocity) without change, which is *inertia*. *Force* is the amount of momentum applied to an element at any instant in time. *Potential energy* is the capability to do work, and *kinetic energy* is the manifestation of that potential. Kinetic energy is effected only on an interaction of some kind. *Entropy* is the inverse measure of the amount of momentum, force, and energy aligned to do work via the arrangement of elements in a system. *Dominant elements* (or *dominant forces*) are situationally the strongest elements or forces operating in a system if and when they are positioned to do work or exert influence (be it constructive or destructive). *Power* is the result of energy being expended or generated during a period of time. Power is not the same thing as efficiency, yet all other things being equal, an increase of efficiency reduces the power needed to do the same work.

Automation, Program, Robots, Computers, Feedback, Cybernetics. *Automation* refers to any machine that is controlled by a set of written or coded instructions stored in some form of memory readable by that machine. Alternatively, the instructions may be the effect of the dynamic configuration—its pattern—of a system, especially in genetics. The set of instructions is the *program*. A *robot* is an automated machine that replicates manual labor. A *computer* is an automated machine that replicates mental computational or logical effort. *Feedback* results from automation that directs a machine to obtain external data and adjust its operation accordingly. *Cybernetics* is the processing of feedback with reference to information the system has on hand and the rules by which it processes that information.

Equilibrium, Static Equilibrium, Dynamic Equilibrium, Homeostasis, Cycles. *Equilibrium* is the general process by which conflicting forces maintain themselves in check or at least keep themselves from running amok. *Static equilibrium* describes a system whereby the operating forces prevent movement of, or flow within, the system. *Dynamic equilibrium*, by contrast, describes the situation wherein movement or flow is controlled but not frozen. *Homeostasis* is the means to achieve and maintain dynamic equilibrium, commonly by using feedback to keep one or more dominant elements or forces in check. *Cycle* describes a process wherein dominance periodically (definitively cyclical) or aperiodically (approximately cyclical) alternates among elements or subunits in a system.

Equation, Inequality, Function.   An *equation* is a statement of equality between two expressions, each of which is a measure of the same attribute. For example, the following equation compares two expressions that measure monetary value:

$$\text{INCOME} + \text{INTEREST} = \text{EXPENSES} + \text{BALANCE}$$

An *inequality* means the two expressions are not equal, for example, the ever popular:

$$\text{EXPENSES} > \text{INCOME}$$

A *mathematical function* is a variation on an equation, wherein known (independent) variables or arguments are grouped in the right-side expression. The values assigned to these known variables determine the value of the single (dependent) variable that appears on the left side. The flow of water at a tap is a function of water pressure, size of the orifice, and so forth:

$$\text{FLOW OF TAP WATER} = \text{FUNCTION OF WATER PRESSURE, ORIFICE, ETC.,} \ or$$
$$\text{FLOW} = f \ (\text{PRESSURE, ORIFICE, } \ldots), or$$
$$z = f \ (x, y, \ldots)$$

Turning Point, Upswing Point, Power Curve, Point of Diminishing Returns, Point of Negative Returns. *Turning point* is a generic term that describes the status of a configuration when it is about to undergo a major and perhaps radical change in operation. An *upswing point* is a turning point whereby a configuration begins to generate significant increase to output or influence as a function of comparatively small input or change. The pattern of configurations that follows this upswing point is called a *power curve*. When the rate of return or output as a function of input begins to diminish, this second turning point is called a *point of diminishing returns*. When the total return or output decreases as a function of input, this third turning point is the *point of negative returns*.

Catastrophe, Critical Point, Crucial Configuration, Critical Mass, Chain Reaction, Point of Attenuation. A *catastrophe* is the result of a configuration reversing entropy by virtue of its own mechanics and configuration. This means that the system continues to produce (without further externally directed input or control) by way of being programmed to take what it needs from its environment or by drawing on internal resources not otherwise available. (In a graph of a catastrophe, the power curve is vertical.) The results may be beneficial or destructive (or both from different perspectives). A *critical point* is a turning point that identifies the time or condition when a catastrophe begins, and *crucial configuration* describes the configuration of a system at the instant that system reaches a critical point. In nuclear reactions, the crucial configuration is called *critical mass*, although that term is widely used in a figurative sense to name other crucial configurations. *Chain reaction* describes a sequence of events, normally where more than one subsequent event, on average, occurs for each preceding event. However, the expression may be used to describe a long sequence of such events with a one-for-one ratio. And in catastrophic phenomena, the point of diminishing returns is renamed the *point of attenuation*.

Breaking Point, Breakdown, Pseudocatastrophe, Culminating Point. A *breaking point* precedes what appears to be a catastrophe but is merely a *breakdown* (or *collapse*). Entropy is not reversed and no work is done. A breakdown is also called a *pseudocatastrophe*. A *culminating point* (or *overstretch*) almost always precedes a breaking point (or a critical point) and describes the condition of a system that is severely if not fatally weakened because it is overcommitted and therefore subject to eventual collapse (or catastrophe if the culminating point precedes a critical point).

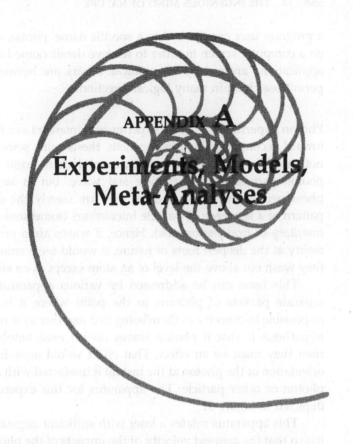

Experiments, Models, Meta-Analyses

This appendix describes 14 experiments, models, and meta-analyses to support confirmation of, or otherwise extend, the physiogenetic thesis, to various systems, to include abstract and sociological systems. *Experiment* means a physical apparatus used to verify or disprove a hypothesis or conjecture. A *model* is an abstract system of ideas and relationships that serves as a generalization with which to study specific systems of various kinds. A *meta-analysis* integrates the results of separate studies and investigations. A quantitative meta-analysis correlates diverse data and/or mathematical models. A qualitative meta-analysis links theories and equivalents by demonstrating logical relationships. *Hypertext* is a computer-based technique, whereby a program user can "point" to a specific name, phrase, or such on a computer screen in order to retrieve detail. Some hypertext applications are strictly hierarchical; others are networked t permit searching in many logical directions.

Photon Experiment   This experiment is intended as a first step toward reconciling the physiogenetic thesis with some of th notions of modern physics, to wit: current doctrine posits that th photon particle is both particle and wave. But in large-scal phenomena, waves do not exist; they are merely the effect c pattern of a sequence of particle interactions (sometimes called *transitory* or *local phenomenon*). Hence, if waves are a primordi: reality at the deepest roots of nature, it would seem strange tha they wash out above the level of an atom except as an effect.

This issue can be addressed by various apparatuses tha separate packets of photons to the point where it is all bt impossible to conceive of them being tied together as waves. Th hypothesis is that if photon waves do *not* exist ontologicall then they must be an effect. That effect would arise from th orientation of the photon at the instant it interacted with anothe photon or other particle. The apparatus for this experiment depicted in Figure 61.

This apparatus rotates a laser with sufficient angular velo ity, so that the *apparent* velocity of the impacts of the photons c a sheet of film is faster than the speed of light by four orders c magnitude. We are not talking about the velocity of a physic: object but the *apparent* velocity by which the trace is recorded o the film from sequential packets of photons. Note that a lase emits a collinear, coherent beam of photons, and quantur mechanics predicts that each packet of photons would impact c the film as a bundle, separated from one another, in th instance, by a centimeter. Furthermore, the energy content c these photons can be calculated as a function of the exposur Hence, if the energy content of the photons is accounted for b dot-shaped exposures on the film, each packet must have tra eled a separate path getting there.

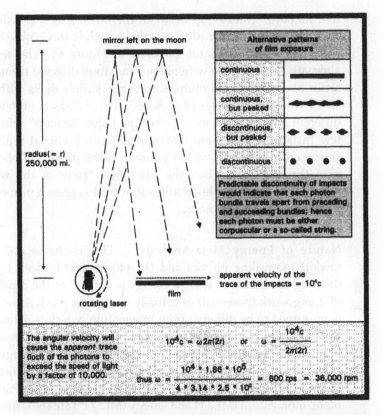

61. Proposed experiment with photons. Photons are commonly presumed to have ontological properties of particles and waves. But if it can be demonstrated that every photon travels freely apart from other photons, of what significance can a wave be except as the orientation in space of the photon at the moment it interacts with another particle?

At least two alternate apparatuses seem feasible. The first alternative is modeled after the experiment that first *accurately* measured the speed of light. This modification consists of two, comparatively tall and narrow, rotating polygonal drums at least 50 miles apart, with one face of each drum having a mirrored surface. The laser beam is aimed just shy of perpendicular so that its path "climbs" up between the mirrors as it is reflected back and forth until an adequate total length of travel obtains, as in the primary apparatus pictured in Figure 61. The second alternative uses very low frequency radiation directed through a series of shutters. The shutters are large, simple disks with one small hole near the edge of each disk. These disks are rotated at different angular velocities so that the holes "line up" only in a few instances, and these instances are synchronized with the supposed radiated "wave" of photons. If the packets of photons pass through only when the presumed "peak of the wave" reaches the aligned holes, that result would suggest a primordial particle nature for photons.

Nature of Energy Meta-Analysis   This is the second step toward reconciling physics and the physiogenetic thesis by addressing the nature of energy. Energy is defined as the capability of doing work (potential) or actually doing that work (kinetic). It is presumed to be different from momentum, which is defined as the product of matter or mass and its velocity. Exchange of momentum is defined as *impulse*, but it is not equal to energy. Yet, energy does not exchange in the absence of particle or particlelike interactions, which also involve impulse in some form. Furthermore, momentum (ML/T) and energy (ML^2/T^2) are the derivative and integral of one another in calculus. Hence, we may conjecture that energy is an extraordinarily useful calculation, but only as a mathematical function of momentum (potential), and exchange of momentum (kinetic). In effect, energy may be an effect.

Bear in mind that the term *energy* did not exist before the 19th century, except to convey a vague notion of efficiency. [The article on energy in the original *Encyclopaedia Brittannica* (1768–1771) was one sentence long.] Newton did *not* use the term—or its calculation—in the *Principia*. He *did* use the notion of force, which is a time–rate measure of momentum. The upshot: It seems very strange that Newton could demonstrate the "System of the World" without reference to energy, if energy is indeed an ontological reality of the universe.

The procedure for this meta-analysis is to point out that every instance of kinetic energy relies on an exchange of momentum in some form, including the value for momentum that can be derived "backward" from photon energy content: $e=hv$. Moreover, it may be possible that the momentum of a photon is a direct function of its spin.

Etymological–Derivations Model: *Logos*   No word or expression can be defined or described except in terms of other words or expressions. The consequence is that all "meaning" is relative. But as suggested by the hierarchical format of *Roget's Thesaurus* (described in Chapter 15, especially Figure 49), all words seem to be hierarchically related and tiered to a handful of generic concepts. Furthermore, this structure is not limited to the English language but has been adapted to numerous other languages. The proposed model *Logos* would use the hypertext technique to network the interrelationships of the etymological roots of 50,000 words. The conjecture is that if a hierarchical structure underwrites all language, *then such a structure would strongly suggest a universal symmetry or ontology.*

Computer Program: OPARIN   This completed computer program is an elementary demonstration of the physiogenetic model applied to the formation of two symbolic sections of genes from symbolic constituent acid molecules. The program does *not* track the location of the molecules. Instead, it shifts each molecule randomly on the screen, then "reads" the screen to determine interactions and/or linkage as applicable. The Microsoft Professional Basic/QuickBasic source code (.txt) for OPARIN can be obtained at www.geocities.com (at the Cape Canaveral neighborhood station 9380). Keep in mind that the program displays on a small, two-dimensional computer screen and takes about 15 minutes per run. By contrast, nature had the whole universe and billions of years; hence, a few corners were cut. Still, the principle is the same.

Physiology Model: BERNARD–CANNON   This proposed program harnesses the hypertext technique to show relationships among significant aspects of human physiology, networked for top-down study from organs down to molecular biology, and in the reverse direction. The intent is to portray graphically how semi-independent subsystems can nevertheless cooperate to an extraordinary degree. In turn, this will support the idea that nature can evoke considerable order out of apparent disorder.

Only representative cells are necessary, and quantitative data can be processed by way of algorithms and databases. This model does not depict growth or binary fission, but the user could reset key parameters to depict physiology at various stages of development. And as might be expected, the programming for this model is complex—like the difficulties Ulysses ran into on the way home from work on one occasion. But keep in mind that thousands of models depicting specific physiological processes are extant. Some of these models can be linked directly; others require modification.

Holographic Computer Apparatus for Systems Analysis

This apparatus tackles the problem of modeling physical systems with a large number of physically independent elements, an obvious tool of research for the physiogenetic thesis. The concept is shown in Figure 62. It can be programmed to "sweep" through the target space thousands of times per second, thus closely approximating actual physical independence of elements or subassemblies under study. And it could be used for *any* system in which excessive degrees of freedom make evaluation by classical mathematics tenuous.

Advanced Physiogenetic Model: *Genesis*   This project combines some of the concepts covered by programs OPARIN and BERNARD–CANNON with the holographic computer apparatus described above. The intent is to come as close as possible to physical reality and away from even the appearance of computer manipulation of data and processes.

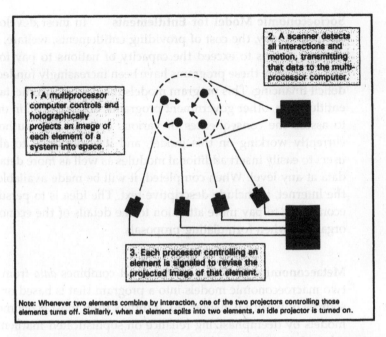

1. A multiprocessor computer controls and holographically projects an image of each element of a system into space.

2. A scanner detects all interactions and motion, transmitting that data to the multiprocessor computer.

3. Each processor controlling an element is signaled to revise the projected image of that element.

Note: Whenever two elements combine by interaction, one of the two projectors controlling those elements turns off. Similarly, when an element splits into two elements, an idle projector is turned on.

62. Holographic programming apparatus. Under normal operating conditions, an ordinary single-processor computer cannot accurately model independent objects in space. But a multiprocessor computer, combined with holography, can be used to do the job.

Macroeconomics Model for Economic Chain Reactions

Economic depressions are the socioeconomic equivalent of a catastrophe, and inevitably they can be described as a chain reaction of economics-based events. This program includes modules (subroutines in computer programs) for the major factors and relationships of macroeconomics. Each module has user-set parameters to be set to reflect economic conditions, government share of gross domestic product, and so forth, any time in economic history. The program would then be tested and revised until it accurately reflected historical evidence, and as such would serve as a bridge between the physiogenetic thesis and its application to sociological organisms.

Socioeconomic Model for Entitlements   In most developed countries today, the cost of providing entitlements, welfare, and so forth, seems to exceed the capacity of nations to pay for it, and in practice these programs have been increasingly funded by deficit financing. This program models the flow of income, taxes, entitlements, other government programs, and so forth, in order to assess the consequences of various options. The author is currently working on this model, and it is designed to allow users to easily insert additional modules as well as more detailed data at any level. When completed, it will be made available on the Internet, to include descriptive text. The idea is to persuade economists to pay more attention to the details of the economic organism when formulating proposals.

Metaeconomics Model   This model combines *data* from the two macroeconomic models into a program that is based on the sources of wealth (Chapter 16). It differs from most econometric models by deemphasizing reliance on sophisticated mathematical models, and instead inserting numerous, user-modified relationship algorithms to be processed currently with the trade-off logic common to fuzzy logic. In short, the emphasis is on the concurrent relationships among systems in lieu of neat sets of equations.

Fuzzy-Logic Model for the Principles of War   As described in Chapter 9, the outcome of battles, and especially of military operations, can be accurately and consistently evaluated on the basis of nine principles of war (which ultimately operate as an application of the physiogenetic model). These principles have logical and sometimes strong mathematical relationships, but they cannot be reduced to a classical mathematical model, and they are to some extent contradictory. But if the technique of fuzzy logic is applied to the application of the principles, the user-set parameters of the program modules could be "played" until the results matched historical fact. As in all such models, the early testing will inevitably point out the need for additional modules and factors.

Geopolitical Analysis Model   This model, which is intended to portray how the sociology of geopolitics comes closest to a pure physical system, is similar to the metaeconomics model described previously, with the factors of national power (described in Chapter 17) substituted for sources of wealth. However, because physical size, configuration, and position of land mass are crucial, the model needs extensive image processing of maps.

Qualitative Meta-Analysis: *The Pinocchio Effect*   Despite the humorous overtones, this is a serious project. The conjecture is: *Every system, or set of related data or information, can be explained by deductive logic emanating from a few simple axioms or assumptions. Any theory purporting to explain the relevant evidence that deviates from these axioms will lead to paradoxes ad absurdum and ad infinitum.* Though the history of science focuses on the success of major theories, the truth of the matter is that virtually every success was preceded by a host of ridiculous, if not ludicrous, concepts. Moreover, once the "correct" theory came to bear, it gained universal scientific (if not popular) acceptance within 30 years and brought debate largely to an end, except on details. To be sure, most of these theories have been subsequently modified, but they still stand largely intact, for example: Harvey's discovery of circulation of the blood, Newton's analysis of gravity, Lavoisier's work in chemistry, and Darwin's theory of evolution.

Demonstrating the validity of this conjecture is simple: a historical analysis of the development of all theories in science, pointing out that once established, *noncontroversial* theory evinces a handful of basic axioms, from which flow deductive logical exposition. The utility is to suggest that whenever current theories grow ludicrous, the assumptions are almost certain to be faulty. And, of course, the thrust here is to forcibly effect a juncture of the simplicity of the physiogenetic model with the baroque notions of modern physics.

Hypertext Knowledge Model   For more than 23 years, the *Propaedia* volume of the 15th edition of the *Encyclopaedia Britannica* has outlined the whole of knowledge, and for more than 44 years, the *Syntopicon* of the *Great Books of the Western World* set has indexed the relationships among significant ideas and concepts developed by arguably the West's most astute thinkers and scientists. Moreover, the *Britannica* is now available on-line and is continuously updated. Lastly, any aspect of its included information is retrievable by a sophisticated indexing system modeled after similar indexes long in use in libraries.

A logical next step would be to combine the *Britannica* with the *Great Books* (including most of the referenced works thereof), then apply the hypertext technique to the point where a user could follow a logical or relational path from any idea or fact, to any other, in any logical direction, backward or forward in time. For obvious reasons, only the Encyclopaedia Britannica Corporation could pursue this, but that corporation and its published works are extensively linked with scholars all over the world.

APPENDIX B
Comparative Systems

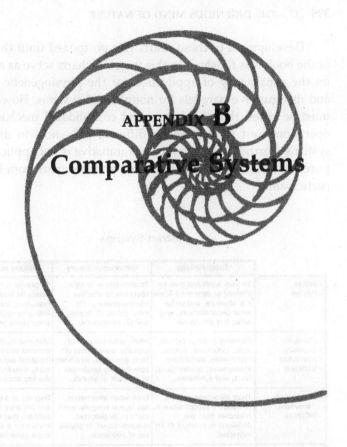

This appendix contains eight **matrix-type** charts outlining the factors and operational characteristics of representative types of systems within the three general categories of systems: abstract, physical, and sociological.

Abstract Systems
Automated Systems
Physics and Engineering
Physiology and Evolution
Psychology and Group Behavior
Economics
Nations and International Relations
Jurisprudence and Ethics

Development of these charts was postponed until the draft of the book was finished. In this way, the charts serve as a check on the consistency of application for the physiogenetic model and its figurative parallels for nonphysical systems. However, it must be noted that the factors and consolidated mechanics of operation exert proportionally different influences in different systems. Furthermore, the factor of initiative is not applicable in pure physical systems, or in abstract systems apart from human participants.

Abstract Systems

	Epistemology	Information theory	Military science
• Intent • Design	To seek truth and thus be enabled to determine when it is absolute, and under what circumstances, and when it is situational.	To eliminate or reduce obstacles to effective communications: (1) mechanical, (2) logical, *and* (3) interpretive.	To prevail in conflict, ideally by leveraging resources, but that failing by relying on sheer mass or clout.
• Elements • Quantity • Attributes • Variance	Elements include: axioms, logic, inductions, theses, hypotheses, deductions, experiments, observations, facts, and syntheses.	Information consists of logically related units of facts and ideas, which are conveyed by sequences of symbols or sounds.	Men and machines demand intense logistical support. As such, mobility is often the key attribute.
• Degrees of freedom • Initiative	Truth eliminates degrees of freedom except when it emanates from the decisions or conduct of an individual.	Even when information has logical integrity, each unit can be distorted, misunderstood, or placed out of sequence.	Degrees of freedom and initiative are so rampant that war is known for the mental fog it generates.
• Crucial arrangements • Interior lines • Dominant elements	Truth radiates by interior lines from axioms into tiers of theorems or equivalent in order to logically explain and relate facts.	Interior lines increase efficiency and often lead to configurations from which dominant ideas and concepts radiate unvexed.	The key principle is mass (concentration), as applied directly or indirectly against dominant objectives.
• Resiliency • Sustainability • Resistance to change	Logically flawless theory may be extended but does not bend. If and when it disagrees with facts, the axioms must be incorrect.	With few exceptions the quality of information tends to degrade rapidly, especially as the amount of it accumulates.	Once bent on war, a nation will pursue it even if the purpose changes or its economy is wrecked.
• Equilibrium • Homeostasis • Cycles • Cybernetics	Inadequate theory points to its own flaws and thus homeostatically impels replacement or revision (sometimes cyclically).	Most systems require extensive quality control measures to reduce the degradation of data in storage and/or its flow.	Absent preponderant mass or deft strategy by one side, most wars devolve into static brawls.
• Power curves • Catastrophe • Culminating points	Good theories spawn momentous discoveries, while poor theories are often artificially maintained despite a lack of utility.	Some words are explosive while pithy phrases have been known to manifest latent human emotions catastrophically.	Aggressors push beyond culminating points; hence they incur devastating losses if not defeat.

Automated Systems

	Hardware	Software	Networks
• Intent • Design	To emulate processing of human thought (or to mimic physical movement) by way of electronics, mechanics, or equivalents.	To codify logic and data so that a computer will implement instructions without intervention except when intended.	To exchange information (data) or programming among two or more systems in a timely, distortion-free manner.
• Elements • Quantity • Attributes • Variance	Computers depend on millions of permutations and combinations of simple circuits, to which robotics adds machinery.	A program consists of a set of logical rules or instructions that tells a computer or robotic machine what to do.	Although configurations of networks vary, the elements are always the nodes and the paths running through them.
• Degrees of freedom • Initiative	Initiative does not apply, and degrees of freedom are usually eliminated (except in parallel arrays of processing units).	This is an application of information theory, in which tens of thousands of error-prone degrees of freedom operate.	The paths a message can take equal the degrees of freedom—an exponential function of the number of nodes.
• Crucial arrangements • Interior lines • Dominant elements	The integrated chip was a crucial configuration that emphasized interior lines, thus setting the stage for the computer revolution.	Programs use modules, structure, and visible interior lines, but crucial algorithms and data exert dominant roles.	Networks use interior lines, in which effective configurations depend on the flow pattern of the heaviest traffic.
• Resiliency • Sustainability • Resistance to change	Hardware is physically stable (excepting purely mechanical components) and lends itself to a wide range of requirements.	Once debugged, the program logic stabilizes, but inevitable changes lead to ad hoc revisions that weaken stability.	Most networks sustain heavy traffic loads by utilizing the nearly infinite number of interior-line patterns.
• Equilibrium • Homeostasis • Cycles • Cybernetics	Hardware must implement the logical cybernetics of software by way of its own systems architecture and processing cycles.	Software can be written to revise or correct itself as a function of the patterns or sequences of interactive usage.	Under heavy load, the control logic for routing messages over the least used paths maintains the network equilibrium.
• Power curves • Catastrophe • Culminating points	Processing technology and data storage offer both an accelerating power curve of efficiency *and* declining cost.	Any error in a program applied to millions of records will typically generate an equal number of errors.	A heavy overload can instantly shut down a network, especially if protocol programs are the least bit faulty.

Physics and Engineering

	Physical and nuclear	States of matter	Engineering
• Intent • Design	To enable the formation of the periodic table of elements from the poles representing the lightest and heaviest elements.	To provide the different stages of dynamics within matter to effect various physical and chemical processes.	To arrange matter and energy sources to achieve a purpose that is impossible or impractical by pure human labor.
• Elements • Quantity • Attributes • Variance	The main variance among unstable nuclei is the ease or difficulty by which they decay, hence the variable half-lives.	States of matter are a function of internal energy levels, as modified by the pressure of the environment.	Engineering can utilize a variety of building materials, providing the configuration is based on elemental strengths.
• Degrees of freedom • Initiative	Variable configurations equate with degrees of freedom, which increase or decrease the rate of decay for a given mass.	Each increase to energy triggers a new quantum equilibrium; hence degrees of freedom are rigorously controlled.	Engineering strives to control degrees of freedom, though the aim is typically to facilitate the exercise of initiative.
• Crucial arrangements • Interior lines • Dominant elements	Radioactive half-life is not a linear function of elements or isotopes but rather the configuration of the nucleus.	Matter changes state by configurations common to all elements, but threshold temperatures vary greatly.	For any specific purpose, only a few configurations work exceptionally well, which in turn become paradigms for variations.
• Resiliency • Sustainability • Resistance to change	Radioactive atoms are inherently unstable, although half-life varies from microseconds to thousands of years.	Atoms gain or lose energy without affecting stability, except plasma, but temperature affects binding properties.	Few projects can tolerate major change, while sustainability and resiliency are usually a matter of design.
• Equilibrium • Homeostasis • Cycles • Cybernetics	The equilibrium is evident in the distribution of elements once the bulk of the fission process has run its course.	Atoms increase or decrease energy content by assuming slightly different patterns of quantum equilibrium.	Cybernetics is not common except for a few control mechanisms that adjust programming based on "experience."
• Power curves • Catastrophe • Culminating points	The catastrophic fission chain reaction is well known, but it depends on configuration of the mass as much as the quantity.	Matter changes state when its energy content overrides the existing pattern of stability (a culminating point).	All man-made and natural systems can be pushed past their breaking point, resulting in destruction by pseudo catastrophe.

Physiology and Evolution

	Physiology and genetics	Pathology	Evolution
• Intent • Design	To enable complex organic compounds and constructs thereof to form into a variety of self-sustaining organisms.	Disease has no purpose per se; it is a failure of the organic design to sustain itself within its environment.	To create organic constructs essential for generating complex organisms and, in turn, to evolve the species.
• Elements • Quantity • Attributes • Variance	Larger organisms can have a quadrillion or more cells, and each cell is built from thousands of largely carbon-based molecules.	Elements can change attributes to the point of disrupting organic processes in biological configurations.	Evolution used only a few subunit paradigms, e.g., polypeptides and cells, though each one has many variations.
• Degrees of freedom • Initiative	Organic chemistry by itself has no initiative, but organisms must control billions of contravening degrees of freedom.	Despite thousands of disorders, most are variations of a few dysfunctional or pathological patterns.	The evolutionary process diverged into millions of species, each of which was a degree of freedom in the environment.
• Crucial arrangements • Interior lines • Dominant elements	The DNA double-helix configuration is the most crucial and far-reaching construct in the whole of natural phenomena.	Disorder disrupts the interior lines necessary for equilibrium, and malignant disorder is a catastrophic disruption.	The cell is the obvious crucial configuration, and most species seem to be variations on a handful of theoretical models.
• Resiliency • Sustainability • Resistance to change	Chromosomes in particular demonstrate strong resistance to change even after replicating themselves a billion-fold. On the other hand, every cell either dies or eliminates itself by cell division (mitosis). As such, resiliency resides in the process.		Only 5 percent of all species escaped the fate of extinction, and many of the survivors live precarious lives.
• Equilibrium • Homeostasis • Cycles • Cybernetics	Organisms intensely rely on various equilibrative and homeostatic mechanics for control, growth, and defense.	Organisms try to remedy disorders by equilibrative dynamics, which can be enhanced by medical practice.	The earth is a meta-organism, its physiology is ecology, and every species is, in effect, a cell living therein.
• Power curves • Catastrophe • Culminating points	Cell reproduction and division are spectacular, constructive catastrophes, as that term is understood in science.	Medical crises often reach a turning point that nudges the body either toward death or rapid recovery.	The history of evolution is one long power curve, but catastrophe beckons whenever species overtax the environment.

Psychology and Group Behavior

	The psyche	Organizations	Cities
• Intent • Design	The meaning of life is a something of a mystery. Perhaps it is related to choices made as man modifies the environment.	To enable a group of persons to achieve goals that its members cannot attain individually, or only partially so.	Originally, to anchor human organization to a locale so as to facilitate trade along routes that intersect at that locale.
• Elements • Quantity • Attributes • Variance	The elements seem to be: (1) the ego, (2) the mind, and (3) the affects or emotions; interacting with the physiological host.	Members are elements, and they are always at variance to one degree or another with the organizational purpose.	Elements are housing, organizations, offices, stores, communication networks, transport, and the ability to trade.
• Degrees of freedom • Initiative	This is the only font of initiative known to man, and it is beset upon by divergent thoughts, feelings and emotions.	Each member represents a degree of freedom, and exercises initiative to the betterment or detriment of the organization.	Cities suppress degrees of freedom and initiative due to a high degree of mutual dependence among residents.
• Crucial arrangements • Interior lines • Dominant elements	Maturation and mental health require integration of the psychical fonts to a point whereby good judgment dominates.	Effectiveness depends on communications, the operation of which is a crucial function of interior lines.	Originally, cities were laid out to facilitate trade and defense, but they usually devolve into hodgepodges.
• Resiliency • Sustainability • Resistance to change	These processes vary tremendously among people. No two can be said to be identical, not even identical twins.	Structure often supplants function as the dominant consideration; hence organizations inure themselves to change.	Because of access to many resources, cities can sustain themselves even as they grow increasingly poor.
• Equilibrium • Homeostasis • Cycles • Cybernetics	Most people seek a form of internal equilibrium and with their environment. Personal learning is analogous to cybernetics.	Organization equilibrium counterbalances the prerogatives of the group ego with the aspirations of individual members.	A high degree of mutual need forces residents to rely extensively on one another, thus creating a passive equilibrium.
• Power curves • Catastrophe • Culminating points	The experience of a crisis occasionally precipitates a catharsis, which is the psychical equivalent of a catastrophe.	Organizations can satisfy member needs up to a point. Reaching beyond that point is overstretch, and can lead to disaster.	Cities expand to a point whereby they can no longer deal successfully with problems and thus begin a slide into decay.

Economics

	Metaeconomics	Macroeconomics	Microeconomics
• Intent • Design	To transform the means of labor, material, technique, and infrastructure into the intermediate ends of sustained wealth/growth.	To employ or control wealth to maintain and improve national economic strength and meet economic needs.	To employ the means of wealth to maintain if not improve profits, service, standard of living, and long-term stability.
• Elements • Quantity • Attributes • Variance	The elements above vary greatly among nations, and the proportionality of those elements is often a key national attribute.	Money is the primary element of both macro- and microeconomics because it is clearly fungible and lends itself to mathematical treatment. But the price of a good or service may or may not coincide with perceived value at any given time.	
• Degrees of freedom • Initiative	If Adam Smith's "invisible hand" thesis is correct, metaeconomics—hence attempts to manipulate national and international economies—have fewer but deeper degrees of freedom than often assumed. As a result, most ad hoc policies fail sooner or later.		Organizations and individuals use initiative more so than nations because their aims are usually narrower.
• Crucial arrangements • Interior lines • Dominant elements	The dominant elements in economics are the assets which generate wealth, but to do so those assets must be appropriately configured and managed. Unfortunately, what is best from a long-term metaeconomic perspective will usually conflict with short-run macroeconomic/microeconomic necessities. As the latter always take precedence, both capitalism and socialism beg failure.		
• Resiliency • Sustainability • Resistance to change	When a nation has ample resources, it gains resiliency. By contrast, shortages force a state into a quasi-beggar status.	Macroeconomics is difficult to change, and most attempts overlook overarching meta-economic processes.	Nations can resist bankruptcy more than people and corporations do because they control massive resources.
• Equilibrium • Homeostasis • Cycles • Cybernetics	Even in sound economies, the evolution of wealth alternates between periods of solid but excess growth, and mild recessions. When the rate of growth exhibits too strong of a power curve, the overstretch often leads to a major recession if not a catastrophic depression.	The cycle of developed nations is an annual budget that may lack discretion due to citizen entitlements.	Economic success often depends on seeking (and using) feedback on the consequences of major decisions.
• Power curves • Catastrophe • Culminating points		Sound economic strategies can lead to wealth, though the power curve is preceded by a period of slow, almost penurious growth. But when an individual or organization leverages resources too far, catastrophic bankruptcy ensues.	

Nations and International Relations

	Nations	Governments	Alliances and war
• Intent • Design	To expand organizational advantages and efficiency to an entire cultural environment, ideally on a permanent basis.	To create infrastructure essential to implement the purposes for which a culture evolved into, or became a nation.	To aggrandize or defend against aggrandizement irrespective if the aggression is military, economic or ideological.
• Elements • Quantity • Attributes • Variance	The elements are ethos, geography, demographics, technology, military power, economic clout, and infrastructure.	Government networks have two elements: citizens and operators. Communications are sporadic and inefficient.	The elements are men, machines, firepower, and logistics in the arena where the conflict is played out.
• Degree of freedom • Initiative	Sovereignty implies initiative, but national decisions are usually constrained by available resources and events.	In governments, the opportunities for initiative are almost inversely proportional to degrees of freedom.	The fluid nature of war ordains many degrees of freedom and initiative, at least until it devolves into a static brawl.
• Crucial arrangements • Interior lines • Dominant elements	The strength of a nation is highly dependent on the configuration of its resources, perhaps more so than the quantity.	Republics offer a near optimum government form due to widespread albeit inefficient, interior lines of communication.	Geography plays a crucial role in alliances and conflicts, but the main determinant of war is clout.
• Resiliency • Survivability • Resistance to change	A cursory review of history suggests that few countries can sustain their international influence on a permanent basis.	Like most organizations, governments seldom have the resiliency to preclude rebellion or revolution in some form. Hence, conflict between the governed and the governors is inevitable to the point whereby power sharing must be rebalanced periodically. If the imbalance grows too skewed, the citizenry will rebel. If the rebellion is suppressed, revolution will likely occur.	Alliances are formed to garner sufficient mass for aggrandizement or to resist the menacing strength of major powers and *their* alliances. Hence nations must depend on a fluid international hegemony to defend and advance their own, more permanent interests.
• Equilibrium • Homeostasis • Cycles • Cybernetics	A nation that endures can successfully balance major conflicting forces that would otherwise tear its fabric to shreds.		
• Power curves • Catastrophe • Culminating points	When nations try to exert too much influence beyond their borders, they inevitably waste resources and in time decay.		Empires collapse when they pass culminating points, Influence weakens tremendously as distance increases.

Jurisprudence and Ethics

	Jurisprudence	The courtroom	Ethics
• Intent • Design	To establish and maintain justice within a nation by way of laws and by procedures to manifest those laws in practice.	To provide a flexible, orderly, and constructive way of resolving the conflicts that inevitably arise within states.	To create the bearings by which organizations, professions, nations, and individuals should conduct themselves.
• Elements • Quantity • Attributes • Variance	The elements are laws and the principles which are used to adjudicate specific cases (gradually modifying the law).	The static elements are procedures and criteria; dynamic, the parties to a conflict plus the evidence and the arguments used.	The elements are similar to law, except the process resides in the mind of each person or group ego.
• Degrees of freedom • Initiative	The process of law needs a door to accommodate change in the polity, but when opened too wide the house risks collapse.	Most options in the courtroom are severely limited to ensure order, enforce due process, and minimize theatrics.	The crux of ethics may be to take personal responsibility for the consequences of acts or the failure to act.
• Crucial arrangements • Interior lines • Dominant elements	Interior lines are often curtailed by conflict in contravening principles, each of which serves the cause of justice.	The arrangement crucial to any case is the picture presented by the evidence and the arguments that tie it together.	Though consequences weigh heavily, most ethical issues are too unsettled to formulate an optimum system.
• Resiliency • Sustainability • Resistance to change	Development of law is dynamic, i.e., the process by which a nation learns to resolve conflict in order to sustain its own structure.	In most English-speaking countries, the procedures of court have not changed much in 500 years, though the criteria have.	Workable systems of ethics must be resilient and resist change, else they degenerate into rationalization.
• Equilibrium • Homeostasis • Cycles • Cybernetics	Some jurists push the envelope; others don't bother to open it, but on the whole this process is an equilibrative dynamic.	Equilibrium is apparent in the exchange of arguments. When justice fails, dissent sooner or later finds its mark.	Similar to development in law, systems of ethics tend to evolve as a synthesis between ideals and experience.
• Power curves • Catastrophe • Culminating points	Short of revolution, catastrophe in jurisprudence is witnessed only in the form of pivotal decisions that affect the polity for generations to come. The drama usually plays itself out in a courtroom, albeit most often in a court of appeals under professional jurists.		Ethical systems gain acceptance or fall into disuse as a function of both logical integrity and practicality.

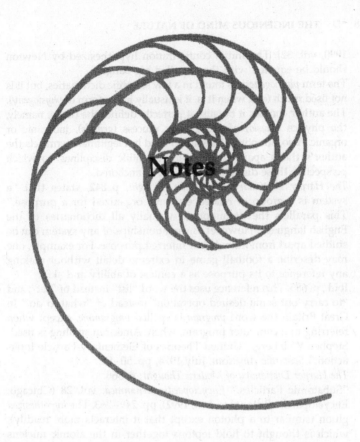

Notes

CHAPTER 1. CONCEPT

1. Philippe Le Corbeiller, "Crystals and the Future of Physics," *Scientific American*, January 1953, p. 52.
2. Alan Bullock and Stephen Trombles, eds., *The Harper Dictionary of Modern Thought*, rev. ed. (New York: Harper & Row, 1988), p. 115. (Sir Alan Bullock was the vice chancellor of Oxford University, Great Britain, at the time this dictionary was recompiled.) As described in the reference, the concept of cellular automata has been applied extensively in biology and the science of chaos, especially via fractals.
3. Isaac Newton, *Optics*, Great Books of the Western World, vol. 34 (Chicago: Encyclopaedia Britannica, Inc., 1952), p. 541 [2nd ed.,

1990, vol. 32]. The initial configuration hypothesized by Newton should be equated with the so-called big bang.

4. The term *physiogenesis* is found in a few scientific dictionaries, but it is not used much (and when it is, it is usually in the form of *physiogeny*). The author chooses it because it perfectly defines the thesis, namely, the physics (*physio*) of the genetic process (*genesis*), inorganic or organic. Also, adopting this seldom-used but legitimate term sets the author's thesis apart from existing scientific disciplines, to which perspective those disciplines can then be anchored.

5. *The Harper Dictionary of Modern Thought*, p. 842, states that "a system is a group of related elements organized for a purpose." This parallels the definition in virtually all dictionaries of the English language. However, the relationships of any system can be studied apart from intended or inferred purpose. For example, one may describe a football game in extreme detail without making any reference to its purpose as a contest of ability and skill.

6. Ibid., p. 685. This reference uses the word "list" instead of "set"; and "to carry out some desired operation" instead of "what to do." In Great Britain the word *program* is spelled *programme*, except when referring to a computer program, where American spelling is used.

7. Stephen Weinberg, "Unified Theories of Elementary-Particle Interaction," *Scientific American*, July 1974, pp. 50–59.

8. *The Harper Dictionary of Modern Thought*, p. 366.

9. "Subatomic Particles," *Encyclopaedia Britannica*, vol. 28 (Chicago: Encyclopaedia Britannica, Inc., 1992), pp. 249–263. The *hypothesized* gluon (similar to a photon except that it interacts more readily), which is thought to hold leptons together in the atomic nucleus (i.e., the strong nuclear binding force), may also have permanence.

10. For example, see John von Neumann, *The Computer and the Brain* (New Haven, Connecticut: Yale University Press, 1958). Literally tens of thousands of references on this subject have been written in the intervening 38 years.

11. For a clear, graphical treatment, see Rick Gore, "The Once and Future Universe," *National Geographic*, June 1983, pp. 704–749, especially pp. 710 and 740.

12. *The Harper Dictionary of Modern Thought*, p. 860. The mathematical difficulties of the three-[or more]-body problem led to the pioneering work of J.H. Poincaré (1854–1912) in topology, which may be considered as the forerunner of chaos science. In any event, equations that address interactions among more than two objects are approximations at best. For example, while pinning down the reaction of A to B, the relationship of A and C, and of B and C may change. This leads to a mathematical "vicious cycle" and is

resolved (without absolute precision) by a technique known as *recursion*. When the number of objects reaches 1000, the computations verge on impossibility. At several million, it is unthinkable.

13. Ibid., pp. 837–838. In this instance, the term *endure* is equivalent to the expression *remain constant*, e.g., conservation of momentum.

14. Physical complexity theory arose, in part, from mathematics, especially the complex manifolds of topology. It is easy to see how the science of chaos, catastrophe theory, and now complexity theory converge on topological manifolds and other fractal patterns in order to decipher the mechanics of biological looms, as it were.

15. *The Harper Dictionary of Modern Thought*, p. 349. This reference is quite clear in pointing out the difficulty faced by general systems theory (GST) coping with the "anarchial and decentralized character" and "irrationalities" of various complex systems. The implication is that GST seeks a form of hierarchical control that does not exist in such systems (which rely more on the give-and-take of equilibrative mechanics than on neat controls).

16. Quoted in James R. Newman, ed., *The World of Mathematics*, vol. 2 (New York: Simon & Schuster, 1956), p. 975.

17. Derek Lovejoy, "The Dialectics of the Tenth Dimension: Some Recent Writings on Science, Philosophy and the Cosmos," *Science & Society*, Summer 1995, pp. 206–222. Also see Gary Taubes, "How to envision (well, sort of) the topology of a ten-dimensional universe," *Discover*, November 1986, pp. 42–48.

18. *The Harper Dictionary of Modern Thought*, p. 365.

19. The third law of thermodynamics states that the amount of available energy in some systems at or very close in temperature to absolute zero may remain constant. Because of the temperature prerequisite, this exception is of little practical interest.

20. The purpose of this system, of course, is to serve as a teaching model. And as described in Chapter 11 of this book, it also served as the model for a hydroelectric power plant that uses excess power at night to move water in a lower reservoir to a higher reservoir in order to generate more electrical power during periods of high usage.

21. This is an oversimplification. Computer "viruses" are commonly classified by their specific mode of operation, to include "Trojan horses." Still, they are all self-sustaining.

22. This is the *Heisenberg uncertainty principle*, as merged into Bohr's *complementarity* dogma. See Niels Bohr, "Causality and Complementarity," in Timothy Ferris, ed., *The World Treasury of Physics, Astronomy, and Mathematics* (Boston: Little, Brown and Company, 1991), pp. 801–807.

23. Ibid. This perspective is known as the *indeterminacy principle*.

24. Ibid. Bohr stated that his theory did *not* mean causality was nonexistent, but that it could be understood only by way of some abstraction that transcended observable, mechanical physics. However, the very need for Bohr to reiterate this point in 1958 strongly indicates that the no-causality idea had taken root. Elsewhere, the nonexistence of causality is popular in some philosophical discourses. See *The Harper Dictionary of Modern Thought*, p. 113.

25. J. Brownoski, *The Ascent of Man* (Boston: Little, Brown and Company, 1973), p. 256. Einstein made this statement numerous times, not always with the same words.

26. Ibid. But see Richard Rhodes, *The Making of the Atomic Bomb* (New York: Simon & Schuster, 1988), p. 133, which quotes Bohr as saying: "nor is it our business to prescribe to God how He should run the world."

27. James Gleik, "Chaos," [excerpt in] Timothy Ferris, ed., *The World Treasury of Physics, Astronomy, and Mathematics*, p. 470. The remark was made by Joseph Ford.

28. Copernicus and Galileo were Roman Catholic; Kepler was Lutheran (and pious enough to persuade church authorities *not* to burn his mother at the stake for alleged witchcraft); Newton and Maxwell were Anglican, though in practice Newton was, in effect, a Unitarian; and Einstein was Jewish, though in terms of theology he was, like Spinoza, more or less a pantheist. Incidentally, Gregor Mendel was an ordained Roman Catholic priest.

29. Isaac Newton, *Mathematical Principles of Natural Philosophy*, Great Books of the Western World, vol. 34, pp. 369–372 [2nd ed., 1990, vol. 32].

30. *The Harper Dictionary of Modern Thought*, pp. 36–37.

31. Ibid. The anthropic principle is really two schools of thought. The *strong anthropic principle* verges on an explicit recognition of a teleological universe, i.e., a design that foreordained evolution of life forms. The *weak anthropic principle* waters this down into a form of "loaded-dice" statistical probability without addressing the matter of teleology.

32. Quoted in Roger Penrose, *The Emperor's New Mind: Concerning Computers, Mind, and the Laws of Physics* (Oxford: Oxford University Press, 1989), p. 2.

33. *The Harper Dictionary of Modern Thought*, pp. 531–532.

34. Originally, Freud admitted a few non-MD psychologists to psychoanalytic training, of whom the most well known was Erich Fromm.

35. See [Baroness] Jane [van Lawick-] Goodall, *The Chimpanzees of Gombe: Patterns of Behavior* (Cambridge, Massachusetts: Belknap Press, 1986). Mrs. Goodall also made this point very clear in her National Geographic Society videotape.
36. Quoted in Gustav Eckstein, *The Body has a Head* (New York: Harper & Row, 1969), p. 253.
37. Brooks Atkinson, ed., *The Selected Writings of Ralph Waldo Emerson* (New York: The Modern Library, 1950), p. 48.

CHAPTER 2. CATEGORIES, STRUCTURE, AND FACTORS

1. *The Harper Dictionary of Modern Thought*, pp. 524, 605.
2. Clifton Fadiman, ed., *The Little, Brown Book of Anecdotes* (Boston: Little, Brown and Company, 1985), p. 556.
3. Patrick Ryan, "Get rid of the people, and the system runs fine," *Smithsonian*, September 1977, p. 140.
4. That Stalin's programs led to between 15 and 20 million deaths prior to World War II is well established. Estimates for China vary between 70 and 80 million. For example, see "Four decades of repression and reform," *U.S. News & World Report*, March 12, 1990, and also "Mao remains bigger than life," *Chicago Tribute*, September 9, 1996, pp. 1–14.
5. *The Little, Brown Book of Anecdotes*, p. 520.
6. Harry G. Summers, Jr., *On Strategy: the Vietnam War in Context* (Carlisle Barracks, Pennsylvania: Strategic Studies Institute, 1981), pp. 59–66.
7. Spinoza, *Ethics*, Great Books of the Western World, vol. 31, p. 422 [2nd ed., 1990, vol. 28, p. 656].
8. Justin Kaplan, ed., *Bartlett's Familiar Quotations* (Boston: Little, Brown and Company, 1992), p. 620.
9. Will and Ariel Durant, *Rousseau and Revolution* (New York: Simon & Schuster, 1967), pp. 439–443.
10. The other six elements (in descending order) are: aluminum, iron, calcium, sodium, potassium, and magnesium. Hydrogen is as plentiful but has a low molecular weight.
11. The author compiled this data from the world data sections of the yearbooks of the *Encyclopaedia Britannica*, and published in *Geopolitics and the Decline of Empire* (Jefferson, North Carolina: McFarland, 1989). See also "A World in the Balance," *U.S. News & World Report*, March 6, 1995, p. 68, which reports that as of 1991, the upper 20 percent of nations controlled 83.7 percent of the world GNP.

12. *Statistical Abstract of the United States 1995* (Washington, D.C.: U.S. Government Printing Office, 1995), tables 527 and 531. Of the 114 million income tax returns in 1992, about 22 percent (of those who filed) paid 75 percent of the individual income taxes.

CHAPTER 3. MECHANICS

1. Will and Ariel Durant, *The Age of Reason Begins* (New York: Simon & Schuster, 1961), p. 503.
2. Jonathan Miller, *The Body in Question* (New York: Random House, 1978), pp. 189–191.
3. Lewis Sorley, "Creighton Abrams and the Active–Reserve Integration in Wartime," *Parameters*, Summer 1991, p. 48.
4. The explanation is that at 8 percent interest/growth per annum, it only takes a small deposit to reach a million dollars at the end of 100 years. In effect, while the first million is drawn down to zero, the small added deposit will compound and replace it. (The first annual withdrawal will reduce the principal by only $450.) By adding that amount to the principal, the new principal will remain at one million dollars essentially forever.
5. Catabolism, of course, is a normal subset of metabolism, but when the body receives insufficient nutrients it will catabolize its own cells. A high-protein diet accelerates this process, and is known as "protein poisoning."
6. For an early article on this either-constructive-or-destructive operation, see E. C. Zeeman, "Catastrophe Theory," *Scientific American*, March 1976, pp. 65–70, 83.
7. For example, see M. Mitchell Waldrop, *Complexity: The Emerging Science at the Edge of Order and Chaos* (New York: Simon & Schuster, 1993).
8. *Bartlett's Familiar Quotations*, p. 739.
9. This is debatable, and depends on various definitions, i.e., types of molecules, sheer numbers, period of time, and, for that matter, exactly what constitutes an organic molecule. But it is fairly clear that the types of molecules normally found in an organism constitute only a small percentage of all possible molecules.
10. *Encyclopaedia Britannica*, 1992, vol. 6, p. 980.
11. Bruce Catton, *Never Call Retreat, The Centennial History of the Civil War*, vol. 3 (Garden City, New York: Doubleday, 1965), p. 404.
12. Morphology inherently changes the internal means of controllability. From greater to lesser controllability amounts to an increase in

entropy. However, a change in the opposite direction does not necessarily reverse entropy, though it may set the stage to do so later.

13. Lewis Thomas, *The Lives of a Cell* (New York: Viking, 1974), p. 5.
14. The other two were his authorship of the Declaration of Independence and the founding of the University of Virginia.
15. Newton, *Optics*, p. 542.

CHAPTER 4. DERIVATIONS AND APPLICATIONS

1. Although Walter Heisenberg developed the *uncertainty principle*, his subsequent 1932 Nobel prize cited it as the *indeterminacy principle*.
2. Niels Bohr balked until Heisenberg anchored the uncertainty principle to his own complementarity idea. See Rhodes, pp. 130–132.
3. Max Planck also balked at the idea of no causality, despite the determinacy-versus-indeterminacy issue. See Max Planck, *Scientific Autobiography and Other Papers*, Great Books of the Western World, 2nd ed., vol. 56, pp. 102–103.
4. James R. Newman, book review of David Bohm's *Causality and Chance in Modern Physics*, *Scientific American*, January 1958, pp. 111–112.
5. "Leon Lederman [interview with D. Teresi]," *Modern Maturity*, June 1994, pp. 60–62.
6. *Webster's Ninth New Collegiate Dictionary* (Springfield, Massachusetts: Merriam-Webster Inc., 1990), p. 1235.
7. *The Oxford Dictionary of Quotations*, 3rd ed. (Oxford: Oxford University Press, 1979), p. 150.
8. Personal letter from David B. Guralnick, May 18, 1976.
9. David B. Guralnick, ed., *Webster's New World Dictionary of the American Language* (New York: The World Publishing Company, 1972), p. 481.
10. "Scientists look into the next century," *Financial Times*, September 14/15, 1996, p. 4.

CHAPTER 5. ROOTS AMONG THE PHYSICAL SCIENCES

1. Planck, p. 81.
2. Newton, *Optics*, p. 529. He wrote "Are not the rays of light very small bodies emitted from shining substances?"

3. Christiaan Huygens, *Treatise on Light*, Great Books of the Western World, vol. 34, pp. 559–563 [2nd ed., 1990, vol. 32]. Huygens admitted that there might be a particle of sorts at the center of the wave.

4. Some perspectives of the big bang do not place it at any computed location; others do, simply by roughly computing "backwards" (in time) the geometric origin of the pattern of the expansion of the universe.

5. Isaac Newton, *Mathematical Principles of Natural Philosophy*, Great Books of the Western World, vol. 34, pp. 8–13 [2nd ed., 1990, vol. 32].

6. "Philosophical Consequences of Relativity," *Encyclopaedia Britannica*, 14th ed., vol. 19 (Chicago: Encyclopaedia Britannica, Inc., 1973), pp. 101–103. This was the last issue of the 14th ed. The article does not appear in the 15th ed. (the *Macropaedia–Micropaedia* format initiated in 1974).

7. Michelson earlier performed a prototype experiment in 1881, but the lack of mechanical finesse was thought to introduce compensating errors larger than the differences the experiment was designed to detect.

8. "Philosophical Consequences of Relativity," p. 102.

9. Hermann Minkowski, "Space and Time," address at the 80th assembly of the German Natural Scientists and Physicians, held at Cologne on September 21, 1908, reprinted in *The Principle of Relativity* (New York: Dover Publications, Inc., 1952), pp. 75–91. Einstein later accepted the tenet of the space-time continuum, becoming one of its most vocal advocates.

10. Albert Einstein, "Space-Time," *Encyclopaedia Britannica*, 1973, vol. 20, pp. 1069–1074.

11. *Encyclopaedia Britannica*, 1992, vol. 18, p. 156. Einstein envisioned determinacy by way of equations of various continuums rather than quantum interactions.

12. For example, see John Horgan, *The End of Science* (Reading, Massachusetts: Addison–Wesley, 1996).

13. Planck was experimenting with radiation emitted from a so-called black box. When the energy emitted did not hew to the predicted continuous wave function, he resolved the matter by presuming radiation escaped in discrete increments or quanta. He did not follow up on this line of reasoning extensively, but Einstein did (at first). See also the next note.

14. Einstein, "Physics and Reality," *The Journal of the Franklin Institute*, vol. 221, no. 3, March 1936. A reprint appears in *Ideas and Opinion* (New York: Crown, 1954). As for Planck's quantum hypothesis, Einstein earlier supported that idea in an article ("On a Heuristic

Viewpoint Concerning the Production and Transformation of Light") published in the same issue of *Annalen der Physik* (1905) that carried his paper on special relativity.

15. James Glanz, "Collisions hint that quarks may not be so indivisible," *Science*, February 9, 1996.

16. *Encyclopaedia Britannica*, vol. 5, 1973, p. 140.

17. Robert Wright, "Can Machines Think?" *Time*, March 25, 1996, pp. 50–58.

18. Joan Baum, *The Calculating Passion of Ada Byron* (Archon Books, 1986), pp. 1, 67–83. A century later, Alan Turing, working at Bletchley Park on the *Colossus* computer, found Ada's notes to be the best available reference on programming.

19. Will and Ariel Durant, *The Age of Napoleon* (New York: Simon & Schuster, 1975), p. 146.

20. For an excellent account, see Richard Hough and Denis Richards, *The Battle of Britain* (New York: W. W. Norton, 1989).

21. "Drs. N. Bohr and L.Onsager on damage uranium atom explosion could cause," *The New York Times*, April 30, 1939, p. 35.

22. Rhodes, p. 544.

CHAPTER 6. INSIGHT FROM THE SOCIAL SCIENCES

1. Durant, *Rousseau and Revolution*, p. 802.

2. Based on a 1990 Harris poll as reported in *Newsweek*, June 3, 1991, pp. 40–42. The comparative figures were 47 percent in 1977 and 79 percent in 1990.

3. In French: "*Il avait été à la peine, c'etait bien raison qu'il fut l'honneur.*"

4. Franz G. Alexander and Sheldon T. Selesnick, *The History of Psychiatry* (New York: Harper & Row, 1966), p. 100.

5. Ibid., p. 14. A few sentences later, the authors wrote: "The integration of brain chemistry with psychology is the principal task which psychiatry is facing in our present era."

6. William Kornblum, *Sociology in a Changing World* (Fort Worth, Texas: Harcourt Brace College Publishers, 1994), pp. 18–23.

7. Abraham Kardiner and Edward Preble, *They Studied Man* (New York: New American Library, 1961), pp. 36, 37, 39, 48–49.

8. Ibid., pp. 54, 55, 57–60.

9. Ibid., pp. 99, 100–102, 107–109.

10. Ibid., pp. 121–124, 128, 134.

11. Ibid., pp. 145, 148, 150–151.

12. Ibid., pp. 169–172, 175.

13. Ibid., pp. 180–182, 186.
14. This may be changing. The study of motivational leadership ha gained some academic respectability, to include the memory c Abraham Lincoln. For example, see Donald T. Phillips, *Lincoln o Leadership* (New York: Warner Books, 1992).
15. *Encyclopaedia Britannica*, 1992, vol. 9, p. 160.
16. The depicted spectrum of governments was derived from th "Great Ideas" in the *Syntopicon* of the *Great Books of the Wester World* set.
17. The term *metaeconomics* was often used in the 19th century t describe "critiques" of economic theory, for example, by Marx an Weber. However, that meaning has fallen into disuse and is seldor found in economic texts (except as historical notes).
18. Broadus Mitchell, *Great Economists and Their Times* (Totowa, Nev Jersey: Littlefield, Adams & Co., 1966), pp. 16–39, 60, 75, 97–114 176, 191, 213–230.
19. George W. Baer, "U.S. Naval Strategy 1890–1945," *Naval Wa College Review*, Winter 1991, pp. 6–33.
20. See Richard B. Morris, *Encyclopedia of American History* (New Yorl Harper & Row, 1961), p. 458. Apparently, the remark was firs made by the director of the U.S. Census (Herman Hollerith) i 1890, then repeated without credit by Turner in 1894.
21. Hawaii was needed as a coaling station for naval control over th Philippines.

CHAPTER 7. TRUTH, LOGIC, AND COMMUNICATIONS

1. *The Harper Dictionary of Modern Thought*, pp. 279, 485, 605.
2. Ibid. In a sense, anything that exists is perforce an element of truth but the root meaning of epistemology is "system," hence th emphasis on relationships.
3. *Encyclopaedia Britannica*, 1992, vol. 27, p. 572. The astronomer wa Urbain-Jean-Joseph Le Verrier.
4. Roy P. Basler, ed., *The Collected Works of Abraham Lincoln*, vol. (New Brunswick, New Jersey: Rutgers University Press, 1953) p. 62. The comment was made within a brief autobiograph Lincoln prepared for John L. Scripps circa August 1860.
5. Ernest Nagel and James R. Newman, "Goedel's Proof," in *Th World of Mathematics*, vol. 3, pp. 1668–1695. His proof may or ma not cover the inability of theorems to prove their own axioms, bu no one has ever achieved that feat in practice.

6. Much has been written about Euclid's postulate (assumption) on parallel lines never meeting no matter how far extended, because it seemed less like a self-evident assumption and more like a theorem to be proved. The attempts to do so led to different (non-Euclidean) geometries. This in no way detracts from Euclid's *Elements* as a deductive model.

7. Rhodes, p. 261. Bohr used the word *idiots* rather than *fools*.

8. C. W. Ceram, ed., *Hands on the Past* (New York: Alfred A. Knopf, 1966), pp. 159–161, 163–164.

9. *The Little, Brown Book of Anecdotes*, p. 555.

10. Robert B. Downs, *Books that Changed the World* (New York: New American Library, 1956), p. 80.

CHAPTER 8. AUTOMATION AND COMPUTER SCIENCE

1. Erwin Schrödinger, *What is Life?: The Physical Aspect of the Living Cell* (Cambridge: Cambridge University Press, 1967), pp. 5, 23. This reference can also be found in the *Great Books of the Western World* set, 2nd ed., 1990, vol. 56, pp. 469–508.

2. Ivars Peterson, "Computing with DNA," *Science News*, July 13, 1996, pp. 26–27.

3. These limited operations become evident with assembly language commands, into which all higher-order languages must be compiled before being run on a computer.

4. This is the National Research and Education Network, funded by the High Performance Computer Act of 1991 for roughly $3 billion. The high-capacity lines will link 12 research stations. About 11,300 other stations will be tied into this network (although those links will transmit at much slower speeds).

5. Some advocates of relational databases claim that they can produce information that was never entered. Well, yes and no. The linkage (logical interior lines) of these databases often suggests relationships that were not previously *recognized*, for essentially the same reason that manual use of a categorical thesaurus often *suggests* new ideas to writers.

6. The author developed these categories in *Strategy, Systems and Integration* (Blue Ridge Summit, Pennsylvania: Tab Books, 1991), pp. 20–21, not realizing at the time that the categorization would prove useful for systems in general.

7. "Use of an Artificial Neural Network for the Diagnosis of Myocardial Infarction," *Annals of Internal Medicine*, December 1991,

pp. 843–848. Dr. William Baxt (University of California, San Diego) wrote a program that when used by emergency-room physicians yielded a 97 percent accuracy rate, whereas physicians who treated patients without the system correctly diagnosed this condition only 78 percent of the time.

8. D.I. Bainbridge, "Computer Aided Diagnosis and Negligence," *Medicine, Science, and the Law*, April 30, 1991, pp. 127–136.

9. Alan Turing, "Can a Machine Think?" in Timothy Ferris, ed., *The World Treasury of Physics, Astronomy, and Mathematics* (Boston: Little, Brown and Company, 1991), pp. 492–519. The original article appeared in *Mind*, 1950.

CHAPTER 9. MILITARY SCIENCE AND GAME THEORY

1. *Bartlett's Familiar Quotations*, p. 371. Wellington once denied he had said it.

2. U.S. Army Field Manual FM 100-5 *Operations*, all editions since 1935.

3. George M. Hall, "Field Expedient for the Principles of War," *Military Review*, March 1983, pp. 34–43. Also, see Stephen J. Kirin, "Synchronization," *Naval War College Review*, Autumn 1996, pp. 7–22. Kirin's article describes the evolving military concept of *synchronization* that is arguably the same thing as the physiogenetic model applied to an abstract system.

4. Samuel Eliot Morrison, *The Two-Ocean War* (Boston: Little, Brown and Company, 1963), pp. 147–163.

5. Carl von Clausewitz, *On War*, ed. and trans. by Michael Howard and Peter Paret (Princeton: Princeton University Press, 1976), pp. 579–581.

6. The author developed these levels of perspective in several articles and then formalized them in *The Fifth Star* (Westport, Connecticut: Praeger, 1994), pp. 139–150.

7. Actually, MacArthur was the most decorated soldier in U.S. history, with the Medal of Honor, three Distinguished Service Crosses, four Distinguished Service Medals, and seven Silver Stars (among other awards). See *Register of Graduates and Former Cadets 1802–1990*, p. 325, published by the Association of Graduates, USMA.

8. Will and Ariel Durant, *The Age of Napoleon* (New York: Simon & Schuster, 1975), p. 748.

9. Dean Acheson, *Present at the Creation* (New York: W.W. Norton, 1969), pp. 451–455.

10. The Maoris gained the respect of the British regular army, who in turn came to detest the conduct of the British settlers. See *History of the English-Speaking Peoples*, The Great Democracies, vol. 4 (New York: Dodd, Mead & Co., 1958), p. 125.

11. Clausewitz, p. 572.

12. U.S. Army Field Manual FM 100-5 *Operations* (Washington, D.C.: U.S. Government Printing Office, 1986), p. 181.

13. Thucydides, *The History of the Peloponnesian War*, Great Books of the Western World, vol. 6, p. 403 [2nd ed., 1990, vol. 5].

14. George M. Hall, "Cycle of Military Technology," *Military Review*, August 1988, pp. 42–48.

CHAPTER 10. PHYSICS AND CHEMISTRY

1. Rhodes, pp. 479–485.

2. Mass and matter are more or less equivalent; whereas weight is a force, in this case the effect of the earth's gravity pulling on the mass of the car.

3. Energy can also be defined tautologically as a time-cross-section of power.

4. Science universally recognizes that kinetic energy results only from motion in some form, and anything that moves possesses momentum by definition. The possible exceptions are the photon and neutrino, which have zero mass but still have energy content.

5. Rhodes, p. 42. Ernest Rutherford developed the schema.

6. Ibid., p. 461.

7. Ibid., p. 702.

8. Ibid., p. 655.

9. *Encyclopaedia Britannica*, 1992, vol. 5, pp. 60–61.

10. The nuclear packing fraction refers to the degree to which an atomic nucleus absorbs or otherwise interacts with free neutrons and other particles—in effect, a measure of stability. On the periodic table, this stability decreases in both directions from silver.

11. Malcolm W. Brown, "Physicists get warmer in search for weird matter close to absolute zero," *The New York Times*, August 23, 1994, p. B5. (Bose–Einstein condensate.)

12. Note that of 259 stable nuclei, 156 (60 percent) have an even number of protons and electrons. Only five have an odd number of both. Of the remaining 98 nuclei, half have an even number of protons and an odd number of neutrons. The other half is the reverse.

13. *Encyclopaedia Britannica*, 1992, vol. 2, p. 849.
14. Nicholas Rashevsky, *Mathematical Biophysics: Physico-Mathematical Foundations of Biology*, 3rd rev. ed., vol. 2 (Chicago: University of Chicago Press, 1960), p. 417.
15. George M. Hall, "Renaissance Warrior," *Army*, December 1988, pp. 50–58.
16. *Medal of Honor Recipients 1863–1973* (Washington, D.C.: U.S. Government Printing Office, 1973), p. 478.

CHAPTER 11. ENGINEERING

1. Bridge models amply illustrate the points of diminishing and negative returns because there comes a length whereby the increased deadweight of an extended beam reduces the load it can take. Trusses and arches substitute semihollow structure for mass in order to reduce weight while remaining able to transfer stress. In turn, these models eventually reach a point of diminishing returns, leaving only the suspension bridge model.
2. The north tower of the Golden Gate Bridge was built adjacent to land, so that the final 1200 feet goes over land, not water.
3. Richard K. Smith, "Not a Success—But a Triumph: 80 Years Since Kitty Hawk," *Naval War College Review*, November–December 1983, pp. 4–20.
4. Ibid.
5. Edward Mead, *Makers of Modern Strategy* (Princeton: Princeton University Press, 1971), pp. 485–503.
6. James Burke, *Connections* (Boston: Little, Brown and Company, 1978), p. 1.
7. Richard S. Westfall, *Never at Rest: A Biography of Isaac Newton* (New York: Cambridge University Press, 1980), p. 92.
8. William S. Ellis, "The Aral: A Soviet Sea Lies Dying," *National Geographic*, February 1990, pp. 73–92.
9. For example, see "Population Wars," *U.S. News & World Report*, September 12, 1994, pp. 54–64.

CHAPTER 12. PHYSIOLOGY AND GENETICS

1. Sharon Begley, "The Cancer Killer," *Newsweek*, December 23, 1996, p. 43.
2. *American Medical Association Encyclopedia of Medicine*, pp. 245–247.

3. A few lesser orders also possess this attribute, for example, large fish such as the common tuna. Some game fish (e.g., the marlin) heat part of their cerebrum and sensory organs. Some dinosaurs may have also possessed this capability, and mammalian temperatures often drop significantly during hibernation.

4. Though a few references state that parthenogenesis occurs in mammals (e.g., W.B. Crow, *A Synopsis of Biology*, 1964, p. 35), there is little evidence to support that statement.

5. Richard Milner, *The Encyclopedia of Evolution* (New York: Facts on File, 1990), p. 77.

6. Will Durant, *Our Oriental Heritage* (New York: Simon & Schuster, 1935), p. 164.

7. See Felix Franks, *Biophysics and Biochemistry at Low Temperatures* (New York: Cambridge University Press, 1985).

8. Lloyd J. Old, "Immunotherapy for Cancer," *Scientific American*, September 1996, pp. 136–143.

9. *American Medical Association Encyclopedia of Medicine*, pp. 59–61.

CHAPTER 13. PATHOLOGY

1. Based on *The American Medical Association Encyclopedia of Medicine*. Other authorities state 3000, 4000, or even 5000 known disorders. Part of the discrepancy arises from subclassifications within the same disorder. The circulatory system diseases and cancers account for approximately 77 percent of all deaths in the United States.

2. Eckstein, pp. 102–103.

3. For example, see "Penicillin From a Screen? Making new drugs 'rationally'—on a computer," *U.S. News & World Report*, September 14, 1992, pp. 58–59.

4. Rhodes, pp. 491, 557, 558.

5. J. L. Lyon, *et al.*, "Cancer incidence among Mormons and non-Mormons in Utah," *Cancer Causes & Control*, March 1994, pp. 149–156; and E. M. Moran, "Epidemiological factors of cancer in California," *Journal of Environmental Pathology, Toxicology & Oncology*, September–October 1992, pp. 303–307. The consensus is that Mormons (and Seventh-Day Adventists) experience 50 percent fewer cancers commonly associated with smoking, and 25 percent fewer cancers of all types. Individual rates correlate with the degree to which Mormon prohibitions (against alcohol, nicotine, or caffeine) are adhered to. These groups also have a 3- to 4-year longer life expectancy than the general population.

6. *American Medical Association Encyclopedia of Medicine*, pp. 478–481.
7. Dimitrios Trichopoulos, Frederick P. Li, and David J. Hunter, "What Causes Cancer?" *Scientific American*, September 1996, p. 87.
8. *American Medical Association Encyclopedia of Medicine*, p. 708.
9. *The New York Times*, September 24, 1993, p. A10. A baby with anencephaly was born in October 1992 at Fairfax Hospital, Fairfax, Virginia. The baby subsequently died of complications from pneumonia during its second year of life.
10. *American Medical Association Encyclopedia of Medicine*, p. 379.
11. Ibid., pp. 844–845.
12. Ibid., p. 60. The data cited was for the year 1982, but current references reiterate the statistic (proportionally). In 1982, 1,974,797 deaths were recorded. Of these, 59,506 were between the ages of 1 and 24 (representing about 3 percent of all deaths). Within the 59,506 deaths, 28,894 were attributed to accidents (about 49 percent of this subgroup).
13. *American Medical Association Encyclopedia of Medicine*, pp. 711–712.
14. Ibid., p. 237.
15. For example, see Stephen S. Morse, *The Evolutionary Biology of Viruses* (New York: Raven Press, 1994).
16. *American Medical Association Encyclopedia of Medicine*, p. 1015.
17. "Does Screening for Prostate Cancer Make Sense?" *Scientific American*, September 1996, pp. 114–115.
18. Erkki Ruoslahti, "How Cancer Spreads," *Scientific American*, September 1996, pp. 72–77.
19. Ibid.
20. Robert A. Weinberg, "How Cancer Arises," *Scientific American*, September 1996, pp. 62–70.
21. "Cancer Research," *Encyclopaedia Britannica*, vol. 4 (Chicago: Encyclopaedia Britannica, Inc., 1973), p. 777. Additionally, Haddow noted that many studies strongly indicated that chemical carcinogens in particular interfere with the synthesis of proteins, which therefore might nullify one or more crucial enzymes (either by chemical combination or by weakening the integral structure of DNA) that are responsible for the host cell's capability of passing on genetic information.
22. Ibid.
23. James R. Watson, *Molecular Biology of the Gene*, 2nd ed. (Menlo Park, California: W. A. Benjamin, Inc., 1970), pp. 598–599.
24. *American Medical Association Encyclopedia of Medicine*, p. 228.
25. Walter C. Willett, Graham A. Colditz, and Nancy E. Mueller, "Strategies for Minimizing Cancer Risk," *Scientific American*, September 1996, pp. 94–95.

26. Old, pp. 136–143.
27. David Sidransky, "Advances in Cancer Detection," *Scientific American*, September 1996, pp. 104–106.
28. *American Medical Association Encyclopedia of Medicine*, pp. 46–48.
29. Michael Specter, "Plunging life expectancy puzzles Russia," *The New York Times*, August 2, 1995, p. A1.
30. The United States does not have clean skirts in this matter. For a brief but excellent discussion of the eugenics craze in the United States during the first few decades of the 20th century, see Liva Baker, *The Justice from Beacon Hill: The Life and Times of Oliver Wendell Holmes* (New York: HarperCollins Publishers, 1991), pp. 590–604.

CHAPTER 14. BLUEPRINT OF EVOLUTION

1. *Encyclopedia of Evolution*, pp. 375–376.
2. Ibid., pp. 387–388.
3. A. I. Oparin, trans. by Sergius Morgulis, *Origin of Life on Earth* (New York: Dover Publications, Inc., 1953), p. 133.
4. *Encyclopaedia Britannica*, 1992, vol. 15, pp. 604–607.
5. The same does not seem to be the case for other musical instruments, probably because they are not as complex. Also, wind instruments self destruct from moisture after 80 years of use. Moreover, their design and tonal attributes have changed in the last 100 years so that even if preserved, old wind instruments can no longer be used by modern orchestras.
6. The *Encyclopedia of Evolution* (New York: Facts on File, 1990) makes no mention of Oparin, and the last issue of the 14th edition of the *Encyclopaedia Britannica* (1973) briefly mentioned him only once. In the 15th edition, however, Oparin now rates a two-column article in the *Micropaedia*, and his work is cited in four other articles.
7. Ibid., p. 77.
8. Ibid., pp. 83–84.
9. *Encyclopaedia Britannica*, 1992, vol. 19, p. 714.
10. J. Travis, "Third Branch of Life Bears its Genes," *Science News*, August 24, 1996, p. 110.
11. *The Harper Dictionary of Modern Thought*, p. 917.
12. *The New York Times*, September 24, 1993, p. A10. Dr. Michael Grodin, head of the Ethics Committee at Boston University School of Medicine, summed up the case with the statement "technology has the life not the patient." The only known ethicist to side with the mother was Dr. Robert M. Veatch, of the Kennedy Institute of Ethics at Georgetown University. He agreed that from a medical

viewpoint, the baby's life should *not* be prolonged, but nevertheless the mother had the right to decide otherwise.

CHAPTER 15. THE PSYCHE, MARRIAGE, AND ORGANIZATIONAL BEHAVIOR

1. Robert L. Chapman, ed., *Roget's International Thesaurus*, 5th ed. (New York: HarperCollins Publishers, 1992), pp. xi–xiii, xvii–xviii.
2. Robert M. Goldenson, *The Encyclopedia of Human Behavior* (Garden City, New York: Doubleday, 1970), p. 375.
3. Ibid., p. 1281.
4. *The American Medical Association Encyclopedia of Medicine*, p. 678.
5. Jane Healy, *Endangered Minds: Why Our Children Don't Think* (New York: Simon & Schuster, 1991). The 20 percent decrease in creativity corresponds almost exactly with the decline in Scholastic Aptitude Test (SAT) results.
6. *The American Medical Association Encyclopedia of Medicine*, p. 678.
7. See *Diagnostic and Statistical Manual of Mental Disorders IV* (Washington, D.C.: American Psychiatric Association, 1994); and the *Pocket Guide to the ICD-10 Classification of Mental and Behavioral Disorders* (World Health Organization, Geneva), published by the American Psychiatric Press. In the latter (p. 257), a new category (F688) has been set up to cover character disorders not elsewhere specified, though the other disorders described are clearly or at least primarily mental disorders. Whether the intent is to create a bin for a few odd disorders or to open Pandora's box to encompass the whole of human behavior is a matter of conjecture, but the potential for it is there.
8. For example, consider the ongoing debate between followers of Freud versus Carl Jung. The latter has its own network called *Friends of Jung.*
9. Goldenson, *The Encyclopedia of Human Behavior*, p. 1153.
10. Ibid., pp. 919–923.
11. *Encyclopaedia Britannica*, 1992, vol. 9, p. 495.
12. Karl Menninger, M.D., *Theory of Psychoanalytic Technique* (New York: Harper, 1964), pp. 75–76.
13. Ibid., p. 75.
14. Sir William Osler, *A Way of Life* (New York: Dover Publications, Inc., 1958), p. 246.
15. "Lincoln, Abraham," *Encyclopaedia Britannica*, vol. 14 (Chicago: Encyclopaedia Britannica, Inc., 1959), pp. 141–143. This article first appeared in 1929.

16. Herman Melville, *Moby Dick*, Great Books of the Western World, vol. 48, p. 136 [2nd ed., 1990, vol. 48, p. 84].

17. Churchill, *Memoirs of the Second World War*, p. 24.

18. Joannie M. Schrof, "A lens on matrimony," *U.S. News & World Report*, February 21, 1994, pp. 66–69, summarizing John Gottman's *Why Marriages Succeed or Fail*.

19. Stanley Milgram, *Obedience to Authority: An Experimental View* (New York: Harper, 1974). The original experiments were reported in *Journal of Abnormal Psychology*, 1963, vol. 67, pp. 317–378. Milgram's experiments were replicated worldwide, including one (without Milgram's prior knowledge) that used animals and actual voltages.

20. T. E. Lawrence, *Seven Pillars of Wisdom* (New York: Doubleday, 1926), p. 1.

21. Bruce Clarke and John G. Hill, *Art and Requirements of Command*, Generalship Study, Technical Report 1–191, vol. II, prepared for the Office of the Director of Special Studies, Office of the Chief of Staff, U.S. Department of the Army, April 1967.

22. Philip Kopper, *Colonial Williamsburg* (New York: Harry W. Abrams, Inc., 1986), pp. 136–155ff.

23. Ibid., p. 164.

24. Adrienne Koch and William Peden, eds., *The Life and Selected Writings of Thomas Jefferson* (New York: Modern Library, 1944), p. 280 [from *Notes on Virginia*].

25. Paul Sniderman and Thomas Piazza, *The Scar of Race* (Cambridge, Massachusetts: Belknap Press, 1993).

26. "Oprah's poverty program stalls," *Chicago Tribune*, August 27, 1996, pp. 1–16.

CHAPTER 16. ECONOMICS

1. This is a variation on a remark made by the columnist Jane Bryant Quinn, namely, that "The chief function of stock-market forecasters is to make astrologers look respectable" (*Newsweek*, October 18, 1993, p. 62).

2. Allan Sloan, "Big Boys with Their Bigger Toys," *Newsweek*, September 11, 1995, p. 58.

3. "The FBI is Shifting 50 Agents to Health Care Fraud Duties in 12 Cities," *The Washington Post*, March 24, 1992, p. H5. The estimate of fraudulent Medicare claims on the part of providers was between $40 and $50 billion annually.

4. As defined in *Dictionary of Economics* (New York: Barnes & Noble, 1970), p. 202, pp. 510–511, gross national product includes (at

market prices) consumer purchases, government purchases, gross private domestic investments, and net export–import of goods and services. The gross domestic product excludes the last category.

5. When a government increases welfare and entitlement programs, it must obtain the necessary revenue by increased taxation and/or borrowing, which takes money out of circulation faster than it would with lower tax rates.

6. *Britannica World Data Annual 1993* (Chicago: Encyclopaedia Britannica, Inc., 1993), p. 727.

7. In 1986, the worldwide gross national product was $15,710 billion. The United States, the Soviet Union, and Japan accounted for $8137 billion (52 percent) of it. France, Germany, the United Kingdom, and Italy accounted for another $2325 billion, so that seven countries controlled 67 percent of the total.

8. The causes of the Great Depression were not quite that simple, yet the complexities do not gainsay the fundamental nature of the chain-reaction collapse. For a short but insightful commentary on the matter, see John Kenneth Gailbraith, "Why the Money Stopped," in *A Sense of History: The Best Writing from the Pages of American Heritage* (Boston: Houghton Mifflin, 1985), pp. 670–680.

9. *Statistical Abstract of the United States: 1995*, table 526, p. 341.

10. James I. Cash, Jr., *Corporate Information Systems Management: Text and Cases*, 2nd ed. (Homewood, Illinois: Richard D. Irwin, Inc., 1988), pp. 146–162.

11. *Statistical Abstract of the United States: 1995*, table 846, p. 543. The data for corporations reporting over $1 million must be further subdivided by reference to the *Fortune* magazine 1000-largest corporations list.

12. Will Durant, *The Age of Faith* (New York: Simon & Schuster, 1950), pp. 614–649.

13. Durant, *The Lessons of History*, pp. 60–61.

14. Frederick R. Stobel, *Upward Dreams, Downward Mobility: The Economic Decline of the American Middle Class* (Savage, Maryland: Rouman & Littlefield Publishers, 1993).

15. *Statistical Abstract of the United States: 1995*, table 534, p. 346.

CHAPTER 17. NATIONS AND GOVERNMENT

1. General Accounting Office Special Report GAO/HR-93-13, *Internal Revenue Service Receivables*, December 1992. The figure for 1992 was $111 billion. Subsequent reports indicate that the annual rate has risen to $130 billion.

2. Edward Gibbon, *The Decline and Fall of the Roman Empire* (chap. 23), Great Books of the Western World, vol. 40 [2nd ed., 1990, vol. 37].

3. Gene Smith, *When the Cheering Stopped: The Last Years of Woodrow Wilson* (New York: William Morrow & Company, 1964), pp. 53–83*ff.*

4. Durant, *Age of Napoleon*, p. 72.

5. At the $4.25 minimum-wage rate (scheduled to be increased), annual earnings equal to 1000 times that wage are $8,840,000. Hundreds of celebrities make more than that.

6. Durant, *Age of Napoleon*, p. 153.

7. Barbara Tuchman, *The March of Folly: From Troy to Vietnam* (London: Michael Joseph, Ltd., 1984), pp. 246–259.

8. The incident occurred when Captain Charles Wilkes, commanding the U.S.S. *San Jacinto*, boarded and searched the British vessel *Trent* on the high seas, taking two Confederate diplomats into custody. Great Britain vaguely threatened to declare war on the Union unless the United States apologized and released the prisoners. Lincoln complied, though the majority of the cabinet and perhaps the Union citizenry were against it.

9. Mostafa Rejai, "The Professional Revolutionary," *Air University Review*, March–April 1980, p. 90.

10. George F. Will, *The Leveling Wind* (New York: Viking, 1994), p. 352.

CHAPTER 18. INTERNATIONAL RELATIONS AND WAR

1. Paul Kennedy, *The Rise and Fall of the Great Powers* (New York: Random House, 1987), pp. 86–89.

2. Hans Morgenthau, "Alliances and National Security," *Perspectives in Defense Management*, Autumn 1973, p. 20.

3. Alfred T. Mahan, *The Influence of Sea Power Upon History, 1660–1783* (Boston: Little, Brown and Company, 1890), pp. 28–88.

4. Frederick H. Hartmann, *The Relations of Nations*, 6th ed. (New York: Macmillan Publishing Company, 1983), pp. 43–64.

5. Maurice Matloff, "The 90-Division Gamble," in Kent Greenfield, ed., *Command Decisions* (Washington, D.C.: U.S. Government Printing Office, 1959), pp. 365–382.

6. For example, see Desmond Flower and James Reeves, eds., *The Taste of Courage* (New York: Harper & Row, 1960), pp. 705–778.

7. *Bartlett's Familiar Quotations*, p. 397. The actual statement was: "We have no eternal allies and we have no perpetual enemies. Our interests are eternal and perpetual, and these interests it is our duty to follow" (Parliament, March 1, 1848).

8. C. de Montesquieu, *The Spirit of Laws*, Great Books of the Western World, vol. 37, p. 60 [2nd ed., 1990, vol. 35].

9. Kennedy, pp. 347–356.

10. The ratio varies with terrain and other circumstances, but 3:1 is generally regarded as a minimum by the military profession because the attacker is much more vulnerable to fire than are defenders.

11. Alan S. Milward, *Economy and Society, 1939–1945* (Berkeley: University of California Press, 1979), p. 74.

12. *Bartlett's Familiar Quotations*, p. 438.

13. Downs, pp. 107–117.

14. This categorization of threats was derived from Donald E. Neuchterlein, "National Interest and Foreign Policy," *Foreign Service Journal*, vol. 54, July 1977, pp. 6–8, 27.

15. Winston S. Churchill, *Memoirs of the Second World War* (Boston: Houghton Mifflin, 1959), p. 5.

16. Captain King did this while commanding the aircraft carrier *Lexington* (1932). In the second instance (1938), Rear Admiral King was in command of a carrier task force. Apparently the Japanese studied the results more thoroughly than did U.S. observers.

17. Churchill, *Memoirs of the Second World War*, p. 12. He said he wrote his memoirs "to show how easily the tragedy of the Second World War could have been prevented."

18. Barbara Tuchman, *The Guns of August* (New York: Macmillan Publishing Company, 1962), pp. 1–15.

19. Clausewitz, *On War*, pp. 105, 139.

20. B. H. Liddell Hart, *Strategy*, 2nd ed. (New York: Praeger, 1967), p. 372.

21. The annual *National Travel Report* is a widely used source because most terrorist crimes are inflicted on victims in transit. The annual total rarely exceeds 2000. Another valid source is the annual State Department report of the subject, which for the year 1993 indicated that worldwide terrorist attacks resulted in 109 deaths and 1393 nonfatal injuries.

22. David Dickson, "Concern grows over China's plans to reduce number of inferior births," *Nature*, January 6, 1994, p. 3.

CHAPTER 19. ETHICS, THEOLOGY, AND JURISPRUDENCE

1. James MacGregor Burns, *Roosevelt: The Soldier of Freedom* (New York: Harcourt Brace Jovanovitch, Inc., 1970), p. 601.

2. The advocacy of civil disobedience traces back to Cicero, St. Aquinas, and John Locke. Thoreau's advocacy implied using it as a form of controlled revolution. Yet in almost all cases, the common theme of these writers is obedience to a superhuman moral law.

3. For an interesting comparison, see David B. McCall, "Profit: spur for solving social ills," *Harvard Business Review*, May–June 1973, pp. 46–54, 180, and Kenneth R. Andrews, "Can the best corporations be made moral?" in the same issue, pp. 57–64.

4. Jerome Lawrence and Robert E. Lee, *Inherit the Wind* (New York: Random House, 1955), p. 82.

5. Edward Tenner, *Why Things Bite Back: Technology and the Revenge of Unintended Consequences* (New York: Alfred A. Knopf, 1966), pp. 26–46, 71–94, 161–183.

6. The author developed this spectrum in *Age of Automation: Technical Genius, Social Dilemma* (Westport, Connecticut: Praeger, 1995), pp. 118–121.

7. An earlier set was Augustus Hopkins Strong, *Systematic Theology* (1886), in three volumes (and which is still used in some seminaries).

8. Will and Ariel Durant, *The Age of Voltaire* (New York: Simon & Schuster, 1965), pp. 787–798.

9. *The Encyclopedia of Evolution*, p. 399.

10. "Senate in Tennessee backs '10 Commandments' posting," *The New York Times*, February 23, 1996, p. A12 (and "Evolution measure killed," March 29, 1996, p. A9).

11. E. N. Da C. Andrade, "Isaac Newton" in *The World of Mathematics*, vol. 1, p. 274. The Archbishop at the time was the Rt. Rev. Thomas Tenison.

12. *Schenck v. U.S. 249 U.S. 47* (1919). Holmes wrote: "The most stringent protection of free speech would not protect a man in falsely shouting fire in a theater and causing a panic." The case involved the right of the government to curtail free speech in time of "clear and present danger," i.e., the U.S. role in World War I.

13. With reference to Holmes's *The Common Law*, see Baker, pp. 246–270.

14. Edward W. Knappman, ed., *Great American Trials* (Boston: New England Publishers Association, 1994). This summarizes 200 cases, dating back to colonial times.

15. The deciding case was that of Dr. John Webster in 1850. Webster, a professor at Harvard, was found guilty of murder based solely on circumstantial but conclusive evidence. He was subsequently hung (*Great American Trials*, pp. 105–108).

16. *The Encyclopedia of Evolution*, p. 399.

17. The camp was *not* benign. Of 140,000 persons sent to Theresienstadt, 88,000 were later shipped to extermination camps and 33,000

perished of malnutrition and disease. Of the 19,000 who survived, most had arrived within the last year of the war. Israel Gretman, ed., *Encyclopedia of the Holocaust*, vol. 4 (New York: Macmillan Publishing Company, 1990), p. 1463.

18. Leo Tolstoy, *War and Peace*, Great Books of the Western World, vol. 51, p. 680.
19. "The other Schindlers," *U.S. News & World Report*, March 21, 1994, pp. 56–64.
20. Max Lerner, ed., *The Mind and Faith of Justice Holmes* (New York: Modern Library, 1943), p. 39.
21. Carl G. Jung, *The Undiscovered Self*, trans. by R.F.C. Hull (New York: New American Library, 1957), p. 73.
22. *Churchill in Memoriam*, ed. *The New York Times* staff (New York: Bantam, 1965), p. 159.

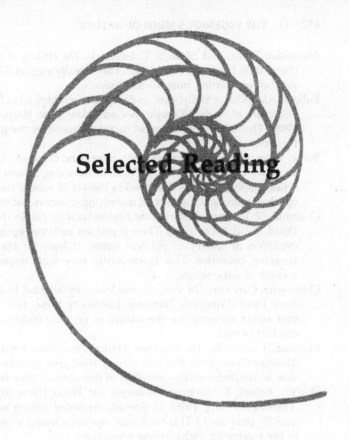

Selected Reading

In developing this book over a period of more than 20 years, I lost track of the tens of thousands of books, articles, and references that contributed to it. On military affairs alone, I read or scanned several thousand articles while contributing 12 of my own. Another problem with references on systems is that most are overwrought with mathematical description and esoteric expressions. The exceptions comprise a comparative handful of general references and these were the most useful. The few listed here were the most helpful to me, exclusive of thousands of forays into the *Encyclopaedia Britannica* and the *Great Books of the Western World* set.

Alexander, Franz, and Sheldon T. Selesnick, *The History of Psychiatry* (New York: Harper & Row, 1966). This is really a general history of psychology starting from ancient times.

Bullock, Alan, Oliver Stallybrass, and Stephen Trombley, eds., *The Harper Dictionary of Modern Thought*, rev. ed. (New York: Harper & Row, 1988). This was the single most useful reference for the purpose of this book.

Burke, James, *Connections* (Boston: Little, Brown and Company, 1978). This book, which was made into a Public Broadcasting System television series, amply illustrates the seamless blanket of science and technology, hence strongly suggests an underlying common systems theory.

Churchill, Sir Winston, *History of the English-Speaking Peoples* (New York: Dodd, Mead & Co., 1962). There is perhaps no better account of the evolution of society in political terms, at least for the English-speaking countries. This is especially true with respect to the concept of nationalism.

Clausewitz, Carl von, *On War*, ed. and trans. by Michael Howard and Peter Paret (Princeton: Princeton University Press, 1976). Still the best single reference on the nature of systems underwriting the conduct of war.

Clayman, Charles B., *The American Medical Association Encyclopedia of Medicine* (New York: Random House, 1989). The practice of medicine is comprehensively described in this authoritative book.

Downs, Robert B., *Books that Changed the World* (New York: New American Library, 1956). (A revised, expanded edition was subsequently published.) This book was especially useful when it came to the mechanics underwriting revolutions.

Durant, Will and Ariel, *The Story of Civilization*, 11 vols. (New York: Simon & Schuster, 1935–1975). Also the one-volume summary volume *The Lessons of History* (1968). It is hard to find a more comprehensive and readable history.

Eckstein, Gustav, *The Body Has a Head* (New York: Harper & Row, 1969). On rereading this book, I realized how influential it had been on my thinking, especially on the logical consequences of anchoring psychical attributes to physiology.

Ferris, Timothy, ed., *The World Treasury of Physics, Astronomy, and Mathematics* (Boston: Little, Brown and Company, 1991), especially John von Neumann, "The Computer and the Brain," pp. 478–491, and Alan Turing, "Can a Machine Think?" pp. 492–519.

Goldenson, Robert M., *The Encyclopedia of Human Behavior*, 2 vols. (Garden City, New York: Doubleday, 1970). Perhaps the most readable comprehensive reference set on psychology and psychiatry ever compiled.

Kardiner, Abram, and Edward Preble, *They Studied Man* (New York: World Publishing, 1961). A very good account of some of the most influential anthropologists in the past two centuries.

Menninger, Karl, *Theory of Psychoanalytic Technique* (New York: Basic Books, 1958). The best, and perhaps the only readable description of the "system" underwriting analysis.

Miller, Richard, *The Encyclopedia of Evolution* (New York: Facts on File, 1990). A good reference for understanding the issue and underlying system of evolution insofar as it has been documented.

Mitchell, Broadus, *Great Economists and Their Times* (Totowa: New Jersey: Littlefield, Adams, 1966). The author could not find a better reference on economics written in nontechnical terms, especially as it applies to political economics.

Newman, James R., ed., *The World of Mathematics*, 4 vols. (New York: Simon & Schuster, 1956). Despite having been published 41 years ago, this anthology contains many papers supporting systems theory.

Wiener, Norbert, *Cybernetic: Or Control and Communications in the Animal and the Machine*, 2nd ed. (New York: John Wiley, 1961). This seminal book (originally published in 1948 by M.I.T. Press) describes the cybernetic process and how it applies to inorganic and organic systems alike. Actually, the book covers information theory in general and is tantamount to a monograph on general systems from that viewpoint.

Wormser, René A., *The Story of the Law*, rev. ed. (New York: Touchstone Books, 1962). Perhaps the best single reference on the development of the law from a systematic perspective.

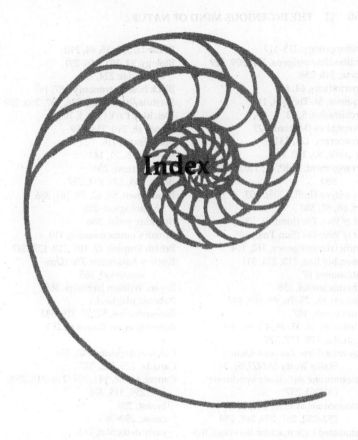

Index

Abortion, 263
Acceleration, 40, 187
Acquired immunity, 240
Adler, Alfred, 114, 289
Aeneid, The (Virgil), 267
Africa, 125, 330
Aid to Families with Dependent Children (AFDC), 305
Alamogordo, 191
Alcohol, 247, 250, 254
Alexander the Great, 168, 346
Alexander, Franz, 113, 114
Alice's Adventures in Wonderland (Carroll), 134, 266
All-or-nothing phenomenon (nervous system), 237

Allergies, 53, 257
Ailuropoda melanoleuca, 287
Alpha particles, 189
American Psychiatric Association, 289
American Revolution, 121, 337
American Scholar, The (Emerson), 25
Amino acids, 271
Amniocentesis, 263
Amoeba, 146, 275, 279
Amyotrophic lateral sclerosis, 257
Analytical engine (Babbage), 103
Analytical Theory of Heat (Fourier), 100
Anaphylactic reaction, 248
Anarchy, 115, 120, 357
Anencephaly, 252, 278
Anthropic principle, 22, 74, 270

Anthropology, 115–117
Antibodies/antigens, 257, 309, 329
Aorta, 246, 258
Aperiodicity, 64, 65
Aquinas, St. Thomas, 110
Archimedes, 5, 43
Areopagitica (Milton), 122
Aristocracy, 121, 331
Aristotle, 93, 110, 326
Arrangement, 8, 40, 41, 120, 124, 197, 380
Ars Magna (Lull, 1305), 103
Art, 69, 87, 380
Art of War, The (Jomini), 106
Art of War, The (Sun Tzu), 105
Artificial intelligence, 162, 164
Assembly line, 213, 214, 311
Astronomy, 87
Atherosclerosis, 258
Atom(s), 66, 75–76, 99, 189, 197
Atom bomb, 107
Attributes, 28, 35, 39, 43, 46, 380
Australia, 125, 177, 327
Australia–New Zealand–United States Treaty (ANZUS), 349
Autoimmune deficiency syndrome (AIDS), 257
Autoimmune system, 241, 247, 252–253, 257, 259, 261, 278
Automated logic machine (Jervons), 103
Automation, 102, 103, 149, 150, 214, 310, 312, 324, 384
Automobile, 6, 69, 210
Awakenings (Sachs), 239
Axioms, 133, 138
Axis powers (World War II), 178

Babbage, Charles, 103
Bacteria, 253, 257, 271
Bacteriophage, 223, 271
Balance of power, 125, 341, 367
Balance of trade, 343
Battle of Britain, 106
Benedict, Ruth, 117
Bernard, Claude, 101, 245
Beta particles, 189
Bipolar disorder, 292
Big bang (cosmology), 12, 22, 74, 96, 218

Billiards, 5, 8, 35, 64, 210
Biology, 94, 101, 116, 201
    molecular, 224, 231
Black hole (astronomy), 87, 193
Blastula/blastocoel, 236, 237, 250, 279
Bletchley Park (U.K.), 105
Blood, 48, 240, 258, 259
Boas, Franz, 116
Bohr, Niels, 21, 142
Bone marrow, 259
Brain, 165, 239, 251, 257
Breakdown, 59, 62, 79, 161, 386
    physiological, 256
Breaking point, 386
British Commonwealth, 110
British Empire, 42, 109, 110, 127, 347
Brothers Karamazov, The (Dostoyevsky), 365
Bryan, William Jennings, 369
Bubonic plague, 61
Bureaucracies, 52, 57, 119, 333
Butterfly effect (Lorenz), 215

Calcium deposits, 246, 258
Canada, 125, 322, 327
Cancer, 61, 88, 141, 239, 240, 247, 253, 259, 315, 376
    breast, 250
    cause, 259–260
    early detection, 261
    morbidity, 58
    prostate, 141, 259
    skin, 254, 260
Cannon, Walter, 48, 102, 112, 245
Capital, 309, 313, 315, 317, 321, 324
Capitalism, 121, 306, 322
Carbon, 53, 68, 201, 218
Carcinogens, 253, 259
Cardiovascular system, 234, 258
Carroll, Lewis (Charles Dodgson), 134, 264
Catalyst, 194
Catastrophe, 35, 57, 59, 146, 203, 356, 386
    cancer, 260
    computer system, 161
    constructive vs. destructive, 58, 60
    economics, 123, 310
    electrical network, 215

Catastrophe (*cont.*)
  overpopulation, 355
  in political science, 121
  and power curves, 60
  revolution, 336, 340
  war, 350
Catastrophe theory, 15, 106
Catherine the Great, 42
Causality, 20, 21, 72, 74, 383
Cause-and-effect, 134, 141, 350
Cavendish, Spencer C., 99
Cell(s), 19, 66, 101, 251, 257, 260
  blood, 227, 230, 232, 253, 259
  brain, 12, 230
  connecting, 232
  epithelial, 232, 253
  glial, 239
  lung, 253
  lymphocyte, 257
  muscle, 232
  nerve, 232
  nonnucleated vs. nucleated, 5, 274, 277
  nucleus, 260
  rudimentary, 268, 271
  structure, 232
Cell differentiation, 146, 277
Cell division (mitosis), 274
Cellular automata, 6
Cellular theory of biology, 101
Center of gravity, 43, 106, 381
Central nervous system, 22
Cerebrum, 239, 251, 284
Chain reaction, 58, 79, 81, 107, 108, 144, 186, 190, 202
  biological, 202, 236–237
  in computer systems, 161
  economic depressions, 317
  fission, 189
  Great Depression (1929), 163, 317
  Malthusian reaction, 356
  network shutdown, 162
  nuclear, 192
  World War I onset, 351
Chambered Nautilus, 4, 195, 219
Chaos, 3, 5, 8, 14, 52, 57, 73, 94, 105–108, 136, 215, 379
  anarchy, 115
  and aperiodicity, 64

Chaos (*cont.*)
  creating order from, 179
  economics, 325, 334
  and gravity, 270
  human behavior, 357
  logical, 135
  revolution, 333, 336
  in war, 341
Chaos: Making a New Science (Gleick), 108
Checks and balances, 78, 329
Chemistry, 19, 74, 99, 102, 186
Chimpanzee, 242, 243, 267, 274
China, 125, 177, 327, 338, 353, 356
Chlorophyllous process, 231
Chordates, 275, 276
Christ, 113, 131
Christina (Queen of Sweden), 47
Chromosomes (DNA), 6, 84, 102, 104, 146, 150, 226, 231, 237, 260, 271
Chrysler Corporation, 118, 320
Church of England, 364
Churchill, Sir Winston, 35, 84, 179, 295, 328, 350, 358, 374
Circuits, 102, 145, 155
Circulation of the Blood, The (Harvey), 48
Circulatory system, 234, 238
Civil Disobedience (Thoreau), 122
Civil War (U.S.), 63, 121, 299, 337
Civilian Conservation Corps, 306
Cladistics (evolution), 237, 274
Clausewitz, Carl von, 106, 178, 351
Clemens, Samuel (Mark Twain), 31
Cnemidophorus (whiptail lizard genus), 225
Collective conscience, 112, 116, 329
Colossus (computer), 104, 105, 149, 154, 166
Common Sense (Paine), 121
Communications, 132, 143–146, 381
  and computers, 155
  international, 42
  line of, 175
Communism, 121, 322, 348
Communist Manifesto (Marx and Engels), 334
Complementarity principle, 72

Complexity theory, 3, 14, 57, 106
Compulsive–obsessive disorders, 292
Computers, 104, 149, 153, 154, 157,
    209, 384
  imaginary model, 151
  robotic, 214
  supercomputer, 343
Concentration, 169, 189, 294, 313
Concert of Europe (1815–1914), 348
Confederacy, 63, 337
Configuration(s), 8, 40, 46, 66, 76, 112,
    168, 171, 238, 380, 383
  aircraft, 212
  automation, 152
  biological, 238
  business, 320
  and chaos, 325
  crucial, 57, 59, 61, 68, 79, 86, 139,
    192, 202, 386
  dynamics, 171
  economics, 309, 319
  genetic and congenital defects,
    250
  geographical, 124
  government, 334
  land, 125
  and magnetism, 218
  metamorphosis, 67
  periodic table, 198
  periodicity, 65
  robotics, 213
  steric hindrence, 227
  strategy and war, 180, 349, 352
Confucianism, 364
Congenital defect, 246, 247, 250, 256,
    258
Connections (Burke), 215
Constancy (concept), 101
Constitution (U.S.), 39, 43, 68, 84, 112,
    301
Constitutional Convention (1787), 44,
    58
Controllability, 51, 81, 383
Copernicus, 22
Corporations, 312–314, 318, 359
Creativity, 119
Creator, 5, 22, 73, 262, 271, 364
Critical mass, 58, 79, 107, 161, 186,
    190, 192, 327, 386

Critical point, 79, 86, 121, 166, 386
  cladistics (evolution), 275
  law, 368
  oncogenesis, 260
  onset of war, 349
  revolution, 333
Cromwell, Oliver, 334
Cryogenics, 240
Cryptography, 145
Crystals, 11, 63
Culminating point, 62, 79, 80, 179,
    377, 386
  ecological, 221, 243
  economics, 315
  political equilibrium, 330
  war, 62, 105, 178
Culture, 111, 115, 116, 326, 365
Current (electrical), 195, 210
Cybernetics, 49, 51, 239, 242, 375, 384
*Cybernetics: Or Control and Communica-
    tion in the Animal and the Ma-
    chine*, 102, 162
Cycle(s), 49, 63, 79, 120, 376, 384
  cellular life, 230
  economics, 313, 315, 316
  physiology, 241
  weapons, 180
Cyst, 248, 258
Cytoplasm, 232, 274

da Vinci, Leonardo, 58
Darwin, Charles, 101, 115, 395
Database, 111, 145, 151, 156, 160
Death, 232, 240, 244, 246, 248
*Decline and Fall of the Roman Empire,
    The* (Gibbon), 109, 125, 328
Definition of life, 278
Deformity, 236, 239, 248
  congenital, 248
  internal structural, 252
  visible, 251
Degenerative disorders, 256
Degrees of freedom, 38, 39, 40, 45, 115,
    159, 383
Democracy, 30, 64, 66, 121, 322, 328, 343
Density, 41, 174, 206
Depression (economic), 124, 308, 315,
    393
Depression (psychiatric), 292

Descartes, 47
Design, 5, 33, 52, 382
Determinism, 21, 75, 76, 266, 383
    actuarial, 21, 266, 267, 383
    biological, 89, 296
    mechanical, 66
    statistical, 21, 77, 181, 266, 383
Diabetes mellitus, 248
Dickens, Charles, 90, 320
Dictatorship, 66, 121, 333, 343
Differentiation (cellular), 237, 238
Digestive system, 240
Dimensional analysis, 186
Diminishing returns, 156, 263, 321
Dinosaurs, 242, 271
Disorder(s), 7, 247–248, 252, 255, 326
Distribution(s), 266, 298
    attributes, 56
    population, 141
    resources, 43
    wealth, 44, 313
Dobzhansky, Theodosius, 270
Doppler effect, 97
Dow Jones average, 163, 312, 317
Dowding, Air Chief Marshal Sir
        Hugh, 107
Duke of Wellington, 167, 174
Durant, Will (and Ariel), 121, 236, 322,
        365
Durkheim, Emile, 112, 116

Earned-income tax credit, 306, 317
Earth, 97, 205, 269
    carrying capacity, 68, 221
    rotation, 97, 219
Earthworm, 274, 277
Eckstein, Gustav, 245
Ecology, 30, 68, 69, 221, 231, 269, 370
Economics, 111, 118, 122, 307, 308, 342,
        397
    business, 118
    corporate, 318
    forces of, 329
Education, 320, 328
Efficacy, 7–8, 33
Efficiency, 54, 82, 155, 381, 382
    database, 156
    energy usage, 215
    physiology, 262

Ego, 285, 287
Einstein, Albert, 21, 22, 97, 98, 138
    quoted, 71, 74, 99, 264
Electromagnetism, 96, 98, 216, 254
Electron(s), 10, 99, 189, 194–195, 198,
        209
Electronics, 145
Elements, 7–8, 32, 35, 132, 379, 380
    atomic, 11
    books, 147
    chemical, 6, 43, 198–199
    dominant, 41, 49, 68, 138, 170, 225,
        321, 383, 385
    linguistics, 144
    mathematical, 140
    of knowledge, 133
Elements (Euclid), 87, 135, 138, 140
Elements of Chemistry (Lavoisier), 99
Elements of Physical Biology (Lotka),
        201
Embolism, 235, 247, 248, 255
Embryo, 19, 237, 274
Emerson, Ralph W., 25, 110, 341
Emotions, 285, 287, 351
Encyclopaedia Britannica, 145, 146, 396
Energy, 11, 18, 20, 28, 76, 78, 187, 188,
        214, 390
    beams, 189
    conservation of, 100
    conversion from/to mass, 62
    electrical, 194, 215
    kinetic, 383, 384
    photons, 388
    potential, 58, 383
    radiant, 189, 191
    transduction, 218, 226, 230, 232
    transformation, 186
    unavailability, 18
Engels, Freidrich, 121
Engineering, 102, 204, 206, 397
    biomedical, 205
    civil, 205
    conveyance, 210
    energy, 215
    genetic, 205
    robotic, 213
    social, 204, 306
England, 103, 328, 334
Enigma (code), 105, 149

Entitlements, 313, 317, 318, 325, 394
Entropic index, 84, 86
Entropy, 18, 52, 57, 60, 78, 85–86, 100,
      101, 119, 297, 303, 383
   communications, 143
   computer programming, 158
   controllability, 60
   financial, 118
   in organizations, 300
   reversal, 20, 59, 61, 79, 81, 83, 86, 87,
      121, 179, 230, 260, 277
   sociological, 304
   in war, 170
Environment, 117, 241, 243, 247, 270,
      275, 300, 353
   and entropy, 60, 61
   influence on conduct, 266
   sociological, 111, 115
Epistemology, 83, 131, 132
Equations, 97, 186, 385
Equilibrative deficiencies (physiol-
      ogy), 252
Equilibrium, 48, 78, 102, 191, 199, 231,
      269, 384
   dynamic, 63, 78, 301, 330, 341, 370,
      384
   economics, 123, 319, 323
   in law, 366
   legal, 366
   in marriage, 296
   organizations, 303
   physiology, 240
   political, 120, 330
   psychological, 284
   static, 384
   in weather, 220
Erikson, Erik, 114, 289
Ether (concept), 95, 101
Ethics, 29, 69, 299, 328, 357–359, 397
   consequences, 362, 363
   corporate, 359
   dilemma, 363
   fixed vs. situational mores, 362
   greatest-good-for-the-greatest-
      number, 363
   individual, 358
   intent, 361–362
   international, 360
   means-to-an-end, 363

Ethics (*cont.*)
   medical, 279
   professional, 360
   systematic exploitation, 361
Ethics (Spinoza), 113, 364
Euclid, 13, 138, 139
Eukaryotes (nucleated cells), 274
Europe, 61, 205, 212, 330, 355
Evidence, 142, 368–369
Evolution, 6, 60, 80, 101, 146, 149, 236,
      237, 264–265, 275, 397
   acceleration, 279
   Darwin's thesis, 101
   misconceptions, 271
   school controversy, 365
   sociological, 115

Factor(s), 28, 46, 124, 245–246, 380
Federal debt, 317, 323
Federal Reserve Board, 307
Feedback, 49, 51, 226
Fetal alcohol syndrome, 250
Fetus, 64, 238, 250
First Principles (Spencer), 116
Fission, 76, 80, 107, 191
   binary, 225, 277, 392
Focal point(s), 41, 42, 53, 106, 170, 299,
      304, 380
Football, 172, 173, 181
Force(s), 42, 187, 188, 216, 383
   angular, 216
   centrifugal, 5
   dominant, 49, 78, 123, 226, 383
   economic, 122
   ideological, 337
   military, 177, 234, 345, 352
   reversal, 225, 227
Fountainhead, The (Rand), 371
Fourier, Joseph, 100
Fractal(s), 4, 66, 108, 380
Frame of reference (motion), 96
France, 109, 113, 337, 348, 351
Frankenstein (fictional character), 38,
      61, 372
Franklin, Benjamin, 109
Fraud, 134, 309, 312
Frazier, James, 116
Free enterprise, 307, 322, 323
French Revolution, 53, 100, 121, 336

Freud, Sigmund, 110, 113, 289
Friction, 49
Fromm, Erich, 114
Fuhrer prinzip, 300
Fuller, Buckminster, 201
Fullerite, 11, 201
Function(s), 112, 115, 197, 275, 297, 385
Fusion, 191
Future of an Illusion, The (Freud), 364
Fuzzy logic, 17, 106, 168, 394

Galileo, 22
Game(s), 29, 30, 124, 167, 169–171, 181
    zero-sum, 181, 193, 314, 315
Gandhi, Mahatma, 328, 329, 339, 347
Gastrointestinal tract, 214, 234
Gastroenterology, 249
General Motors Corporation, 118, 301,
    320
General systems theory, 3, 94
*General Theory of Employment, Interest
    and Money, The* (Keynes), 124
Genes, 235, 250–251, 274, 277
Genetic engineering, 205
Genetics, 25, 74, 80, 101, 149, 288, 384
    central dogma, 235, 271
    as destiny, 266
Genetics and the Origin of the Species
    (Dobzhansky), 270
Genome project, 260
Geococcyx californianus (roadrunner),
    173
Geographical Pivot of History, The
    (Mackinder), 125
Geography, 124, 342
Geometry, 28, 35, 189, 198
Geopolitics, 124, 125, 127, 327, 342,
    370, 395
Germany, 107, 123, 127, 214, 263, 293,
    330, 348, 351
Gesell, Arnold, 114, 289
Gestalt (synonym for configuration), 8
Gestation, 250, 252, 274
Giantism, 236
Gibbon, Edward, 109, 125, 328
Global warming, 221
Glorious Revolution (1688), 336
God and Golem, Inc. (Wiener), 162
Goedel, Kurt, 139

Goethe, 350
Goldbach, Christian, 13
Gossamer (human-powered aircraft),
    213
Government, 121, 322, 328, 332, 367
Grand Inquisitor, The (in *The Brothers
    Karamazov*), 365
Graphite, 11, 201
Gravity, 5, 19, 74, 98, 208, 211, 213,
    218, 226, 270
Great Books of the Western World, 135,
    298, 396
Great Britain, 103, 107, 112, 122, 125,
    126, 174, 322, 334, 337, 339, 348,
    351, 364
Great Depression (U.S.), 306
Great Pyramids, 19, 205
Greely, Major General Adolphus
    Washington, 203
Gross domestic product, 44, 312–315,
    322, 343, 393
Gross national product, 312, 343
Grotius, Hugo, 370
Group behavior, 115, 397
Group ego, 116, 299, 329, 330

Haddow, Sir Alexander (quoted), 259
Half-life, 189–190
Hart, B. H. Liddell, 352
Harvey, William, 48, 395
Hawking, Stephen, 257
Health care, 34, 322, 363
Heart, 101, 234, 240, 246, 252, 255, 258
Heisenberg, Walter, 72
Hemophilia, 250, 252
Heredity, 101, 146
Hermaphroditism, 236, 252
Hippocampus erectus, 236
History, 112, 177, 298, 309
Hitler, Adolph, 125, 178, 179, 293, 294,
    300, 345, 349, 370
Holmes, Justice Oliver W., Jr., 366,
    373, 377
Holocaust, 34, 110, 373
Homeostasis, 48, 66, 78, 102, 112, 277,
    375, 384
    biology, 231
    compartmentalization, 277

Homeostasis (*cont.*)
  and cybernetics, 82
  economics, 123, 314, 323
  government, 331
  jurisprudence, 366
  organizations, 303
  physiology, 240, 242
  political, 115, 120
  sociological, 360
Homo sapiens, 5, 38, 75, 210, 219, 230,
    232, 243, 247, 255, 278, 286
Hooke, Robert, 101
Horney, Karen, 114, 289
Horstmann, A. F., 100
Human behavior, 111, 113, 116, 120,
    127, 357, 361, 366
Human immunodeficiency virus
    (HIV), 257
Huygens, Christiaan, 95
Hydraulics, 5, 197, 209
Hydrocephalus, 258
Hydrogen, 31, 198
Hydrogen-ion activity, 84
Hypertext (defined), 387

Iacocca, Lee, 118, 320
Iatrogenesis, 241
Image processing, 147
Implosion technique (nuclear weap-
    ons), 191
Impulse (physics), 390
Income tax, 44
Indeterminacy, 20, 266
India, 122, 125, 327
Industrial Revolution, 334
Inequality (mathematics), 385
Inertia, 188, 383
Infection, 251, 255, 257
Influence of Sea Power Upon History 1660–
    1783, The (Mahan), 125, 342
Information theory, 101, 143, 145, 147
Infrastructure, 125, 126, 311
  economics, 310
  political, 342, 343, 345, 346
  revolutions, 338
Initiative, 5, 23, 28, 29, 37, 40, 60, 76,
    82–83, 89, 116, 117, 164, 169,
    266, 279, 285, 326, 360, 371, 372,
    383, 398

Insanity, 239
Instincts, 115
Institutes of the Christian Religion
    (Calvin), 364
Instructions (programming), 150, 384
Integration(s), 160–161, 292
Intelligence, 287
Interactionalism, 23, 24, 284
Interdependence (economic), 370
Interferometer, 96
Interior lines, 66, 135, 147, 381
  authority, 331, 336, 338
  cities, 304
  communication, 298
  critical mass, 190
  database, 156
  of evidence, 369
  geopolitics, 345
  law, 135
  logic, 133, 138, 158
  nations, 336
  organizations, 303, 319
  political infrastructure, 354
  programming, 158
  psychological, 293
  sociological, 333
  war, 177
International Business Machines, Inc.
    (IBM), 105, 319
International Red Cross, 370
International relations, 341, 344, 370,
    397
Introduction to the Study of Experimen-
    tal Medicine (Bernard), 101
Invisible hand (economics concept),
    123, 310
Irish Republican Army (IRA), 355
Irreversible differentiation, 260
Isotopes, 189, 198
Israel, 295, 364

Jacquard, Joseph Marie, 104, 213
James, William, 37, 110, 113, 114
Japan, 125, 127, 327
Jefferson, Thomas, 68, 110, 304
Jervons, Stanley, 103
Jews (persecution of), 350, 364
Johnson, Samuel, 88
Jomini, Baron de, 106

Jumping (computer program), 151, 157
Jung, Carl, 114, 289, 373
Jurisprudence, 42, 82, 366–368, 397

Kant, Immanuel (quoted), 80
Kennedy, John F., 107
Kepler, Johannes, 22, 27
Keynes, John Maynard, 124
Kidneys, 234, 258, 277
Kinematics, 28, 97, 187, 209
Kinetics, 21, 74, 94, 187, 209
King, Fleet Admiral Ernest J., 349
Korean war, 329, 353
Kroeber, Alfred, 116
Kuwait, 329, 349

Labor, 310–312, 324, 384
Lamarckism, 269
Languages, 144
Lavoisier, Antoine, 99, 100, 395
Law, 29, 120
    canon, 365
    civil, 365
    common, 368
    constitutional, 365, 367, 369
    criminal, 365
    distinction between law and fact, 368
    international, 360, 365, 368, 370
Lawrence, Colonel T. E. (quoted), 300
Leadership, 118, 303, 310, 329, 338
Leeuwenhoek, Antonie van, 101
Lagrange, Joseph, 100
Length, 186, 238
Lenin, Vladimer, 34
Lessons of History, The (Durant), 322
Leukemia, 259
Levels of perspective (in war), 69, 172
Leverage, 125, 208, 353
    and controllability, 81
    geopolitics, 346
    revolution, 338
Lewin, Kurt, 114, 289
Libraries, 146, 147
Life expectancy, 244, 262
Light, 95, 97, 254, 388
Lincoln, Abraham, 138, 146, 283, 294, 299, 331, 337, 340

Line of regression, 141, 180, 314
Linearity, 8, 12, 181, 379
Linguistics, 132, 143–145, 147
Linkage, 134–135, 160, 381
    database, 156
    economics, 321
List, Friedrich, 123
Liver, 87, 234
Locomotion (biological), 230, 238, 277, 278
Logic, 115, 134, 144, 156, 157, 180
Logistics (biological), 227, 232, 233, 255, 277
Logos, 132, 285, 376
Looms (automated), 103–104, 150, 209, 213
Lorentz, Hendrik Antoon, 97
Lorenz, Edward, 108
Los Alamos, 107, 186
Lotka, Alfred J., 201
Lou Gehrig's disease (ALS), 257
Lovelace, Augusta Ada, 103
Lull, Ramon, 103
Lungs, 234, 277
Lymph system, 234, 259

MacArthur, General Douglas, 173, 175
Machine, 31, 150, 213, 246, 379
    economic, 310
    vs. human response, 166
    perpetual motion, 217
    physiological, 247
Machine language, 152, 154
Mackinder, Sir Halford, 125, 347
Macroeconomics, 122, 308, 312, 314, 318
Magna Carta, 328
Magnetism, 216
Mahan, Admiral Alfred T., 125
Malinowski, Bronislaw, 116
Malnutrition, 246, 288, 355, 356
Malthus, Thomas, 221
Malthusian reaction, 68, 80, 220, 308
Manifesto of the Communist Party (Marx and Engels), 121
Mao Zedong, 34, 338
Marshall, Alfred, 123
Marshall, General George C., 328
Marx, Karl, 33, 111, 333

Maslow, Abraham, 114
Mass, 52, 62, 76, 187, 190, 192, 382
    as clout, 175, 321
    electron, 198
    geographical, 124, 126
    land, 44, 219, 343, 345, 346
    physics, 8, 186, 187
    principle of, 106
    relativistic, 98
*Mathematical Biophysics: Physico-Mathe-
    matical Foundations of Biology*, 201
Mathematics, 12, 28, 74, 82, 139, 140,
    143, 150, 180, 209
Matter, 195, 215, 238
Maturity (psychological), 26, 289
Maxwell, James Clerk, 16, 22, 96, 97
May, Rollo, 114
Mechanical advantage, 208, 210
Mechanics, 52, 94, 100
    biochemical, 274
    cerebral, 279
    physiological, 287
Medicare, 325
Medicine (profession), 72, 243, 247,
    248, 263
Meiosis, 225
Membranes, 227, 230, 231
Mendel , Gregor, 101, 201
Mendeleyev, Dimitry, 9, 100, 133, 197
Menninger, Karl, 114, 289, 293
Meta-analysis (defined), 387
Metaeconomics, 122, 308, 309
Metabolism, 56, 133
Metastasis, 239, 259, 261
Methanococcus jannischii, 274
Michelangelo, 58
Michelson, A. A., 96
Michelson–Morley experiment, 13, 96,
    97
Microeconomics, 122, 308, 318, 321
    consumer-unit, 122, 320
    corporate, 318, 319
Micrographia (Hooke), 101
Microorganisms, 5, 58, 66, 231, 253,
    274, 355
Milgram, Stanley, 299
Mill, John Stuart, 110, 111, 122, 326
Mind–body issue, 22, 24
Minkowski, Hermann, 98

Mitchell, General William "Billy," 203,
    212, 349
Mitochondria, 230
Mitosis, 225, 277
Moby Dick (Melville), 294
Model(s), 181, 155, 234, 381, 393
    decision support, 165
    deductive reasoning, 138
    epistemogenetic, 83
    linear, 108
    mathematical, 98, 108
    physiogenetic, 10, 12, 71, 75, 82, 88,
        222, 227, 237, 391, 398
    psychogenetic, 290
    sociogenetic, 83
Molecular Biology of the Gene (Watson),
    224
Molecular weight, 198
Molecule(s), 5, 31, 66, 76, 194, 196
Momentum, 10, 140, 187, 209, 216,
    238, 383, 390
Monarchy, 121, 331, 332
Money, 118, 122, 308, 312, 321
Monism, 23, 24
Monopolies, 310, 318
Monopterus albus, 236
More, Sir Thomas, 358
Morganthau, Hans, 342
Morley, E. W., 96
Mormons, 247
Morphology, 119, 213, 252, 381
Moscow, 172, 179
Motion, 96, 238
Mozart, Wolfgang Amadeus, 264
Muscular system, 231, 278
Muslim fanaticism, 348
Mutation, 236, 250
    cumulative, 227
    gene, 251
    random, 235, 271
    stray, 237
My Years with General Motors (Sloan),
    118
Myocardial infarction, 235, 255

Nagasaki, 191
Napoleon, 105, 106, 125, 167, 168,
    172–175, 178, 179, 345, 348
Narcissism, 284

Nation, 309, 326, 329, 397
National economy, 69, 318
National Geographic Society, 203
National power, 342–343
National purpose (in war), 69, 351
National System of Political Economy, The (List), 123
Nationalism, 112, 329, 330
Natural law (universality), 94
Natural resources, 44, 221, 342
Nature, 5–6, 60, 77, 161, 219, 234, 242, 255, 268, 387
    actuarial determinism, 267
    evolution, 237
    methodology, 150
Neoplastic transformation, 224
Neptune (planet), 133
Nervous system, 231, 239, 251, 279
Network(s), 145, 214, 381
    computer, 41, 155, 160
    neural, 231
    telephone, 155
    utilities, 214
Neuron(s), 12, 227, 237, 239
Neutron(s), 10, 99, 190, 191, 198
New York Times, The, 107
New Zealand, 125, 177
Newton, Sir Isaac, 22, 95, 216–217, 365
    quoted, 6, 68, 185
Nonlinearity, 8, 12, 379
Normandy (France), 173, 334
North Atlantic Treaty Organization (NATO), 348, 349
North Korea, 175, 353
North Vietnam, 336
Nuclear binding force, 10, 198
Nuclear decay force, 10
Nuclear packing-fraction, 193
Nuclear reactions, 58
Nuclear weapons, 61, 191, 253, 371
Nucleus
    atom, 199
    cell, 225–227, 232, 274
Nutrients, 20, 227, 277, 278, 286

Obedience to authority, 299, 328, 336, 339
Oedipus complex, 284

Oligarchy, 121, 331
On Liberty (Mill), 122
On the Law of War and Peace (Grotius), 370
On the Principle of Population (Malthus), 221
On War (Clausewitz), 106
Oncogenesis, 260, 261
Oparin, A. I., 267, 270
Operations, 69, 140–141, 382
Operations research, 105, 106
Order, 5, 379
    from chaos, 3
    and complexity, 57
    and disorder, 60
Organism, 5, 60, 223, 246, 286
    development, 146, 375
    economics, 325
    multicellular, 229–231, 235, 268, 277
    prototype, 274
    single-cell, 229, 253
    zoological, 6, 10, 222, 230
Organizations, 119, 288, 300, 302
    autonomic control, 303
    balance-of-power model, 301
    semi-independent satellite model, 301
    structure vs. function, 297
Origin of Life on Earth, The (Oparin), 267, 270
Origin of the Species, The (Darwin), 115
Osler, Sir William, 293
Out of Chaos (Halle), 108
Overstretch, 316, 386
Oxygen, 43, 234, 286

Palliative care, 248
Panama Canal, 127, 342
Pancreas, 234, 277
Paradigm(s), 99, 140, 210, 382
Paranoia, 292
Parasites, 240, 253
Parasympathetic subsystem (physiology), 239
Paris, 109, 336
Parliament (G.B.), 109, 123, 364
Parthenogenesis, 58, 225
Pasteur, Louis (quoted), 25
Pathology, 244, 245

Pattern(s), 7–8, 17, 64, 103
Pearl Harbor, 180, 346, 349
Peck, M. Scott, 289
Pediatrics, 114
Peloponnesian war, 179
Penicillin, 248
Peptides, 268, 271
Periodic table, 9, 100, 198–200
Periodicity, 65, 197
Perpetual motion, 20
Peter III, Emperor (Russia), 42
pH scale, 84
Pharmaceuticals, 246, 253, 256
Philip of Macedon, 350
Philippines, 127
Philosophy, 114, 134
Photons, 11, 20, 189, 388
Physics, 98, 186, 245, 366, 397
  nuclear/subatomic, 16, 75, 100
Physiogenesis, 7, 72, 81, 85, 228, 387
Physiology, 6, 36, 233, 244, 255, 301,
    397
Piaget, Jean, 114, 289
Pineal gland, 279
Pinocchio Effect, 395
Planck, Max, 94, 99, 197, 292, 293
Plasma (blood), 195
Plasma (state of matter), 195
Plate tectonics, 219
Plato, 110
Point of attenuation, 60, 80, 386
Point of diminishing returns, 54, 56,
    304, 385
  aircraft design, 211
  economics, 123
  life expectancy, 262
Point of negative returns, 385
Poisons, 240, 253
Population, 309, 313, 342
  excess, 220, 221
  world, 44, 326, 355, 371
Positron emission tomography (PET),
    245
Power, 215, 383
  economic, 124
  electrical, 216
  focal point, 331
  military, 124
  national, 126, 301, 342, 343

Power (*cont.*)
  naval, 126
  nuclear, 191
  physics, 187, 189, 215
  political, 328
Power curve, 54–58, 63, 385
  catastrophe, 59, 386
  computers, 155, 156
  and efficiency, 82
  pharmaceuticals, 253
  physical health, 262
  strategy, 175, 176
Pressure, 197, 209, 213
Preventive medicine, 252, 363
Priestly, J. B., 99
Prime numbers, 13
Principia Mathematica (Russell and
    Whitehead), 139
*Principia [Philosophiae Naturalis ...
    Mathematica]* (Newton), 94, 114,
    390
Principles of Economics (Marshall), 123
*Principles of Political Economy and Taxa-
    tion, The* (Ricardo), 123
Principles of Psychology, The (James),
    113
Principles of war, 106, 168, 171, 180,
    181, 318, 394
Probability, 141, 266, 383
Productivity, 122, 124, 311
Profit, 44, 119, 318–319
Program/programming, 8–9, 21, 159,
    384
  acquired, 239
  autoimmune system, 51
  automation, 102
  computer, 75, 154, 157–158, 160
  and cybernetics, 51
  decision support, 181
  branching, 157
  modular, 158, 160
  robotics, 213
  self-modifying, 164
  three-dimensional, 157
  in various dimensions, 159
Prokaryotes (nonnucleated cells), 274
Propaedia (Encyclopaedia Britannica),
    145, 146, 396
Proteins, 234, 253, 271

Proton(s), 10, 99, 198
Protozoan, 225, 274
Pseudocatastrophe, 59, 62, 386
Psyche, 29, 41, 75, 88, 114, 284–287, 352
   subelements, 284, 285, 288
Psychiatric dysfunctions, 113, 288
Psychiatry, 113, 114, 134, 248, 288
Psychoanalysis, 114, 134, 293
Psychology, 113–115, 146, 397
Psychopathology, 289
Psychosis, 292
Psychotherapy, 24
Puberty, 241, 287
Purpose, 8, 231, 382
Pythagoras, 138

Quantum mechanics/elec-
   tromechanics, 23, 94, 99
Quarks, 99
Quartz (crystal), 63

Radiation, 96–97, 254
Radioactivity, 79
Radionuclide scanning, 245
Rand, Ayn, 371
Rashevsky, Nicholas, 201
Rebellion, 115, 117, 310, 333, 370
Recession, 124, 315
Red Queen dilemma (evolution the-
   ory), 266
Regression analysis, 141
Reign of Terror (France), 361
Relativity, 97–98
Renaissance, 58
Representative Government (Mill), 122
Reproduction, 65, 80, 146, 230
   asexual, 235
   parthenogenetic, 235, 260
   sexual, 235
Resiliency, 52, 82, 208, 221, 354
Resistance, 210, 227
Revolution, 29, 115, 117, 121, 303, 310,
   330, 333, 335, 336, 351, 370
Revolutionary War (U.S.), 121, 337
Ricardo, David, 123
Rickettsia, 240, 253
Rights (legal), 120, 301, 358, 360, 366
Rise and Fall of the Great Powers, The
   (Kennedy), 345

RNA, 231, 235, 271
Robot/robotics, 102, 151, 213, 384
Roget's Thesaurus, 145, 285, 286, 391
Roman Catholic Church, 364
Roman Empire, 110, 125, 126, 322, 334
Roosevelt, Franklin D., 340, 358
Roosevelt, Theodore, 323, 340
Rosetta Stone, 144
Royal Society, 101, 103
Russell, Bertrand, 139
Russia, 42, 100, 125–127, 178, 262, 322,
   327, 345, 348

Sachs, Oliver, 239
Schindler's List (film), 373
Schizophrenia, 292
Schleiden, Mathias J., 101
Schrödinger, Erwin, 102, 150
Schwann, Theodor, 101
Science, 101, 131, 134, 334
   computer, 146, 375
   information, 147
   medical, 262
   military, 105, 170
   physical, 108, 113, 116
   political, 120
   social, 117
   traditional, 3, 16, 94, 108
Scientific American, 260
Sclerosis, 235, 248
Scopes "monkey" trial, 369
Seamless blanket of nature, 14
Second Piano Concerto (Rachmaninoff),
   226
Septal defect, 252
Serendipity, 8
Sexual drive, 114, 287
Shakespeare, 267
Shaw, George Bernard, 365
Shay's rebellion (1792), 333
Sheen, Bishop Fulton, 134
Shining Path (Peruvian terrorists), 355
Short stature (dwarfism), 251
Silent Spring (Carson), 356
Silicon, 43, 201
Situs inverses, 252
Sloan, Alfred P., 118, 320
Smith, Adam, 110, 122, 310
Social Security, 317, 321, 325

Socialism, 121, 306, 322, 323, 325, 332
Socioeconomic status, 304, 306
Sociology, 115, 120, 146
Solar system, 66, 218
Somalia, 172, 330, 347
South America, 221
Sovereignty, 314, 327, 330, 331, 337, 368
Soviet Union, 125, 127, 178, 220, 330
Space, 41, 74, 96, 108, 382
Space–time continuum, 98
Spanish–American War, 127
Sparta, 179, 350
Spencer, Herbert, 116
Sphere of influence, 382
Spinal cord, 257
Spinoza, Benedict, 109–110, 131–132
St. Joan (Joan of Arc), 103, 112, 377
Stabilization, 205
Stalin, Josef, 34, 349
Standard deviation, 36, 140
State (nation), 120, 347, 359, 368
States of matter, 194, 196
Statistical Abstract of the United States, 317
Statistics, 36, 116, 140–141, 266
Steady-state doctrine (biology), 101
Stereoisomers, 11
Steric hindrance, 227
Stock market, 163, 315, 321
Story of Civilization (Durant), 121
Stowe, Harriet Beecher, 146
Stradivari, Antonio, 265, 270
Strategy, 53, 135, 167, 176, 180, 380
Stress, 240, 256
Structure, 112, 201
Subassemblies, 6, 36, 267–268, 376, 380
Subsystem(s), 36, 242, 368, 380
Suicide (Durkheim), 116
Sullivan, Harry Stack, 114, 289
Sun Tzu, 105
Superego, 287
Superpower, 126, 346
Supranational authority, 326, 329, 341, 370, 371
Supreme Court (U.S.), 370
Survival instinct, 243
Survival of the fittest, 115
Sustainability, 52, 82

Sweden, 47, 322, 364
Symbiosis, 230, 237, 251
Symmetry, 10, 14, 17, 16, 28, 69, 186, 194, 380, 391
Sympathetic subsystem (physiology), 239
Synapses, 12, 239
Syndactyly, 251
Synergism, 12, 41, 158, 379
Syntopicon (Great Books of the Western World), 135
System of the World, The (in Newton's *Principia*), 135
Systematic Theology (Tillich), 363
Systemic lupus erythematosus (SLE), 241
Systems, 55, 78, 80, 115, 120, 132, 302, 379
    abstract, 28, 40, 52, 86, 89, 115, 122, 380, 397
    autoimmune, 51
    automated, 160, 214, 397
    biological, 254
    categories, 26, 27, 32
    circulatory, 214
    closed, 18, 101
    computer, 36, 52, 154
    cultural, 117
    database, 156
    deterministic, 383
    dynamics, 106
    economic, 121, 314
    and entropy, 60, 87, 100
    ethical, 357, 365
    forms, 27, 64
    free-enterprise, 323
    hierarchical form, 29, 30, 64, 81, 138, 156, 381
    immuno-response, 231
    integral form, 29, 31, 64, 81, 370, 381
    judicial, 37, 357, 359
    logical relationships, 40
    management information, 157
    mathematical, 43
    mechanical, 254, 255
    morphology, 30, 67
    nonlinear, 105
    operating characteristics, 380
    organic, 301

Systems (*cont.*)
  organizational, 118
  physical, 19, 39, 40, 115, 124, 194,
    230, 233, 238, 262, 268, 360, 375,
    380, 392, 397
  political, 43
  relational form, 29, 30, 64, 81, 147,
    381
  serendipitous, 33, 382
  Sociological, 16, 29, 39, 40, 42, 43, 52,
    56, 71, 86, 89, 112, 127, 166, 203,
    307, 359, 362, 365, 371, 380, 397
  solar, 5, 22
  structured, 57
  theological, 357
  theory, 136, 155, 304, 398
Systems analysis, 106, 107

Tactics, 25, 69, 105, 106, 167, 173, 180,
    354
Taxation, 123, 309, 314, 317
Tay–Sachs disease, 252
Taylor, Frederick, 118, 214
Technology, 111, 310, 311, 323, 334, 342
  automation, 145, 311
  progression, 310
  and tactics, 180
Teleology, 22
Temperature, 194–196, 235, 240
Template, 237, 381
Terrorism/terrorists, 353, 354, 371
Theater of operations/war, 69, 173,
    178
Theology, 29, 363, 365
Theory, 99, 135
  game, 180
  information, 132
  mathematical, 116
  organizational, 118, 120
  phlogiston, 99
  quantum, 99
  relativity, 97
  systems, 135, 237, 325
Theory of Games and Economic Behavior
    (von Neumann), 124
Theresienstadt, 370
Thermodynamics, 18–19, 74, 100
Third Reich, 61
Thoreau, Henry, 338

Threats (national) , 331, 347–348
Three-body problem, 12
Thrombosis, 235, 248
Thucydides, 179
Tiananmen Square, 25, 42
Time, 75, 186, 238, 382
Time–motion technique, 214
Tissue (biological), 102
Titanic (ship), 31–32, 45, 208, 297
Tocqueville, Alexis de, 126, 346
Tolstoy, Leo, 110, 372
Tooth-to-tail ratio (military), 234
Torque, 205, 226
Toxins, 248, 250, 253
Toynbee, Arnold, 112
Trade balance, 313
Transaction(s), 314, 382
Transformation (neoplastic), 259, 261
Transformation equations, 97
Treasury (U.S.), 325
Treaties, 341, 371
Treatise on Electricity and Magnetism
    (Maxwell), 96
Trent affair, 337
Trigger effect, 215
Troy (ancient), 304
Truman, Harry S, 353
Trust funds (government), 317
Truth, 131, 133, 135, 140, 145
Tumor(s), 227, 248, 258–259
Turing, Alan, 166
Turner, Frederick Jackson, 127
Turning point(s), 54, 55, 62, 178, 385
Twain, Mark (Samuel Clemens), 146,
    219
Tyler, Edward, 116

U.S. Army War College, 52, 295
Umbilical cord, 238
Uncertainty principle, 17
Uncle Tom's Cabin (Stowe), 121, 146
Unemployment, 315, 317, 324, 335
Union (U.S. Civil War), 299, 337
United Kingdom, 334
United Nations, 177, 329
United States, 34, 44, 112, 254–255,
    260, 295, 322–323
  and English system of law, 110
  geographic advantages, 125

United States (*cont.*)
 and geopolitics, 127
 and Hawaii, 127
 and Mahan's thesis, 126
 Pearl Harbor, 180
 pivotal legal cases, 369
 Revolutionary War, 337, 349
 stabalized boundaries, 125
 superpower, 126, 346
 world wars, 107, 173, 349
Upswing point, 54, 56, 156, 385
Uranium, 190
Uterus, 237, 238

Vaccines, 253, 355
Variance, 36, 39, 140
Velocity, 8, 35, 40, 97, 187, 188
Venereal diseases, 253
Ventricles, 252
Versailles Peace Treaty (World War I),
 349
Vertebrates, 39, 238, 274
Vietnam (conflict), 34, 53, 69
Virus, 61, 240, 246, 253, 257, 259, 355
 computer, 19
Vital force (concept), 101
Voltage, 210
Voltaire, 365
von Neumann, John, 124

War, 30, 69, 105, 167, 168, 170, 171,
 351, 353, 370
 absolute vs. limited war, 170

War (*cont.*)
 and chaos, 341
 civil, 330, 334, 336, 337, 349, 351, 371
 economic, 309, 347
 guerrilla, 337, 354
 international, 337, 351
 potential, 350
War and Peace (Tolstoy), 372
Washington, D.C., 162, 214, 299, 343,
 353
Washington, General George, 303
Waves, 17, 145
Wealth, 56, 122, 308–309, 309, 311, 325,
 343
Wealth of Nations, The (Smith), 122, 310
Welfare, 309, 313, 317, 322, 394
Western Europe, 125
What is Life? (Schrödinger), 102
Whitehead, Alfred North, 139
*Why Things Bite Back: Technology and
 the Revenge of Unintended Conse-
 quences*, 360
Wiener, Norbert, 102
Wisdom of the Body, The (Cannon), 48,
 102
Work, 51, 102, 188, 384
World War I, 25, 175, 212, 295, 349, 351
World War II, 61, 84, 105, 336, 373

Yugoslavia, 327

Zone of negative returns, 56, 254
Zygote, 19, 236, 260

www.ingramcontent.com/pod-product-compliance
Ingram Content Group UK Ltd.
Pitfield, Milton Keynes, MK11 3LW, UK
UKHW022304280225
455674UK00001B/166

9 780738 205847